Application of CFD Simulations to Marine Hydrodynamic Problems

Application of CFD Simulations to Marine Hydrodynamic Problems

Guest Editors

Peng Du
Abdellatif Ouahsine

MDPI

Basel • Beijing • Wuhan • Barcelona • Belgrade • Novi Sad • Cluj • Manchester

Guest Editors
Peng Du
School of Marine Science and Technology
Northwestern Polytechnical University
Xi'an
China

Abdellatif Ouahsine
Laboratoire Roberval
Université de Technology de Compiègne
Compiègne
France

Editorial Office
MDPI AG
Grosspeteranlage 5
4052 Basel, Switzerland

This is a reprint of the Special Issue, published open access by the journal *Journal of Marine Science and Engineering* (ISSN 2077-1312), freely accessible at: https://www.mdpi.com/journal/jmse/special_issues/H62A50YHCA.

For citation purposes, cite each article independently as indicated on the article page online and as indicated below:

Lastname, A.A.; Lastname, B.B. Article Title. *Journal Name* **Year**, *Volume Number*, Page Range.

ISBN 978-3-7258-5893-4 (Hbk)
ISBN 978-3-7258-5894-1 (PDF)
https://doi.org/10.3390/books978-3-7258-5894-1

Contents

About the Editors vii

Peng Du and Abdellatif Ouahsine
Application of CFD Simulations to Marine Hydrodynamic Problems
Reprinted from: *J. Mar. Sci. Eng.* **2025**, *13*, 1212, https://doi.org/10.3390/jmse13071212 1

Mo Chen, Nan Zhang, Hailang Sun and Xuan Zhang
Large Eddy Simulation of the Flow around a Generic Submarine under Straight-Ahead and 10° Yaw Conditions
Reprinted from: *J. Mar. Sci. Eng.* **2023**, *11*, 2286, https://doi.org/10.3390/jmse11122286 4

Miao Zhang, Haibao Hu, Abdellatif Ouahsine, Peng Du, Xiao Huang and Luo Xie
Numerical Study of the Force Characteristics and Flow Field Patterns of a Cylinder in the Internal Solitary Wave
Reprinted from: *J. Mar. Sci. Eng.* **2024**, *12*, 906, https://doi.org/10.3390/jmse12060906 26

Xuanquan Wang, Suwei Xiao, Xinchun Wang and Debo Qi
Numerical Simulation and Analysis of Added Mass for the Underwater Variable Speed Motion of Small Objects
Reprinted from: *J. Mar. Sci. Eng.* **2024**, *12*, 686, https://doi.org/10.3390/jmse12040686 48

Xiaolong Chen, Xuan Che, Yin Zhou, Changfeng Tian and Xinfeng Li
A Numerical Simulation Study and Effectiveness Evaluation on the Flow Field Effect of Trapezoidal Artificial Reefs in Different Layouts
Reprinted from: *J. Mar. Sci. Eng.* **2024**, *12*, 3, https://doi.org/10.3390/jmse12010003 69

Siddharth Gupta, Atul Sharma, Amit Agrawal, Mark C. Thompson and Kerry Hourigan
Role of Shape and Kinematics in the Hydrodynamics of a Fish-like Oscillating Hydrofoil
Reprinted from: *J. Mar. Sci. Eng.* **2023**, *11*, 1923, https://doi.org/10.3390/jmse11101923 84

Chen Yang, Ke Zeng, Jilong Chu, Shuxia Bu and Zhang Zhu
Computational Study on Influence Factors and Vortical Structures in Static Drift Tests
Reprinted from: *J. Mar. Sci. Eng.* **2024**, *12*, 789, https://doi.org/10.3390/jmse12050789 101

Héctor Rubén Díaz Ojeda, Sebastian Oyuela, Roberto Sosa, Alejandro Otero and Francisco Pérez Arribas
Fishing Vessel Bulbous Bow Hydrodynamics—A Numerical Reverse Design Approach
Reprinted from: *J. Mar. Sci. Eng.* **2024**, *12*, 436, https://doi.org/10.3390/jmse12030436 122

Yu Zhang and Dongqin Li
Parametric Design of a New Float-Type Wave Energy Generator and Numerical Simulation of Its Hydrodynamic Performance
Reprinted from: *J. Mar. Sci. Eng.* **2023**, *11*, 2192, https://doi.org/10.3390/jmse11112192 140

Young Jun Kim, Maxime Canard, Benjamin Bouscasse, Guillaume Ducrozet, David Le Touzé and Young-Myung Choi
High-Order Spectral Irregular Wave Generation Procedure in Experimental and Computational Fluid Dynamics Numerical Wave Tanks, with Application in a Physical Wave Tank and in Open-Source Field Operation and Manipulation
Reprinted from: *J. Mar. Sci. Eng.* **2024**, *12*, 227, https://doi.org/10.3390/jmse12020227 165

About the Editors

Peng Du

Peng Du obtained his bachelor's and master's degrees from Northwestern Polytechnical University in 2012 and 2015. His received the Ph.D. from the CNRS Roberval Laboratory in France. He is currently an Associate Professor and doctoral supervisor at the School of Marine Science and Technology, Northwestern Polytechnical University. He also serves as the deputy director of the Shaanxi Provincial Key Laboratory, the deputy secretary-general of the University's Science Association, etc. Dr. Peng Du's research focuses on the environmental hydrodynamics of marine structures. He has published over 70 journal papers, applied for more than 40 invention patents, and registered over 30 software copyrights. He serves as an Editorial Board Member for the *Journal of Hydrodynamics* and the *Chinese Journal of Ship Research*, a Young Editorial Board Member for *Engineering Reports* and *Marine Engineering*, and a reviewer for more than 30 journals. He has been invited to give plenary/keynote speeches at numerous academic conferences and has led several national and provincial-level projects.

Abdellatif Ouahsine

Abdellatif Ouahsine is a full and distinguished Professor at the University of Technology of Compiègne (UTC), Alliance Sorbonne University (France). He directed the Joint Computational Hydraulics Laboratory (LHN) of UTC and the Ministry of Sustainable Development and Energy (CEREMA) for fifteen years. He was also President of the Board of Directors in restricted formation of the UTC (2013–2021). Prof. Ouahsine's research focuses on computational fluid dynamics and the development of numerical tools for hydrodynamic and environmental processes in oceanic and fluvial environments. His concern is the study of concrete problems linked to the numerical modeling of the transport and propagation of surface waves in an environmental area (marine, fluvial, etc.) and to the numerical treatment of applied fluid–structure interactions to naval hydrodynamics (resistance to the advancement of ships in confined environments, bow waves, maneuverability, rolling, squatting, etc.), which also cover the modeling of water–sediment–boat interactions in confined environments. Part of his applied research concerns ship transport in inland waterways, which is linked to the problem of clean, economical, and safe boats.

Journal of
Marine Science and Engineering

MDPI

Editorial

Application of CFD Simulations to Marine Hydrodynamic Problems

Peng Du [1,2,*] and Abdellatif Ouahsine [3]

[1] School of Marine Science and Technology, Northwestern Polytechnical University (NPU), 127 Youyi Road, Beilin, Xi'an 710072, China
[2] Hanjiang National Laboratory, Wuhan 430060, China
[3] Laboratoire Roberval, Sorbonne Université, Université de Technology de Compiègne, Centre de Recherches Royallieu, CS 60319, CEDEX, 60203 Compiègne, France
* Correspondence: dupeng@nwpu.edu.cn

Growing global energy demand and the increasing exploration of the ocean have brought about significant challenges and opportunities in the field of marine engineering and fluid mechanics [1]. This Special Issue features nine cutting-edge research articles that focus on key areas such as submarine flow fields, ship navigation characteristics, wave energy utilization, artificial reefs, the dynamic stability of pontoons, and fish-shaped oscillating wings. Our aim is to provide a platform for researchers, engineers, and academics to exchange ideas and gain new insights, thereby advancing research and technological progress in this field. The topics of the articles featured in this Special Issue can be classified into three categories: underwater structures, surface structures, and surface waves.

The hydrodynamics of underwater structures, including submarines, hydrofoil, reefs, and other simplified structures, have been investigated in several studies. In [2], an investigation into the flow field characteristics of a new submarine model under both 0 and 10° yaw conditions was carried out. By employing Large Eddy Simulation (LES) technology, complex flow phenomena were accurately captured, and the interaction laws between different appendages and the evolution of the wake were revealed. This research offers crucial theoretical support for enhancing the hydrodynamic design of submarines. The hydrodynamic behavior of cylinders under the action of internal solitary waves is analyzed in [3]. Through numerical simulations, the flow field distribution characteristics and force characteristics of the cylinders are determined, offering critical data support for the design of safe marine structures situated in complex ocean environments. The authors of [4] conduct research on the added mass of underwater objects in variable-speed motion using a numerical simulation approach combined with regression analysis and parameter separation analysis. The study investigates the sources and variation patterns of added-mass forces for irregularly shaped small objects, providing new insights to inform the mechanical analysis and design optimization of underwater vehicles. The flow field effects of trapezoidal artificial reefs are evaluated in a study that combines numerical simulations with field surveys [5]. The study comprehensively assesses the flow field effects, stability, and ecological benefits of different reef layouts under various flow velocities, serving as a valuable reference for the rational design and deployment of artificial reefs. The authors of [6], in a study on fish-like oscillating hydrofoils, investigate the effects of hydrofoil geometric parameters and kinematic parameters on propulsive performance. By analyzing the near-wake structure, the study establishes a correlation between engineering parameters and flow separation, providing key insights for the development of biomimetic propulsion technologies.

The hydrodynamics of surface structures, including ships and wave energy converter, have been investigated in several studies. Ship drift trials were conducted in [7], and the Orthogonal Experiment Method (OEM) was used to analyze the effects of mesh size, turbulence models, and time steps. The reliability of the simulation results was verified, and the Detached Eddy Simulation (DES) method successfully captured the vortex structures around the ship's hull, highlighting the potential of numerical simulation in ship design. In the field of ship design, a study on the hydrodynamics of bulbous bows for fishing vessels presents a novel numerical reverse design approach [8]. Based on the prototype of an Argentine trawler, this research utilizes the OpenFOAM platform for numerical simulations and combines towing tank experiments for validation. It systematically evaluates the resistance performance of vessels with and without bulbous bows, offering a fresh perspective for the design optimization of small-scale fishing vessels. Research on wave energy utilization is gaining prominence due to the increasing focus on clean and abundant ocean energy resources. The authors of [9] explore parameterized design and employ numerical wave tank technology to analyze the motion response and power output characteristics of a new pontoon-type wave energy converter in regular-wave environments. This work lays a foundation for the engineering application of wave energy conversion technologies.

In the field of wave simulation technology, ref. [10] proposes a high-order spectral method for irregular-wave generation and calibration. By integrating physical wave tank experiments with numerical simulations, this method enhances the accuracy of wave simulations and provides an efficient and reliable tool for research on wave–structure interaction.

The nine studies featured in this Special Issue showcase innovative research from diverse perspectives in the field of marine engineering and fluid mechanics. They not only advance the development of theories and technologies in this field, but also offer valuable guidance for practical engineering applications. We hope that these research findings will inspire readers, promote collaboration between academia and industry, and foster the exchange of ideas to address the complex challenges in marine engineering and contribute to sustainable development.

Funding: This research was funded by the National Natural Science Foundation of China (Nos. 52471344, 52201380), the Open Fund Project of Hanjiang National Laboratory (No. KF2024046), and the Joint Training Fund Project of Hanjiang National Laboratory (No. HJLJ20240304).

Acknowledgments: The authors wish to thank all of the contributors to this Special Issue, as well as the *JMSE* editorial staff, without whom the publication of this Special Issue would not have been possible.

Conflicts of Interest: The authors declare no conflicts of interest.

List of Contributions

1. Chen, M.; Zhang, N.; Sun, H.; Zhang, X. Large Eddy Simulation of the Flow around a Generic Submarine under Straight-Ahead and 10° Yaw Conditions. *J. Mar. Sci. Eng.* **2023**, *11*, 2286. https://doi.org/10.3390/jmse11122286.
2. Zhang, M.; Hu, H.; Ouahsine, A.; Du, P.; Huang, X.; Xie, L. Numerical Study of the Force Characteristics and Flow Field Patterns of a Cylinder in the Internal Solitary Wave. *J. Mar. Sci. Eng.* **2024**, *12*, 906. https://doi.org/10.3390/jmse12060906.

3. Wang, X.; Xiao, S.; Wang, X.; Qi, D. Numerical Simulation and Analysis of Added Mass for the Underwater Variable Speed Motion of Small Objects. *J. Mar. Sci. Eng.* **2024**, *12*, 686. https://doi.org/10.3390/jmse12040686.
4. Chen, X.; Che, X.; Zhou, Y.; Tian, C.; Li, X. A Numerical Simulation Study and Effectiveness Evaluation on the Flow Field Effect of Trapezoidal Artifcial Reefs in Different Layouts. *J. Mar. Sci. Eng.* **2024**, *12*, 3. https://doi.org/10.3390/jmse12010003.
5. Gupta, S.; Sharma, A.; Agrawal, A.; Thompson, M; Hourgian, K. Role of Shape and Kinematics in the Hydrodynamics of a Fish-like Oscillating Hydrofoil. *J. Mar. Sci. Eng.* **2023**, *11*, 1923. https://doi.org/10.3390/jmse11101923.
6. Yang, C.; Zeng, K.; Chu, J.; Bu, S.; Zhu, Z. Computational Study on Influence Factors and Vortical Structures in Static Drift Tests. *J. Mar. Sci. Eng.* **2024**, *12*, 789. https://doi.org/10.3390/jmse12050789.
7. Díaz Ojeda, H.R.; Oyuela, S.; Sosa, R.; Otero, A.D.; Pérez Arribas, F. Fishing Vessel Bulbous Bow Hydrodynamics—A Numerical Reverse Design Approach. *J. Mar. Sci. Eng.* **2024**, *12*, 436. https://doi.org/10.3390/jmse12030436.
8. Zhang, Y.; Li, D. Parametric Design of a New Float-Type Wave Energy Generator and Numerical Simulation of Its Hydrodynamic Performance. *J. Mar. Sci. Eng.* **2023**, *11*, 2192. https://doi.org/10.3390/jmse11112192.
9. Kim, Y.J.; Canard, M.; Bouscasse, B.; Ducrozet, G.; Le Touzé, D.; Choi, Y.-M. High-Order Spectral Irregular Wave Generation Procedure in Experimental and Computational Fluid Dynamics Numerical Wave Tanks, with Application in a Physical Wave Tank and in Open-Source Field Operation and Manipulation. *J. Mar. Sci. Eng.* **2024**, *12*, 227. https://doi.org/10.3390/jmse12020227.

References

1. Zhang, M.; Hu, H.; Guo, B.; Liang, Q.; Zhang, F.; Chen, X.; Xie, Z.; Du, P. Predicting shear stress distribution on structural surfaces under internal solitary wave loading: A deep learning perspective. *Phys. Fluids* **2024**, *36*, 035153. [CrossRef]
2. Chen, M.; Zhang, N.; Sun, H.; Zhang, X. Large Eddy Simulation of the Flow around a Generic Submarine under Straight-Ahead and 10° Yaw Conditions. *J. Mar. Sci. Eng.* **2023**, *11*, 2286. [CrossRef]
3. Zhang, M.; Hu, H.; Ouahsine, A.; Du, P.; Huang, X.; Xie, L. Numerical Study of the Force Characteristics and Flow Field Patterns of a Cylinder in the Internal Solitary Wave. *J. Mar. Sci. Eng.* **2024**, *12*, 906. [CrossRef]
4. Wang, X.; Xiao, S.; Wang, X.; Qi, D. Numerical Simulation and Analysis of Added Mass for the Underwater Variable Speed Motion of Small Objects. *J. Mar. Sci. Eng.* **2024**, *12*, 686. [CrossRef]
5. Chen, X.; Che, X.; Zhou, Y.; Tian, C.; Li, X. A Numerical Simulation Study and Effectiveness Evaluation on the Flow Field Effect of Trapezoidal Artifcial Reefs in Different Layouts. *J. Mar. Sci. Eng.* **2024**, *12*, 3. [CrossRef]
6. Gupta, S.; Sharma, A.; Agrawal, A.; Thompson, M.; Hourgian, K. Role of Shape and Kinematics in the Hydrodynamics of a Fish-like Oscillating Hydrofoil. *J. Mar. Sci. Eng.* **2023**, *11*, 1923. [CrossRef]
7. Yang, C.; Zeng, K.; Chu, J.; Bu, S.; Zhu, Z. Computational Study on Influence Factors and Vortical Structures in Static Drift Tests. *J. Mar. Sci. Eng.* **2024**, *12*, 789. [CrossRef]
8. Ojeda, H.R.D.; Oyuela, S.; Sosa, R.; Otero, A.D.; Arribas, F.P. Fishing Vessel Bulbous Bow Hydrodynamics—A Numerical Reverse Design Approach. *J. Mar. Sci. Eng.* **2024**, *12*, 436. [CrossRef]
9. Zhang, Y.; Li, D. Parametric Design of a New Float-Type Wave Energy Generator and Numerical Simulation of Its Hydrodynamic Performance. *J. Mar. Sci. Eng.* **2023**, *11*, 2192. [CrossRef]
10. Kim, Y.J.; Canard, M.; Bouscasse, B.; Ducrozet, G.; Le Touzé, D.; Choi, Y.-M. High-Order Spectral Irregular Wave Generation Procedure in Experimental and Computational Fluid Dynamics Numerical Wave Tanks, with Application in a Physical Wave Tank and in Open-Source Field Operation and Manipulation. *J. Mar. Sci. Eng.* **2024**, *12*, 227. [CrossRef]

Journal of
Marine Science and Engineering

MDPI

Article

Large Eddy Simulation of the Flow around a Generic Submarine under Straight-Ahead and 10° Yaw Conditions

Mo Chen [1,2,*], Nan Zhang [1,2], Hailang Sun [1,2] and Xuan Zhang [1,2]

1 Department of Hydrodynamics, China Ship Scientific Research Centre, Wuxi 214082, China; zn_nan@sina.com (N.Z.); sunhailang@cssrc.com.cn (H.S.); zhangxuan0302@126.com (X.Z.)
2 National Key Laboratory of Hydrodynamics, China Ship Scientific Research Centre, Wuxi 214082, China
* Correspondence: chenmo@cssrc.com.cn

Abstract: Aiming towards a better understanding of the flow field around a fully appended Joubert BB2 submarine model, and in order to complement the experimental investigations of the wake of the hydroplanes and sail, large eddy simulation (LES) with the dynamic Smagorinsky model was conducted. Three sets of grids with a maximum grid number of up to 228 million were designed to perform the LES simulation for the Joubert BB2 under 10° yaw conditions, with a freestream Reynolds number based on the local freestream velocity and a hull length of $Re_L = 2.2 \times 10^7$. Comparisons of the wake of the cruciform appendage were made with experiments to verify the computational accuracy and to examine the influence of the spatial resolution. A satisfactory result was more representative of the experiments with the improvement in grid spatial resolution. The evolution characteristics of three co-rotating vortices originating from the cruciform appendage under the most refined grid arrangement are further described in detail under straight-ahead and 10° yaw conditions. The comparison results show that, in the core-flow region, the resultant velocity, vorticity magnitude, and TKE were stronger and the wake was more complicated under 10° yaw conditions. Tip vortex tracking under 10° yaw conditions exhibited significant three-dimensional characteristics as the wake developed downstream.

Keywords: submarine; LES; yaw; straight-ahead

1. Introduction

Submarines will generate vortex structures with various scales and forms that develop downstream during navigation. Generalized appendages, such as sails, hydroplanes, rudders, decks, and flood holes, create coherent vortex structures, like the horseshoe vortex, tip vortex, Karman vortex, discrete vortex, etc., leading to intense flow vibration. The development of these vortex structures also influences the maneuverability and deteriorates the wake of the platform. The vortex oscillation and its induced pressure fluctuation excite the hull, resulting in flow-induced noise, which seriously influences the stealth [1–3]. The investigation of the flow physics of a submarine can help us to better understand maneuverability limitations and flow-induced noise sources.

A certain amount of the available literature has described the flow around a submarine, within which experimentation plays an important role, especially in early studies. Fu utilized a PIV (particle image velocimetry) system to characterize the flow field around a sting-mounted captive ONR (Office of Naval Research) Body-1 submarine model in a steady turn [4]. Jimenez and Ashok utilized a hot-wire system and an SPIV (stereo particle image velocimetry) system to measure the flow field around an axisymmetric DARPA (Defense Advanced Research Projects Agency) SUBOFF (Submarine Technology Program Office) model, respectively, and flow field experimental databases with different Reynolds numbers, pitches, and yaw angles were obtained [5–10]. Similar research works on the SUBOFF model were conducted by Khan [11] and Shokrallah [12].

Another DSTO (Defense Science and Technology Organization) generic submarine model with a large deck, sail, and an X-form rudder arrangement [13,14], which can provide a useful representation of a conventional submarine, has been widely discussed in the available literature. The flow field measurements of the fully appended model both under straight-ahead conditions and during a 10° side-slip were conducted by the Defense Science and Technology Group, with data collected using pressure probes, PIV, and flow visualization of the wool-tuft streamers [15–20]. It provided a benchmark prototype with which to validate and improve numerical simulations of submarine wakes.

Since a new vertical sail and two horizontal hydroplanes, known as "Joubert BB2", were recently designed to improve stability and control characteristics [21,22], the wind-tunnel experiment of the model with 10∘ yaw was conducted again, with China-clay visualization and a high-resolution SPIV system [23–25]. The wake of this cruciform appendage can be made available to assist validation studies.

With the development of computational fluid dynamic (CFD) methods, the numerical simulation of the flow field around a fully appended submarine model can complement experimental investigations and create opportunities to advance the understanding of the flows, as the experiments only obtain limited flow field information. Careful verification and validation studies should be conducted with experimental data.

Four main CFD methods have been developed for predicting the flow field, and direct numerical simulation (DNS) is not practical in engineering predictions because of its enormous demand for computing resources. Reynolds-averaged Navier–Stokes (RANS), DES (detached eddy simulation), and large eddy simulation (LES) are gradually becoming dominant in computing the flow field around underwater vehicles. For the fully appended hull, however, the RANS seems to lack prediction accuracy somewhat, particularly for the second-order statistical moments and the local flow topology, and the DES and its variants lack good representation for the development of wall turbulence. The development of LES can provide very useful insights into the complicated transient nature of the flow, including unsteady wake flow, flow-induced noise, and vibrations [16,26–28].

The LES method was first proposed by Smagorinsky [29], and the applicability of predicting the flow field around submarines has been verified by many researchers. Shi evaluated the predictive capabilities of LES on the flows around a bare hull model by comparing it with experiments, and a good agreement was observed for the pressure and the skin-friction coefficients on the body and the streamwise velocity in the wake [30]. Zhou performed LES computations of the SUBOFF submarine model to investigate the wake characteristics of an underwater vehicle with and without a propeller [31]. Rosa performed LES simulations of the same model, with a comparison between towed and self-propelled cases, for boundary layer development over both the hull and wake flow [32]. Then, LES simulations that focus on the effects of the Reynolds number on the structure of the boundary layer in the stern area, as well as the near wake, were conducted. Results showed that the influence of the junction vortices on the first- and second-order statistics of most of the stern boundary layer is weakly dependent on the Reynolds number [33]. The computational ability of LES for the surrounding flow of submarines under maneuvering conditions has been verified.

Zhang studied the numerical prediction approach for the hydrodynamic force and noise of a SUBOFF submarine appended by an AU5-65 propeller using LES and Powell vortex theory [34], and expanded LES to the field of submarine noise prediction and more complex maneuvering conditions, such as submarine propeller interaction and crashback [35–37]. The results further validated the numerical prediction ability of LES. Kroll performed an LES of the flow over the SUBOFF submarine appended by the DTMB-4381 propeller in forward mode, and crashback, mean flow fields, and propeller load statistics showed good agreement with experiments and previous simulations [38].

In this paper, we focus on the evolution characteristics of three co-rotating vortices originating from the sails and hydroplanes and the development trend of the far wake flow, and provide a comparison of them under straight-ahead and 10∘ yaw conditions.

Three sets of grids with a maximum grid number of up to 228 million were designed to perform LES simulation for the fully appended Joubert BB2 submarine model for 10° yaw. A comparison of the wake of the cruciform appendage was made with Lee's work [24], to verify the grid convergence and computational accuracy of LES in terms of predicting the flow field around the submarine. Simulations were conducted at a freestream Reynolds number based on the local freestream velocity and the hull length of $\mathrm{Re}_L = 2.2 \times 10^7$, which is higher than those in previous work and more representative of a full-scale submarine. Then, the characteristics of the flow around the submarine are described in detail under straight-ahead and 10° yaw conditions. These computations further elucidate the structure of the flow around the fully appended Joubert BB2 submarine model, and provide an effective complement to the experimental investigations.

The remainder of the paper is organized as follows: Section 1 provides an introduction to the Joubert BB2 submarine model and some information on the SPIV experiments in Lee's work [24]. In Section 3, the numerical methodology is explained, including the LES method, computational domain, boundary condition, and the grid arrangements. Section 4 presents the results and discussion, including the validation of the numerical approach and analysis of the evolution of the flow. Section 5 presents the conclusions.

2. The Joubert BB2 Submarine Model

Figure 1 shows the Joubert BB2 submarine model with a length of $L = 3.826$ m and a model scale of 1:18.348. The hull was designed as an axisymmetric body of revolution with a length-to-diameter ratio $L/2r_m = 7.3$, and the bow profile is derived from an NACA0018 curve [39], splined to allow the rise in pressure to occur further aft [20]. The cross-section of the sail is based on an NACA0022 with a height of $0.080L$ and a chord length of $0.157L$. Two horizontal NACA0015 hydroplanes are assembled on the sail, and the combined span and the root chord were designed to be $0.117L$ and $0.033L$, respectively. The stern control plane consists of four rudders in an X configuration, and the tailing edge is $0.075L$ away from the end of the hull. Further detailed descriptions can be found in Joubert [13,14], Bettle [21], and Overpelt [22].

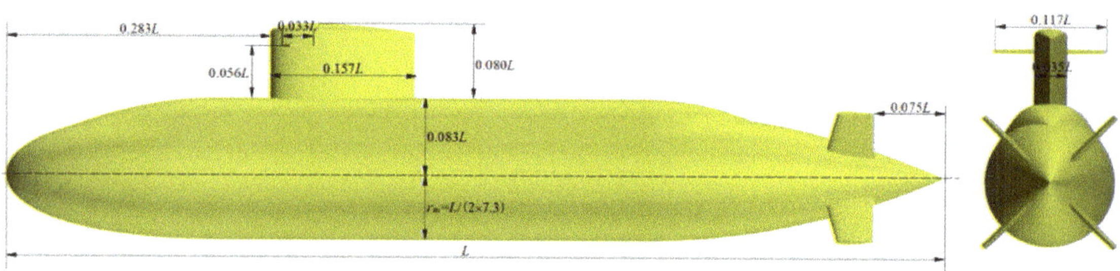

Figure 1. The Joubert BB2 geometry.

In Lee's work [24], the experimental results are discussed in a wind axis coordinate system, with the x-axis direction defined as the direction of freestream, as shown in Figure 2, which is also adopted in the paper for a more direct comparison. The velocity field is obtained with an SPIV measurement system at three selected model-length locations, $x/L = 0.511$, $x/L = 0.650$, and $x/L = 0.815$, and the normal direction of these experimental planes is parallel to the free-stream direction. For further details on SPIV setup and measurements, see Lee [24]. It should be noted that the model-length Reynolds numbers were chosen to be $\mathrm{Re}_L = 4 \times 10^6$ and $\mathrm{Re}_L = 8 \times 10^6$ in those experiments, which are lower than those of this paper. The conclusion has been drawn that similar experimental results for the velocity field, turbulence kinetic energy (TKE), cross-stream Reynolds stress, etc., are shown with different Re_L values. According to previous research results on the wake field of submarine models, when the Re_L increases from 1×10^7 to 3.5×10^7, the dimensionless mean velocity changes between 3% and 10%; the change in Re_L only influences the amplitude of the fluid variables, not the peak and valley characteristics, and thus the reciprocal validation of computations and experiments is still reliable.

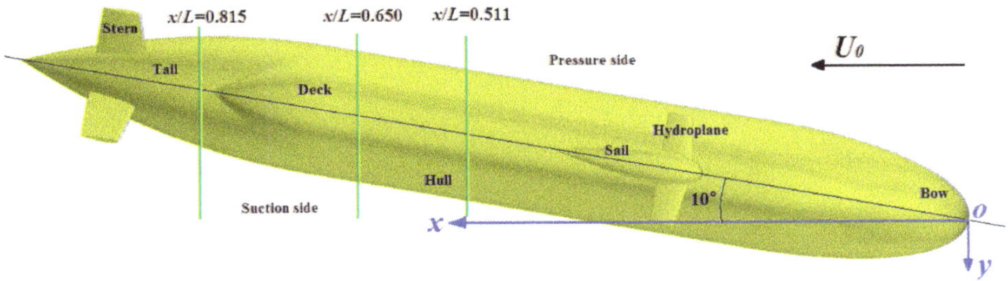

Figure 2. The schematic diagram of SPIV measurement and the wind axis coordinate system.

3. Numerical Methodology

3.1. Large Eddy Simulation

The commercial CFD code StarCCM+ by Siemens PLM, based on the finite volume method, was utilized to conduct the LES simulations presented in the paper. The filtered Navier–Stokes equations are as follows:

$$\frac{\partial \rho}{\partial t} + \nabla \cdot (\rho \widetilde{\mathbf{v}}) = 0 \tag{1}$$

$$\frac{\partial}{\partial t}(\rho \widetilde{\mathbf{v}}) + \nabla \cdot (\rho \widetilde{\mathbf{v}} \otimes \widetilde{\mathbf{v}}) = -\nabla \cdot \widetilde{p}\mathbf{I} + \nabla \cdot (\widetilde{\mathbf{T}} + \mathbf{T}_{SGS}) + \mathbf{f}_b. \tag{2}$$

Here, ρ is the density, $\widetilde{\mathbf{v}}$ is the filtered velocity, $\widetilde{p}$ is the filtered pressure, $\mathbf{I}$ is the identity tensor, and $\mathbf{f}_b$ is the resultant of the body forces. $\widetilde{\mathbf{T}}$ is the filtered stress tensor due to molecular viscosity, and $\mathbf{T}_{SGS} = 2\mu_t\mathbf{S} - \frac{2}{3}(\mu_t\nabla \cdot \widetilde{\mathbf{v}})\mathbf{I}$ is the sub-grid-scale stress, where $\mathbf{S}$ is the strain rate tensor and is computed from the resolved velocity field. The sub-grid scale turbulent viscosity μ_t must be described by a sub-grid scale model that accounts for the effects of small eddies on the resolved flow, and the dynamic Smagorinsky model proposed by Germano [40] and modified by Lilly [41] is used here. This approach has shown good performance for a variety of complex marine flows of fully appended submarines [31,37,42–44].

As for the discretization of the governing equations, time integration was performed using a semi-implicit, second-order, two-point backward differencing scheme. Convective fluxes were reconstructed using multi-dimensional, cell-limited linear interpolation, whereas diffusive fluxes were reconstructed using a combination of central-difference approximations and gradient face interpolation to minimize the non-orthogonality error. The coupling between the velocity and pressure was achieved by means of the classic PISO (Pressure Implicit with Splitting of Operators) algorithm, and the algebraic multigrid method was employed to accelerate the solution convergence.

3.2. Computational Domain and Boundary Condition

Figure 3 shows the cylindrical computational domain; the domain has a length of $4L$ and a radius of L, and extends from L upstream of the front of the hull to $2L$ downstream of the stern of the hull, to model a fully developed wake. Free-stream boundary conditions are imposed at the inflow and radial boundaries, and a pressure-outlet boundary condition is imposed at the outflow boundary. The no-slip wall treatment is used for shear stress specification.

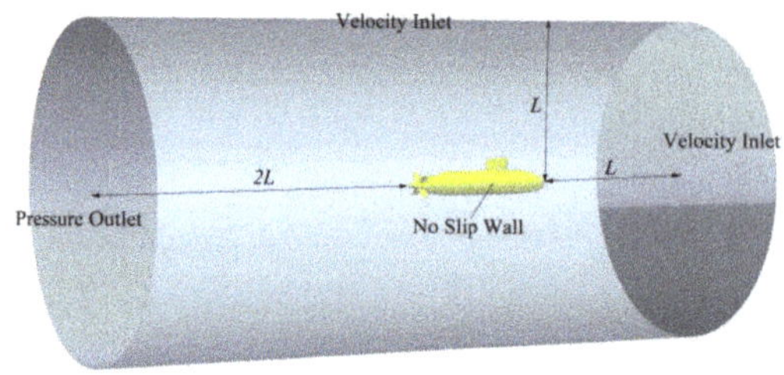

Figure 3. The computational domain.

3.3. Computation Mesh

In this subsection, three sets of grids are designed for the convergence study. The simulation domain is discretized using the unstructured hex-dominant grid (trimmer mesher, Octree grid), and a prism layer mesher is used for the generation of a boundary layer mesh with the dimensionless wall distance $y^+ \approx 1$ for all grid sets. The refinement ratio of grids in three directions is $\sqrt{2}$, as recommended by the International Towing Tank Conference [45]. The corresponding grid numbers are 3.30×10^7, 8.62×10^7, and 2.28×10^8, respectively, and the three grid codes are *G1* to *G3*, respectively. All numerical simulations are carried out by parallel processing in CSSRC (China Ship Scientific Research Centre) with 50 nodes (2400 processors). The time needed to finish a simulation with 228 million cells is about 30 days.

Figure 4 shows a diagram of the basic grid arrangement and the intuitive comparison of grid spatial resolution around the hydroplanes and sail under different grid sets. For capturing the flow field more accurately, a local volumetric control block was established around the submarine to control the surrounding grid size. Care was taken to resolve the initial vortex formation and roll-up of the free shear layer, and to avoid the rapid dissipation of the vortex structure in the wake of the sail, hydroplanes, and X-rudder; local volumetric control blocks were also established around these appendages for more refined volume mesh control. There is an angle of approximately 5.5 degrees between the blocks and the longitudinal axis of the hull in calculations under 10° yaw conditions.

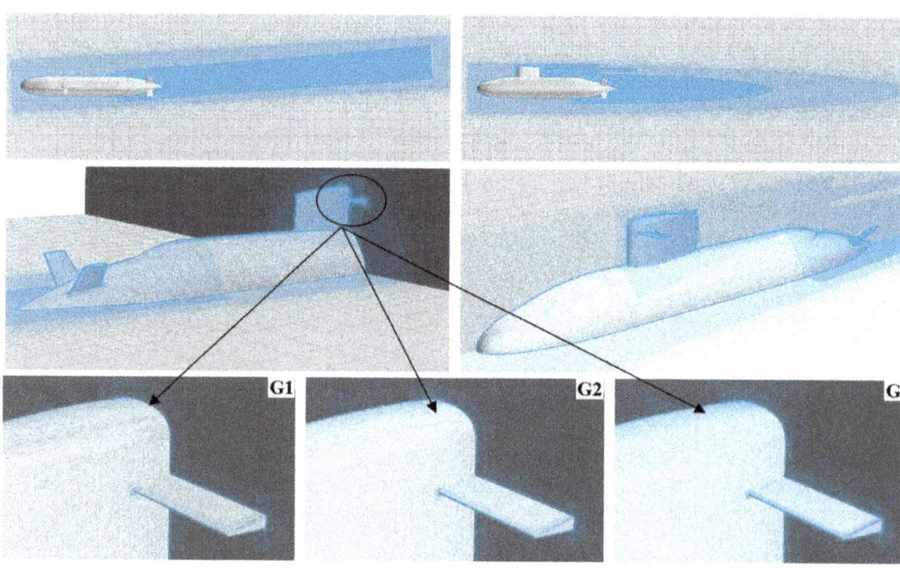

Figure 4. Grid arrangements for computation.

4. Results and Discussion

In this section, we initially present results for the qualitative and quantitative comparison between the numerical results and experimental results for 10° yaw with three sets of grids, and then under the most refined grid arrangement. The detailed numerical investigation of the flow around the submarine was conducted under straight-ahead and 10° yaw conditions.

4.1. Validation of the Numerical Approach

Figures 5–7 provide a direct comparison of the wake of the cruciform appendage, including the mean resultant velocity $<\overline{U}_{xyz}>/U_{\infty}$, the vertical component of velocity $<\overline{U}_{z}>/U_{\infty}$, and the cross-stream Reynolds stress $<u_y u_z>/U_{\infty}^2$, where U_{∞} denotes the freestream velocity. The experimental results with $\mathrm{Re}_L = 4 \times 10^6$ and $\mathrm{Re}_L = 8 \times 10^6$ are denoted by Exp. 1 and Exp. 2, respectively. It can be observed that the sail-tip vortex formed by the rolling up of the wake can be presented by numerical calculations under all sets of grids. The region of the core flow, defined as a region of vortex flow from its center to its radial location of maximum swirl, is quite similar to the experimental results, while the range of the core flow and low-velocity region presents minor differences under different sets of grids. For the vertical component of velocity, in particular, as the number of grids increases, the distribution of high/low-velocity regions becomes more concentrated and obvious, and the numerical results become closer to the experiments. In addition, the numerical dissipation decreases as the grid number increases, and the capture of the cross-stream Reynolds stress seems to be more refined.

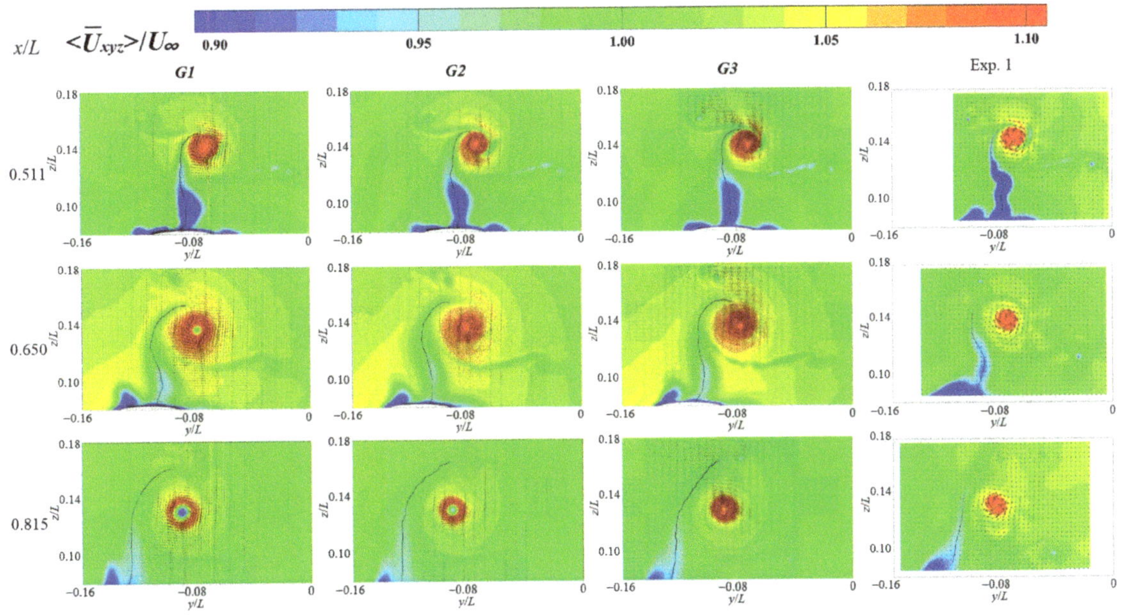

Figure 5. Qualitative comparison of the mean resultant velocity.

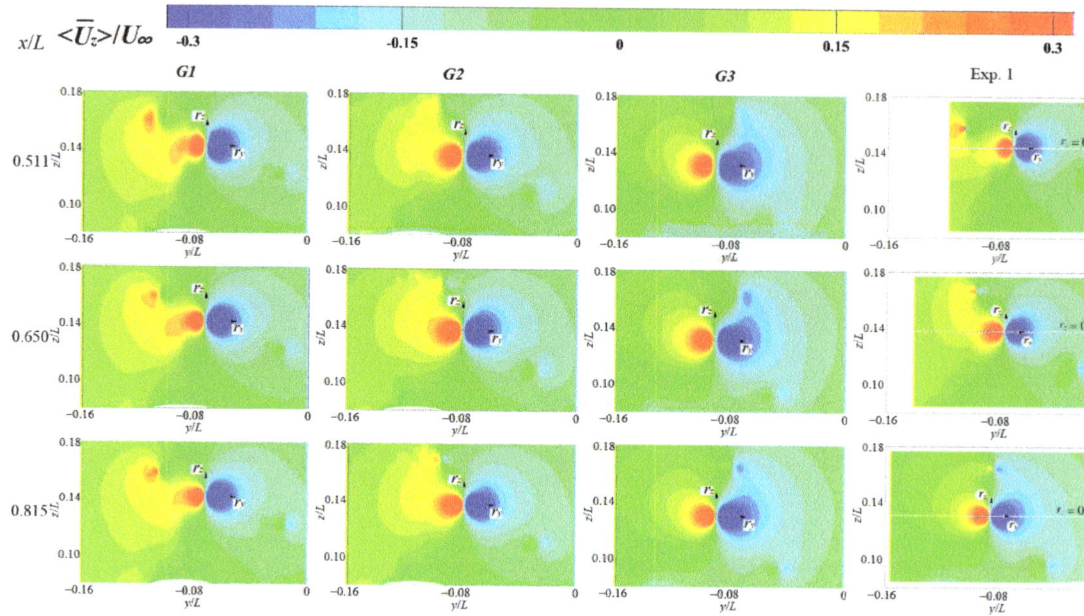

Figure 6. Qualitative comparison of the vertical component of velocity.

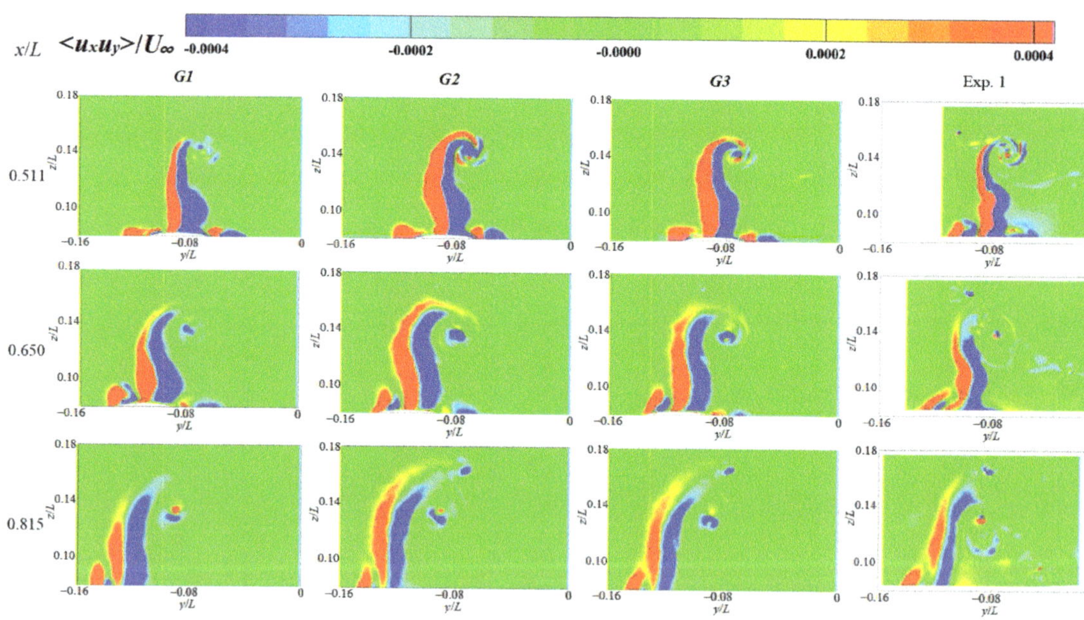

Figure 7. Qualitative comparison of t velocity cross-stream Reynolds stress.

Figure 8 shows a comparison of the mean resultant velocity figures $< \overline{U}_{xyz} > /U_\infty$, as functions of radial distance r_y/L from the vortex center, in the horizontal profiles through the sail-tip vortex. The mean resultant velocity under different grid sets exhibits a considerable difference, especially for the region of the core flow, which presents a corresponding increase with the grid number's continuous growth. The core-flow velocity obtained from numerical calculations exhibits a significant difference at different axial distances from the bow, but is quite close for the experiments. The core flow obtained in the first set of grids (*G1*) has a mean velocity $< \overline{U}_{xyz} > /U_\infty \approx 1.12$ and $< \overline{U}_{xyz} > /U_\infty \approx 0.91$ at $x/L = 0.511$ and $x/L = 0.815$, respectively, with an axial descent rate of 18.8%. It corresponds to $< \overline{U}_{xyz} > /U_\infty \approx 1.26$ and $< \overline{U}_{xyz} > /U_\infty \approx 1.11$, respectively, in *G3*, with an axial descent rate of 11.9%. This indicates that numerical attenuation inevitably exists in the LES simulations with the dynamic Smagorinsky model of the core flow, which is not advantageous compared to experiments, and can be partly eliminated by the improvement in the spatial resolution of the grid.

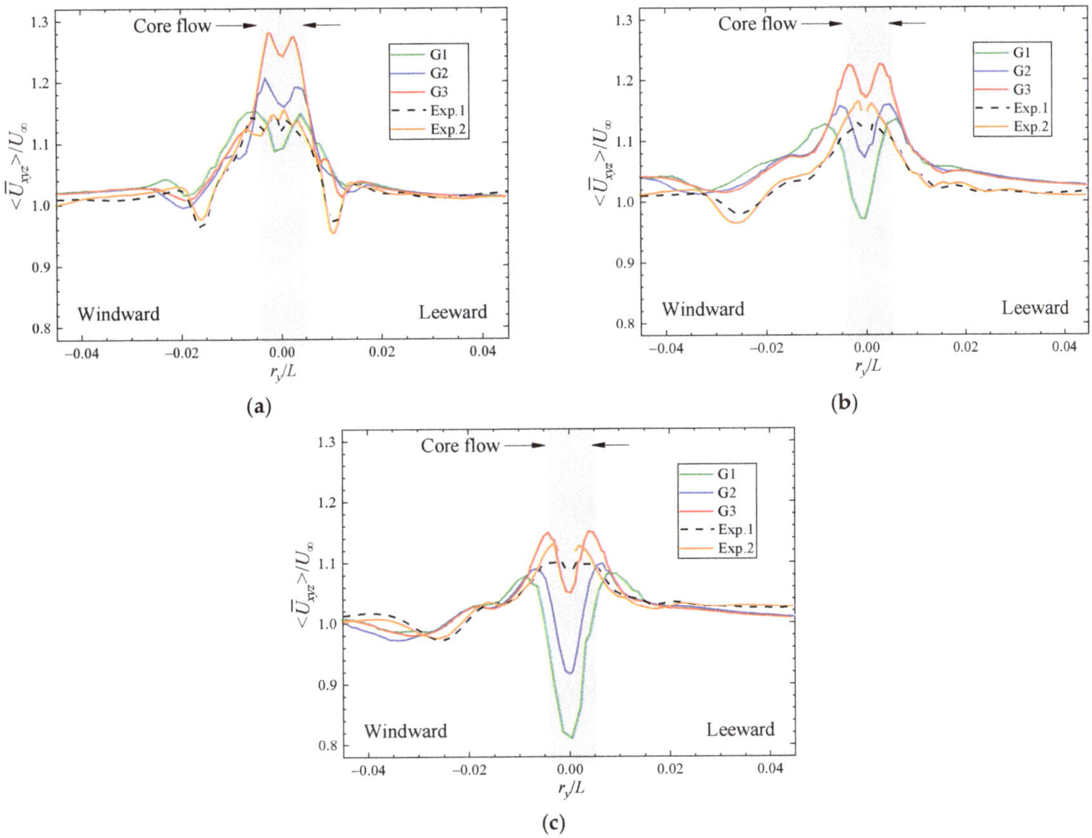

Figure 8. (**a**) $x/L = 0.511$; (**b**) $x/L = 0.650$; (**c**) $x/L = 0.815$. Comparison of the mean resultant velocity under different sets of grids.

Figure 9 shows a comparison of the vertical component of velocity $< \overline{U}_z > /U_\infty$. It can be seen that the numerical results become gradually closer to the experimental values as the grid number increases. For the prediction of the peak and valley values, especially, the results obtained in *G3* are quite close to those of Exp. 2, with a relatively higher Reynolds number. Overall, the relative errors of the peak and valley values in *G3* are 10.6% and 4.3%, respectively, which becomes 13.2% and 14.0% in *G2*, and 21.6% and 21.1% in *G1*. It can be concluded that, with the refinement of the grids, the extreme values can be more accurately captured in numerical simulations due to the improvement in spatial resolution.

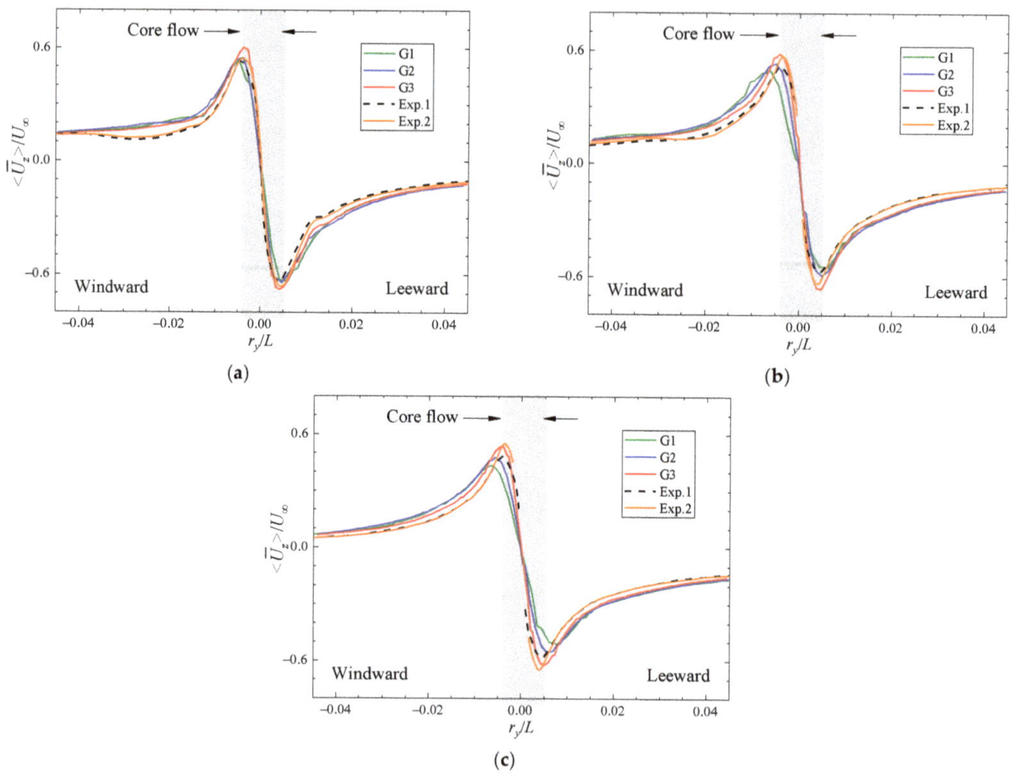

Figure 9. (**a**) x/L = 0.511; (**b**) x/L = 0.650; (**c**) x/L = 0.815. Comparison of the vertical component of velocity under different sets of grids.

Figure 10 provides a plot of the comparison of the turbulence kinetic energy k/U_∞^2 for the sail-tip vortex. The distribution of TKE along the radial distance r_y/L for all grid sets is consistent with the experimental results, and the simulations of *G3* are more representative of the experiments, with the relative error of the peak value in the region of core flow being relatively smaller.

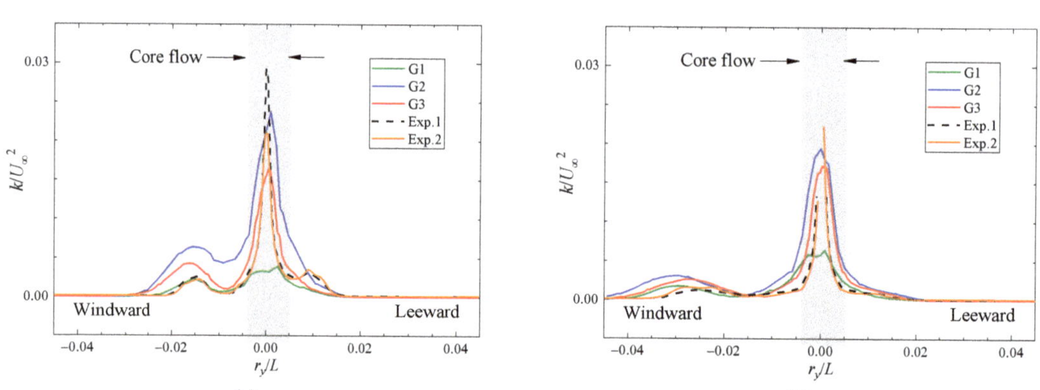

Figure 10. *Cont.*

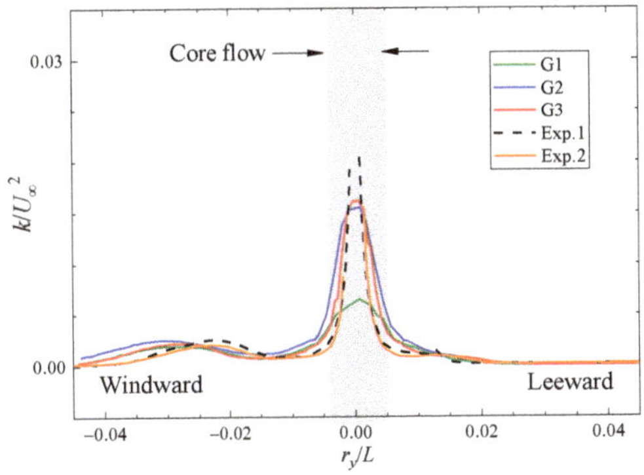

(**c**)

Figure 10. (**a**) x/L = 0.511; (**b**) x/L = 0.650; (**c**) x/L = 0.815. Comparison of the turbulence kinetic energy under different sets of grids.

Figure 11 summarizes the comparison of the centerline of the sail wake, which is defined by tracing the boundary $<u_x u_y> = 0$ between the positive and the negative Reynolds stresses. Overall, the sail-wake centerline shifts very slightly windward by no more than 0.01L, and by refining the grids from *G1* to *G3*, it shifts leeward and closer to the experiments. The centers of the vortices on the upper hull are listed in Table 1, and their relative errors to Exp. 2 are listed in Table 2. In the near sail-wake region (x/L = 0.511), the numerical simulations for all grid sets can define the centers of the vortices well, but as they evolve downstream, the advantages of refining the grids gradually become reflected. For the case of *G3*, the maximum horizontal and vertical relative errors of the vertex centers at x/L = 0.815 are 7.2% and 1.5%, which is quite satisfactory considering the actual engineering prediction needs.

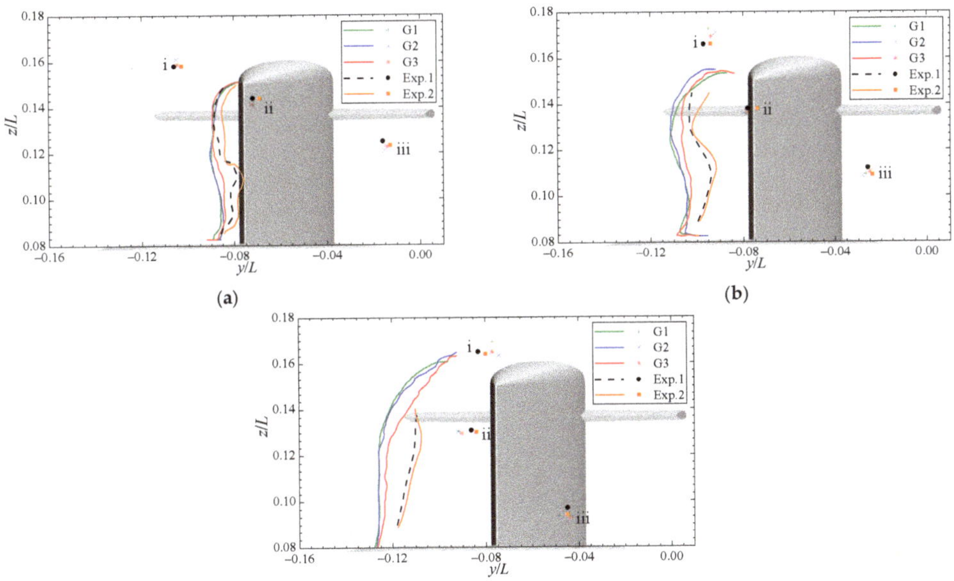

(**c**)

Figure 11. (**a**) x/L = 0.511; (**b**) x/L = 0.650; (**c**) x/L = 0.815. Comparison of the centerline of the sail wake under different sets of grids.

Table 1. Summary of the centers of the vortices on the upper hull.

Measurement Plane	Grid Scheme	Sail Tip		Hydroplanes			
				Windward		Leeward	
		y/L	z/L	y/L	z/L	y/L	z/L
$x/L = 0.511$	G1	−0.072	0.141	−0.108	0.161	−0.013	0.123
	G2	−0.073	0.141	−0.105	0.161	−0.015	0.121
	G3	−0.072	0.141	−0.105	0.159	−0.014	0.123
$x/L = 0.650$	G1	−0.079	0.136	−0.095	0.173	−0.027	0.109
	G2	−0.080	0.136	−0.092	0.171	−0.028	0.109
	G3	−0.079	0.136	−0.094	0.169	−0.025	0.110
$x/L = 0.815$	G1	−0.092	0.130	−0.077	0.169	−0.048	0.094
	G2	−0.092	0.130	−0.074	0.163	−0.048	0.092
	G3	−0.090	0.130	−0.077	0.165	−0.044	0.093

Table 2. Summary of the relative errors to Exp. 2 of the centers of the vortices.

Measurement Plane	Grid Scheme	Sail Tip		Hydroplanes			
				Windward		Leeward	
		y/L (%)	z/L (%)	y/L (%)	z/L (%)	y/L (%)	z/L (%)
$x/L = 0.511$	G1	4.3	−1.9	5.3	1.6	2.7	0.3
	G2	5.3	−1.8	2.2	1.9	9.2	−1.6
	G3	5.0	−1.9	1.8	0.5	6.7	−0.3
$x/L = 0.650$	G1	6.5	−1.3	1.0	4.2	12.0	0.4
	G2	7.7	−1.4	−2.3	3.0	9.6	−0.4
	G3	6.1	−1.1	−0.1	2.0	4.8	1.2
$x/L = 0.815$	G1	9.0	0.4	−3.7	3.3	7.5	−0.5
	G2	9.2	0.3	−7.0	−0.4	6.2	−1.8
	G3	7.2	−0.3	−3.5	0.6	−2.5	−1.5

4.2. Analysis of the Evolution of the Flow

An overall conclusion can be drawn that the results of the LES simulations are more representative of the experiments as the grid spatial resolution is improved, and then under the most refined grid arrangement, the evolution of the flow under straight-ahead and 10° yaw conditions is further analyzed.

Figure 12 presents an overall view of the flow past Joubert BB2 under straight-ahead conditions in terms of the second invariant of the velocity gradient Q, colored by the mean resultant velocity $< \overline{U}_{xyz} > / U_{\infty}$. At the immediate front of the junctions of the sail root and the deck, the flow rolls up into a horseshoe vortex system surrounding the sail, the legs of which develop downstream following the deck. Because of the adverse pressure gradient at the trailing edge of the sail, the flow gradually develops into turbulence, and the side vortices are formed and interact with the horseshoe vortex at the root. The side vortices over the sail cap are transported along the sail edge and then merge with the sail-tip vortices and dissipate rapidly. The flow over the outer edge of the hydroplanes induces a pair of hydroplane-tip vortices with opposite circulation, which develop downstream independently, then dissipates and disappears at a distance of approximately one submarine length from the stern. Further, horseshoe vortex, tip vortex, and wake vortex systems can be observed around the X-rudders. Between the two upper rudders, they interact with the vortices sweeping down and sideways over the end of the deck, complicating the flow into the propeller disk, which is unstable and the main source of propeller hydrodynamic noise. Downstream, far away from the hull, all the tip vortices dissipate and really only the wake vortices, after complicated interaction, dynamically evolve, with the energy gradually weakening.

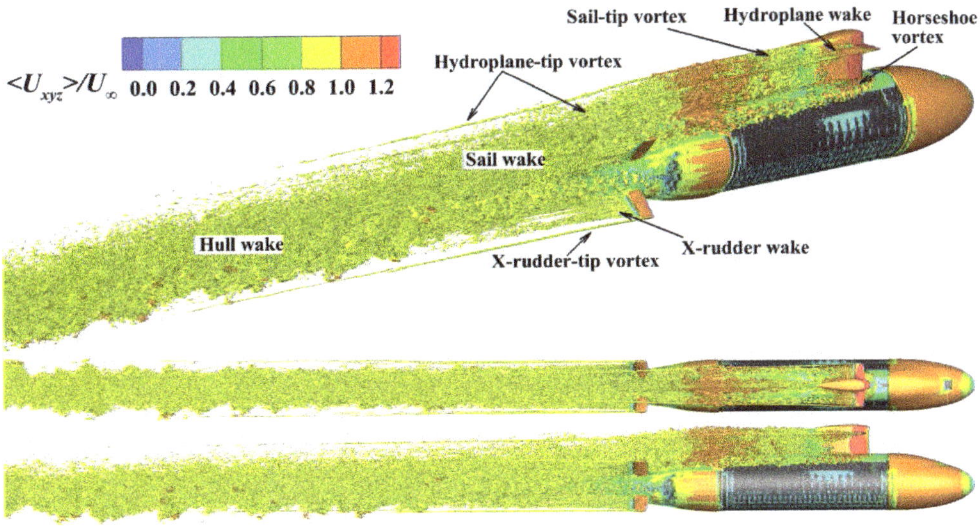

Figure 12. Development of vortex systems past Joubert BB2 under straight-ahead conditions.

Similarly, Figure 13 presents various vortex systems past Joubert BB2 under 10° yaw conditions, from perspectives of oblique, top, and side views. It can be clearly seen that the vortex systems are more complicated. The side vortices on the leeward side of the sail occur further forward, and the sail-tip vortices are relatively strong enough to develop far downstream. The same is true for the hydroplane-tip vortices, but they are not clearly observed downstream because of strong interactions with the sail wake. Figure 14 shows the development of the trajectories of the cores of tip vortices originating from the cruciform appendage, including the port and starboard hydroplanes and the vertical sail, under 10° yaw conditions. Referring to Akkermans's work [46], tip vortex tracking under 10° yaw conditions exhibits significant three-dimensional characteristics as the wake develops downstream. The clockwise rotating sail-tip vortices maintain an axial angle of approximately 8 degrees with the hull and develop downstream and leeward, and are almost stable vertically after experiencing a brief downwash immediately behind the sail. The position of the port hydroplane-tip vortices fluctuates widely, the horizontal and vertical coordinates of which reach their maximum values approximately at $x/L = 1.1$ and $x/L = 0.7$, respectively, and then experience a sharp drop. Overall, the port hydroplane-tip vortices develop and revolve around the sail-tip vortices. The development of the starboard hydroplane-tip vortices is relatively stable, and their core keeps moving towards the leeward side, with the vertical position gradually rising away from the hull after passing through a valley at approximately $x/L = 1.1$, due to the repulsive interaction of the hull wake.

Under 10° yaw conditions, another obvious feature that distinguishes the straight-ahead conditions is the flow separation on the leeward side of the middle hull. The upper and lower vortex system can be clearly seen, and the former eventually interacts with the horseshoe vortex system originating from the sail, while the latter merges into the wake between the upper and lower rudders. The wake of the submarine becomes quite complicated, and the flow behind the stern is dominated by the mixing of various component vortex systems, including the tilted horseshoe vortex system, the upper and lower hull vortices, the tip vortices, and the wake of the sail, hydroplanes, X-rudders, and hull.

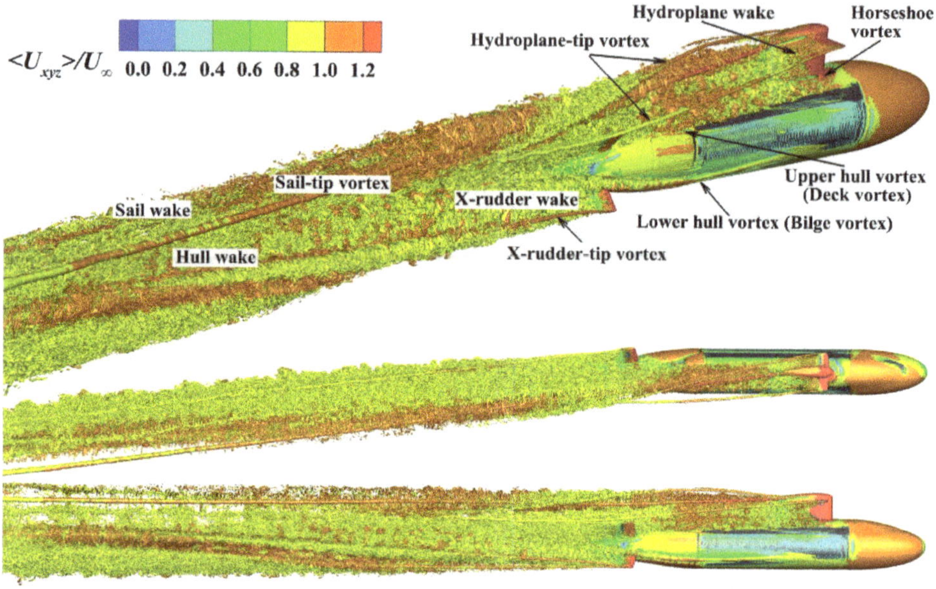

Figure 13. Development of vortex systems past Joubert BB2 under 10° yaw conditions.

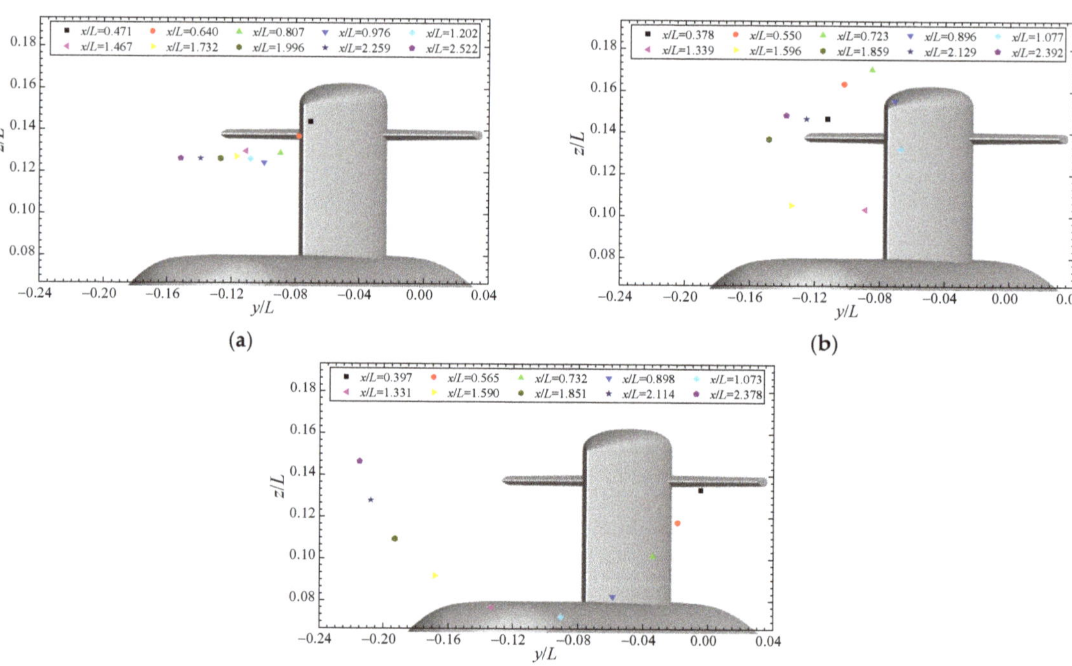

Figure 14. (**a**) The sail-tip vortices; (**b**) The port hydroplane-tip vortices; (**c**) The starboard hydroplane-tip vortices. Trajectories of the cores of tip vortices under 10° yaw conditions.

Figure 15 shows the evolution of the mean vorticity magnitude $< \overline{\omega}_{xyz} > r_m / U_\infty$ along the submarine axial, under straight-ahead and 10° yaw conditions, where $r_m = L/(2 \times 7.3)$. The momentum and energy are transported by the development of vortex systems, where the vortex systems generated by the sail, hydroplanes, deck, hull, and X-rudder pass present an increase in vorticity, especially for the tip vortices, horseshoe vortices, wake vortices, and generated hull side vortices, possibly. It can be seen that the instabilities of the horseshoe-vortex system and its interaction with the hull boundary layer cause the legs to break up and develop connected vortex loops, which results in the transport of momentum across the hull and influences the distribution of the vorticity, as mentioned by Fureby [20]. As the vortex structure gradually dissipates downstream, the vorticity gradually decreases, with the sail-tip vortices and hydroplane-tip vortices being particularly prominent. For the case of straight-ahead conditions, the obvious vorticity near the sail induced by the tip vortices of the cruciform appendage quickly decays, quite significantly far away from the stern for the case with 10° yaw. Increased vorticity is also found around and behind the hull under 10° yaw conditions because of considerable interaction between the flow and the hull. Therefore, the flow-induced noise of submarines under maneuvering conditions has always been a research highlight in the international hydrodynamics field. As the vortex systems develop from a concentrated distribution in the near wake region to a dispersed mode in the far field, in both cases, the vorticity gradually weakens while the coverage range increases, due to energy conservation.

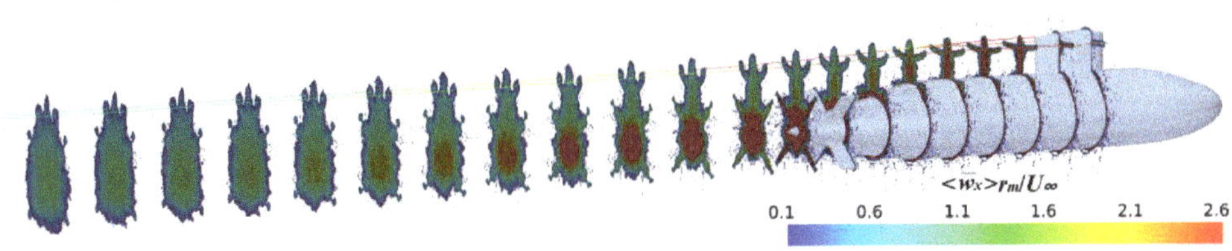

(**a**)

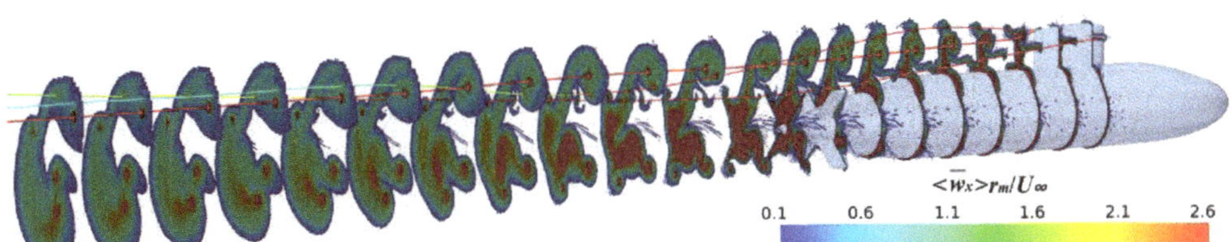

(**b**)

Figure 15. (**a**) Straight-ahead; (**b**) 10° yaw. Evolution of the mean vorticity magnitude.

Figure 16 shows the evolution of the turbulence kinetic energy k/U_∞^2 along the submarine axial, under straight-ahead and 10° yaw conditions. The evolution of TKE is quite similar to that of the mean vorticity magnitude, where the vorticity is concentrated, the momentum is intense, and the TKE is significant. For the case with 10° yaw, it can be obviously seen that the TKE induced by the sail-tip vortices is stronger than that induced by the hydroplane-tip, with the same pattern as the vorticity followed. In addition, the TKE in the wake is strongly influenced by the X-rudder under both conditions, which exacerbates the velocity fluctuations of the flow and induces additional propeller noise; thus, the hydrodynamic design of the X-rudder is also a research highlight.

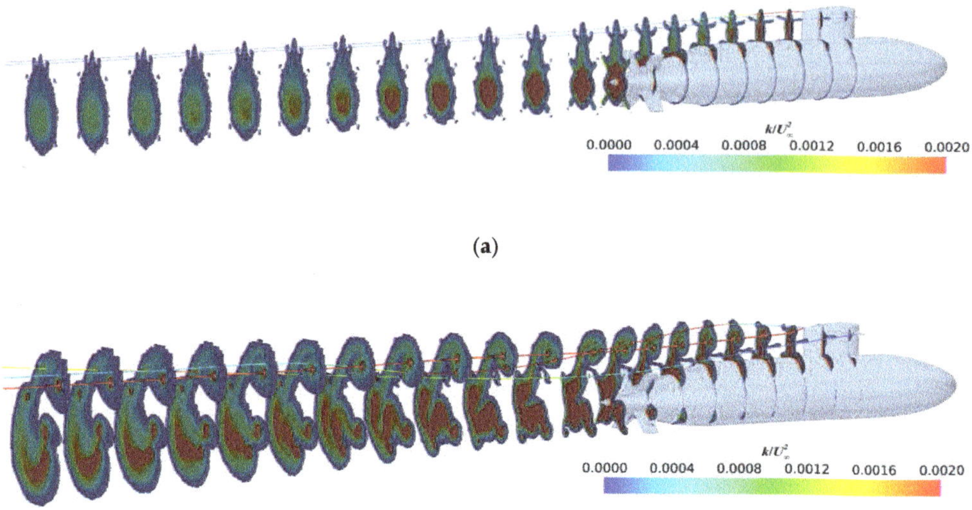

(a)

(b)

Figure 16. (**a**) Straight-ahead; (**b**) 10° yaw. Evolution of the turbulence kinetic energy.

Figures 17–19 show a comparison of the mean resultant velocity $<\overline{U}_{xyz}>/U_\infty$, the mean vorticity magnitude $<\overline{\omega}_{xyz}>r_m/U_\infty$, and the turbulence kinetic energy k/U_∞^2, respectively, for the model-length locations x/L = 0.484, under straight-ahead and 10° yaw conditions. A pair of sail-tip vortices with opposite circulation can be found in straight-ahead conditions, and in the core-flow region, the mean resultant velocity, the mean vorticity magnitude, and the turbulence kinetic energy were considerably smaller than those under 10° yaw conditions.

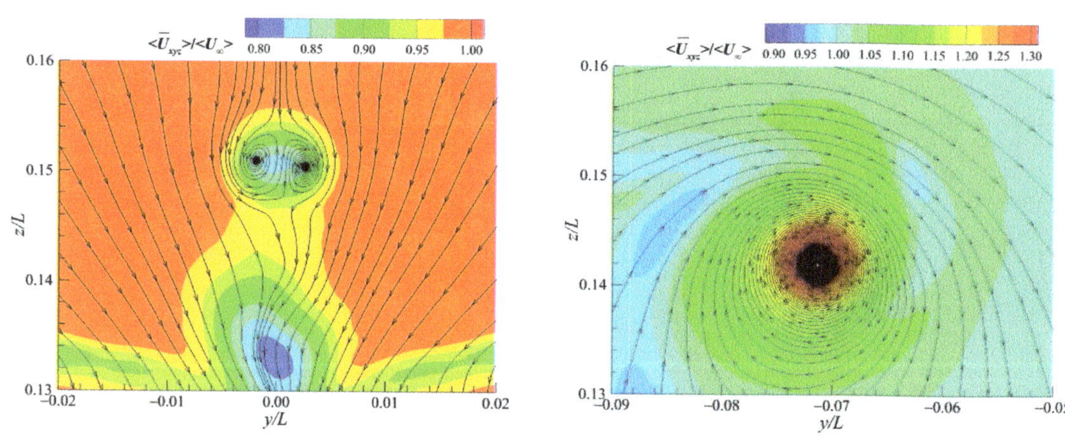

(a)

(b)

Figure 17. (**a**) Straight-ahead; (**b**) 10° yaw. Comparison of the mean resultant velocity under straight-ahead and 10° yaw conditions.

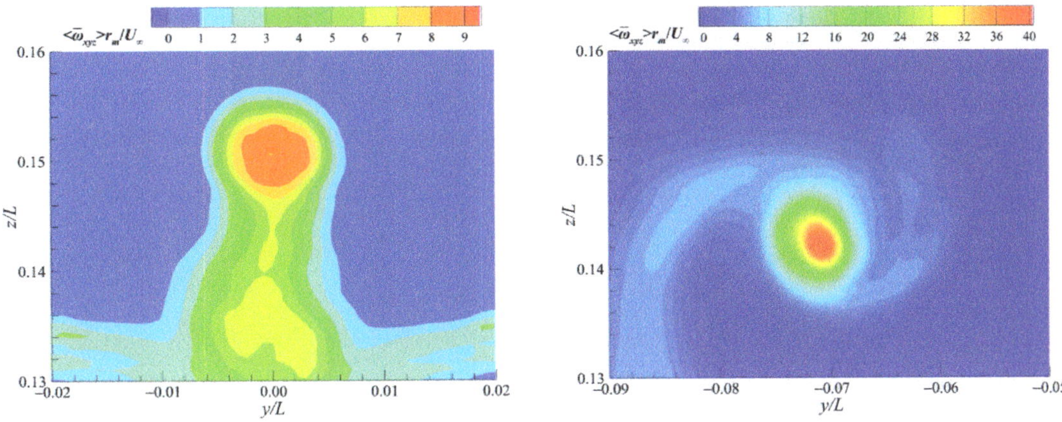

(a) (b)

Figure 18. (**a**) Straight-ahead; (**b**) 10° yaw. Comparison of the mean vorticity magnitude under straight-ahead and 10° yaw conditions.

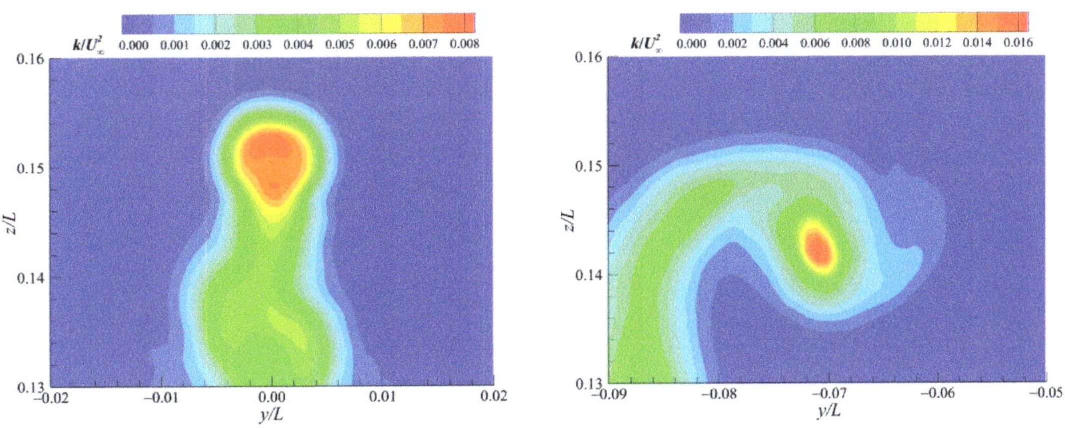

(a) (b)

Figure 19. (**a**) Straight-ahead; (**b**) 10° yaw. Comparison of the turbulence kinetic energy under straight-ahead and 10° yaw conditions.

In view of the fact that the momentum and energy transported by the sail-tip vortices under 10° yaw conditions were predominant, the flow characteristics of the sail-tip vortices are further studied below. Figure 20 provides a comparison of the velocity for the sail-tip vortex under different longitudinal locations, including mean resultant velocity $<\overline{U}_{xyz}>/U_\infty$ and the three-dimensional component of velocity $<\overline{U}_x>/U_\infty$, $<\overline{U}_y>/U_\infty$, and $<\overline{U}_z>/U_\infty$. It can be seen intuitively that, as the wake of the sail-tip vortices develops downstream, the mean resultant velocity, streamwise velocity, and horizontal velocity show a gradually decreasing trend, and the fluctuation of the vertical velocity between peak and valley values gradually weakens. In the near-wake region, where $x/L \leq 0.807$, the core flow exhibits a high-velocity characteristic, while in the far-wake region, the velocity is smaller than the freestream velocity.

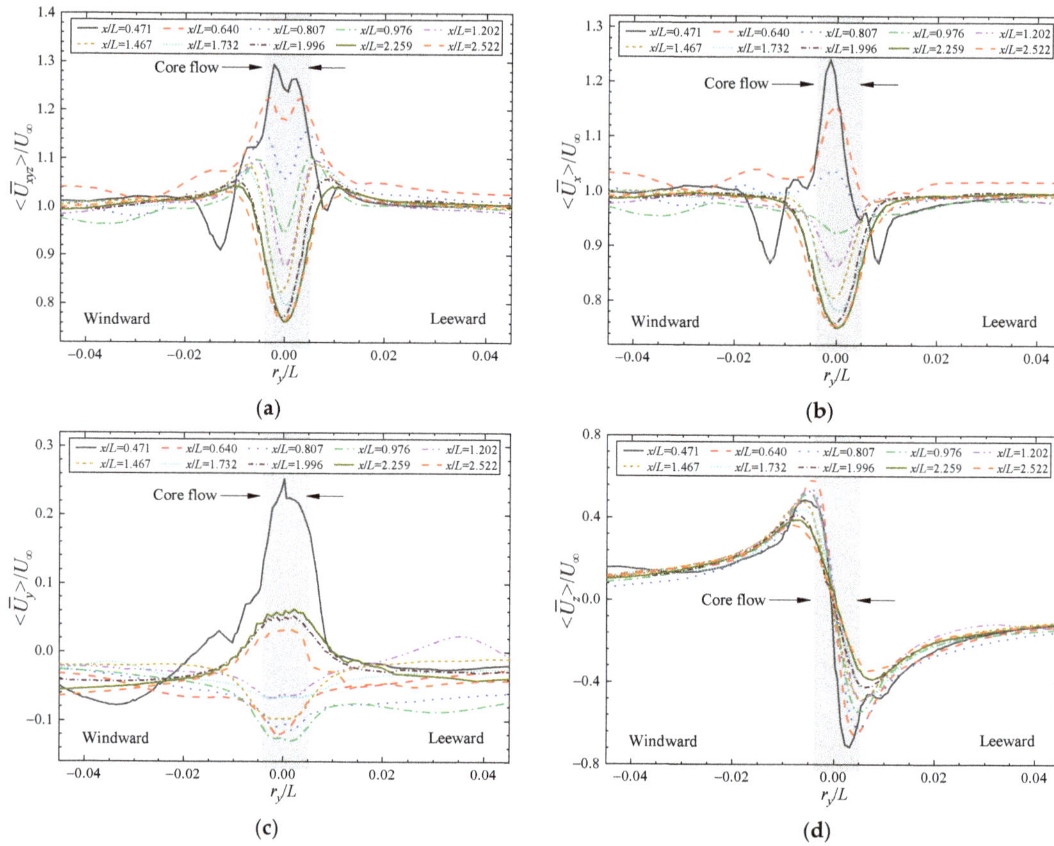

Figure 20. (**a**) The mean resultant velocity; (**b**) The mean streamwise velocity; (**c**) The mean horizontal velocity; (**d**) The mean vertical velocity. Comparison of the mean velocity under different streamwise locations.

Figure 21 shows a comparison of the mean vorticity under different streamwise locations. The dominant component of the mean vorticity magnitude $< \overline{\omega}_{xyz} > r_m/U_\infty$ is the streamwise mean vorticity $< \overline{\omega}_x > r_m/U_\infty$, which indicates that the sail-tip vortex rotates rather faster around the x-axis compared to the other two. The core-flow vorticity weakens rapidly as the wake develops downstream, while the decay rate gradually slows down. An interesting phenomenon shows that the distribution of the peaks and valleys of the vertical mean vorticity $< \overline{\omega}_z > r_m/U_\infty$ changes with the evolution of the flow. In the near wake region, where $x/L \leq 0.807$, the peaks are located on the windward side and the valleys are located on the leeward side, while in the far wake region, the positions of the two are exactly opposite.

Figure 22 shows a comparison of the turbulence kinetic energy under different streamwise locations. The roll-up from the sail is accompanied by a downwash of the sail-tip vortex (see Figure 14a), and the normal stress in the vertical component $<u_x u_x>$ is the strongest contribution to the TKE, as mentioned by Lee [24]. Because of the significant cross-stream, the normal stress in the horizontal component accounts for the second strongest contribution, and the normal stress in the streamwise component is the weakest. The strongest TKE does not occur immediately behind the sail, but approximately in the range of x/L equal to 0.6 to 1.0, where the downwash of the sail-tip vortex is quite intense.

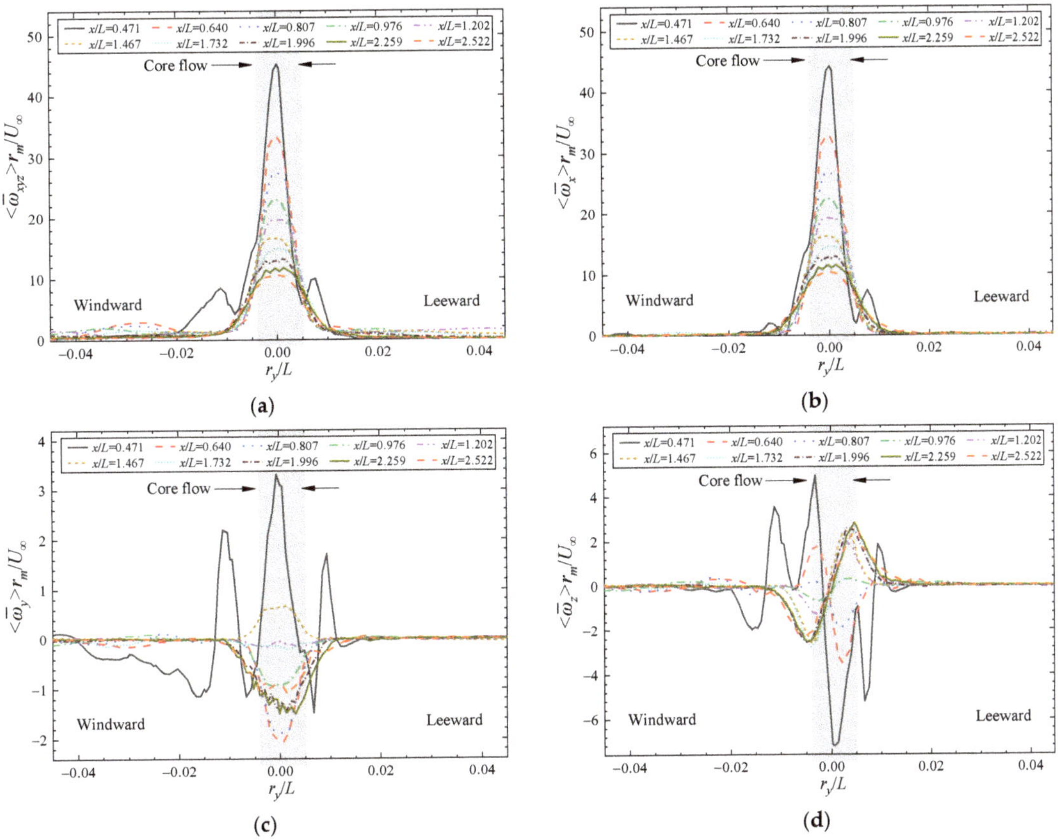

Figure 21. (**a**) The mean vorticity magnitude; (**b**) The mean streamwise velocity; (**c**) The mean horizontal velocity; (**d**) The mean vertical velocity. Comparison of the mean vorticity under different streamwise locations.

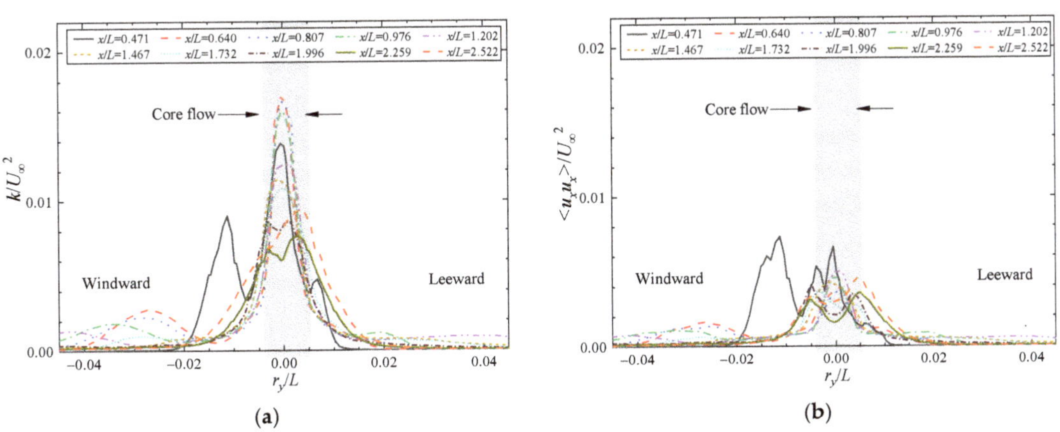

Figure 22. *Cont.*

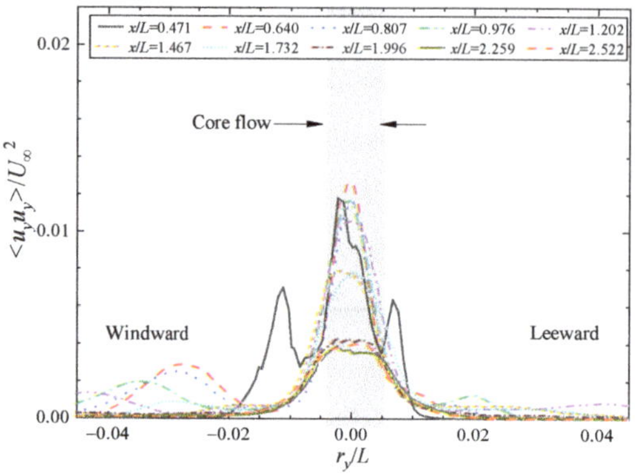

(c)

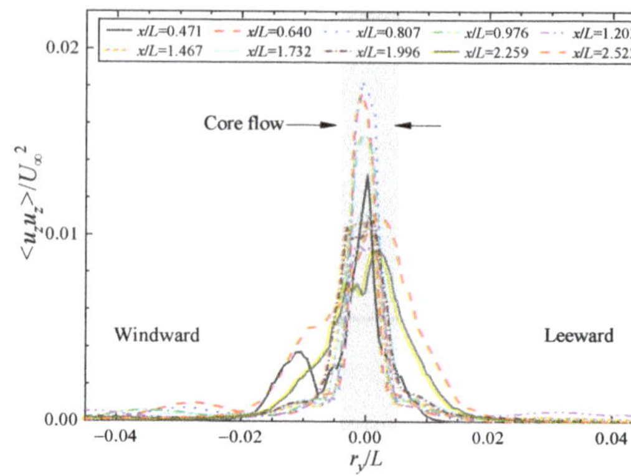

(d)

Figure 22. (**a**) The turbulence kinetic energy; (**b**) The normal stress in the streamwise component; (**c**) The normal stress in the horizontal component; (**d**) The normal stress in the vertical component. Comparison of the turbulence kinetic energy under different streamwise locations.

5. Conclusions

In the paper, large eddy simulation with the dynamic Smagorinsky model was conducted to investigate the flow field around a fully appended Joubert BB2 submarine model under straight-ahead and 10° yaw conditions. Conclusions acquired from the analysis of computed results can be summarized as follows.

(1) Three sets of grids under 10° yaw conditions were designed to examine the grid convergence and computational accuracy. As the grid number increased, the numerical dissipation decreased, and the capture of the vertical component of velocity, the cross-stream Reynolds stress, seemed to be more refined, especially for the extreme values, which were more representative of the experiments, with the relative error of the peak value in the region of core flow being relatively smaller. The numerical attenuation of the mean resultant velocity of the core flow inevitably existed in the LES simulations with the dynamic Smagorinsky model, which could be eliminated through the improvement in the grid's spatial resolution. In the near sail-wake region, the numerical simulations for all grid sets could define the centers of the vortices well, but as they evolve downstream, the advantages of refining the grids are gradually reflected. Overall, through qualitative and quantitative comparison with experiments under 10° yaw conditions, the computational accuracy was verified and results shown are more representative of the experiments with the improvement in grid spatial resolution.

(2) A comparison of the evolution of the flow under straight-ahead and 10° yaw conditions shows that in the core-flow region, the resultant velocity, vorticity magnitude, and TKE under straight-ahead conditions were somewhat smaller than those under 10° yaw conditions. The side-vortices on the leeward side of the sail occurred further forward, and the sail-tip and hydroplane-tip vortices were strong enough, relatively, to develop far downstream under 10° yaw conditions. Another obvious feature that distinguishes the straight-ahead conditions is the flow separation on the leeward side of the middle hull; the wake of the submarine becomes quite complicated, and the flow behind the stern is dominated by the mixing of various component vortex systems, including the tilted horseshoe-vortex system, the upper and lower hull vortices, the tip vortices, and the wake of the sail, hydroplanes, X-rudders, and the hull under 10° yaw conditions. But downstream, far away from the hull under straight-ahead conditions, all the tip vortices dissipated, and only the wake vortices, after complicated interaction, dynamically evolved, with the energy gradually weakening.

(3) The tip vortex tracking under 10° yaw conditions exhibited significant three-dimensional characteristics compared to those under straight-ahead conditions. Under10° yaw conditions, sail-tip vortex tracking maintained an axial angle of approximately 8 degrees with the hull, and was almost stable vertically after experiencing a downwash immediately behind the sail. The port hydroplane-tip vortices developed and spiraled around the sail-tip vortices, while the core of the starboard hydroplane-tip vortices kept moving towards the leeward side, with the vertical position gradually rising away from the hull after passing through a valley at approximately $x/L = 1.1$, due to the repulsive interaction of the hull wake.

(4) The resultant velocity, vorticity magnitude, and TKE showed a gradually decreasing trend as the wake of the cruciform appendage developed downstream. Under 10° yaw conditions, the core-flow exhibited a high-velocity characteristic, in which the peaks of mean vorticity magnitude were located on the windward side in the near wake region, while in the far wake region, the velocity was smaller than the freestream velocity and the valleys of mean vorticity magnitude were located on the windward side. The strongest TKE did not occur immediately behind the sail, but approximately in the range of x/L equal to 0.6 to 1.0, where the downwash of the sail-tip vortex was quite intense.

Author Contributions: M.C.: Conceptualization, Methodology, Software, Writing—original draft. N.Z.: Con-ceptualization, Methodology, Data curation, Validation, Supervision. H.S.: Visualization, Investigation. X.Z.: Writing—review and editing. All authors have read and agreed to the published version of the manuscript.

Funding: Support from the Foundation of Equipment Pre-research Area with grant no. 50907010101 is acknowledged.

Institutional Review Board Statement: Not applicable.

Informed Consent Statement: Not applicable.

Data Availability Statement: Data sharing is not applicable to this article.

Conflicts of Interest: The authors declare that they have no known competing financial interests or personal relationships that could have appeared to influence the work reported in this paper.

Nomenclature

CFD	Computational fluid dynamics	SUBOFF	Submarine Technology Program Office
CSSRC	China Ship Scientific Research Centre	$\widetilde{\mathbf{T}}$	Filtered stress tensor
DARPA	Defense Advanced Research Projects Agency	$\mathbf{T}_{SGS}$	Sub-grid-scale stress
DES	Detached eddy simulation	TKE, k	Turbulence kinetic energy, m^2/s^2
DNS	Direct numerical simulation	$< \overline{U}_x >$	Mean streamwise velocity, m/s
DSTO	Defense Science and Technology Organization	$<u_x u_x>$	Normal stress in the streamwise component, m^2/s^2
Exp. 1	Experimental results with $Re_L = 4 \times 10^6$	$< \overline{U}_{xyz} >$	Mean resultant velocity, m/s
Exp. 2	Experimental results with $Re_L = 8 \times 10^6$	$< \overline{U}_y >$	Mean horizontal velocity, m/s
$\mathbf{f}_b$	Resultant of the body forces	$<u_y u_y>$	Normal stress in the horizontal component, m^2/s^2
$G1$	Most coarse grid set	$<u_y u_z>$	Cross-stream Reynolds stress, m^2/s^2
$G2$	Median grid set	$< \overline{U}_z >$	Mean vertical velocity, m/s
$G3$	Most refined grid set	$<u_z u_z>$	Normal stress in the vertical component, m^2/s^2
$\mathbf{I}$	Identity tensor	U_∞	Freestream velocity, m/s
L	Model length, m	$\widetilde{\mathbf{v}}$	Filtered velocity

LES	Large eddy simulation	x	Streamwise coordinate, m
o	Coordinate origin	y	Horizontal coordinate, m
ONR	Office of Naval Research	y^+	Dimensionless wall distance
$\widetilde{p}$	Filtered pressure	z	Vertical coordinate, m
PISO	Pressure implicit with splitting of operators		
PIV	Particle image velocimetry	***Greek symbols***	
RANS	Reynolds averaged Navier–Stokes	μ_t	Sub-grid scale turbulent viscosity
Re_L	Reynolds number	ρ	Density
r_m	Model radius, m	$< \overline{\omega}_x >$	Mean streamwise vorticity,/s
r_y	Radial distance, m	$< \overline{\omega}_{xyz} >$	Mean vorticity magnitude,/s
S	Strain rate tensor	$< \overline{\omega}_y >$	Mean horizontal vorticity,/s
SPIV	Stereo particle image velocimetry	$< \overline{\omega}_z >$	Mean vertical vorticity,/s

References

1. Liu, Y.W.; Li, Y.L.; Shang, D.J. The generation mechanism of the flow-induced noise from a sail hull on the scaled submarine model. *Appl. Sci.* **2019**, *9*, 106. [CrossRef]
2. Deng, R.; Zhang, Z.Z.; Pang, F.Z.; Tiecheng, W.; Wanzhen, L. Investigating the sound power level of a simplified underwater vehicle induced by flow separation. *Ocean Eng.* **2020**, *204*, 107286.
3. Rocca, A.; Cianferra, M.; Broglia, R.; Armenio, V. Computational hydroacoustic analysis of the BB2 submarine using the advective Ffowcs Williams and Hawkings equation with Wall-Modeled LES. *Appl. Ocean Res.* **2022**, *129*, 103360. [CrossRef]
4. Fu, T.C.; Atsavapranee, P.; Hess, D.E. PIV measurements of the cross-flow wake of a turning submarine model (ONR Body-1). In Proceedings of the 24th Symposium on Naval Hydrodynamics, Fukuoka, Japan, 8–13 July 2002.
5. Jimenez, J.M.; Hultmark, M.; Smits, M. The intermediate wake of a body of revolution at high Reynolds numbers. *J. Fluid Mech.* **2010**, *659*, 516–539. [CrossRef]
6. Ashok, A.; Smits, A.J. The turbulent wake of a submarine model at varying pitch and yaw angle. In Proceedings of the 65th Annual Meeting of the APS Division of Fluid Dynamics, San Diego, CA, USA, 18–20 November 2012.
7. Ashok, A.; Smits, A.J. The turbulent wake of a submarine model in pitch and yaw. In Proceedings of the 51st AIAA Aerospace Sciences Meeting including the New Horizons Forum and Aerospace Exposition, Grapevine, TX, USA, 7–10 January 2013.
8. Ashok, A.; Buren, T.V.; Smits, A.J. Asymmetries in the high Reynolds number wake of a submarine model in pitch. In Proceedings of the 67th Annual Meeting of the APS Division of Fluid Dynamics, San Francisco, CA, USA, 23–25 November 2014.
9. Ashok, A.; Buren, T.V.; Smits, A.J. The structure of the wake generated by a submarine model in yaw. *Exp. Fluids* **2015**, *56*, 123. [CrossRef]
10. Ashok, A.; Buren, T.V.; Smits, A.J. Asymmetries in the wake of a submarine model in pitch. *J. Fluid Mech.* **2015**, *774*, 416–442. [CrossRef]
11. Khan, M.K.; Korulla, M.; Nagarajan, V.; Sha, O.P. Surface pressure measurements on a generic submarine hull form at high angles of incidence. *J. Mar. Sci. Technol.* **2022**, *27*, 677–694. [CrossRef]
12. Shokrallah, M.B.; Alireza, S.; Mojtaba, D.M. Experimental study of the submarine's wake improvement by displacement of stern planes. *Ships Offshore Struct.* **2022**, *17*, 2103–2115.
13. Joubert, P.N. *Some Aspects of Submarine Design Part 1: Hydrodynamics*; Technical Report DSTO-TR-1622; Australia Defence Science and Technology Organization: Canberra, Australia, 2004.
14. Joubert, P.N. *Some Aspects of Submarine Design Part 2: Shape of a Submarine 2026*; Technical Report DSTO-TR-1920; Australia Defence Science and Technology Organization: Canberra, Australia, 2006.
15. Kumar, C.; Manovski, P.; Giacobello, M. Particle image velocimetry measurements on a generic submarine hull form. In Proceedings of the 18th Australasian Fluid Mechanics Conference, Launceston, Australia, 3–7 December 2012.
16. Anderson, B.; Chapuis, M.; Erm, L.; Fureby, C.; Giacobello, M.; Henbest, S.; Jones, D.; Jones, M.; Kumar, C.; Liefvendahl, M.; et al. Experimental and computational investigation of a generic conventional submarine hull form. In Proceedings of the 29th Symposium on Naval Hydrodynamics, Gothenburg, Sweden, 26–31 August 2012.
17. Jones, M.B.; Erm, L.P.; Valiyf, A.; Henbest, S.M. *Skin-Friction Measurements on a Model Submarine*; Technical Report DSTO-TR-2898; Australia Defence Science and Technology Organization: Canberra, Australia, 2013.
18. Lee, S.K. Topology of the flow around a conventional submarine hull. In Proceedings of the 19th Australasian Fluid Mechanics Conference, Melbourne, Australia, 8–11 December 2014.
19. Manovski, P.; Giacobello, M.; Jacquemin, P. *Smoke Flow Visualization and Particle Image Velocimetry Measurements over a Generic Submarine Model*; Technical Report DSTO-TR-2944; Australia Defence Science and Technology Organization: Canberra, Australia, 2014.
20. Fureby, C.; Anderson, B.; Clarke, D.; Erm, L.; Henbest, S.; Giacobello, M.; Jones, D.; Nguyen, M.; Johansson, M.; Jones, M.; et al. Experimental and numerical study of a generic conventional submarine at 10° yaw. *Ocean Eng.* **2016**, *116*, 1–20. [CrossRef]
21. Bettle, M.C. Validating design methods for sizing submarine tailfns. In Proceedings of the Warship 2014: Naval Submarines and UUV's, Bath, UK, 18–19 June 2014.

22. Overpelt, B.; Nienhuis, B.; Anderson, B. Free running maneuvering model tests on a modern generic SSK Class submarine (BB2). In Proceedings of the Pacific International Maritime Conference, Sidney, Australia, 8–6 October 2015.
23. Lee, S.K. Longitudinal development of flow-separation lines on slender bodies in translation. *J. Fluid Mech.* **2018**, *837*, 627–639. [CrossRef]
24. Lee, S.K.; Manovski, P.; Kumar, C. Wake of a cruciform appendage on a generic submarine at 10° yaw. *J. Mar. Sci. Technol.* **2019**, *25*, 787–799. [CrossRef]
25. Lee, S.K.; Jones, M.B. Surface-pressure pattern of separating flows over inclined slender bodies. *Phys. Fluids* **2020**, *32*, 095123. [CrossRef]
26. Alin, N.; Bensow, R.; Fureby, C.; Huuva, T.; Svennberg, U. Current capabilities of RANS, DES and LES for submarine flow simulations. *J. Ship. Res.* **2010**, *54*, 184–196. [CrossRef]
27. Bensow, R.E.; Larson, M.G. Towards a novel residual based subgrid modelling based on multiscale techniques. In Proceedings of the 9th Numerical Towing Tank Symposium, Le Croisic, France, 1–3 October 2006.
28. Zhang, N.; Wang, X.; Xie, H.; Li, Y. Research on numerical simulation approach for flow induced noise and the influence of the acoustic integral surface. *J. Ship Mech.* **2016**, *20*, 892–908. (In Chinese)
29. Smagorinsky, J. General circulation experiments with the primitive equations. I. the basic experiment. *Month. Wea. Rev.* **1963**, *91*, 99–164. [CrossRef]
30. Shi, B.; Yang, X.; Jin, G.; He, G.; Wang, S. Wall-modeling for large-eddy simulation of flows around an axisymmetric body using the diffuse-interface immersed boundary method. *Appl. Math. Mech.* **2019**, *40*, 305–320. [CrossRef]
31. Zhou, Z.; Li, Z.B.; Yang, X.L.; Wang, S.; Xu, D. Investigation of the wake characteristics of an underwater vehicle with and without a propeller. *Ocean Eng.* **2022**, *266*, 113107. [CrossRef]
32. Posa, A.; Balaras, E. Large-eddy simulations of a notional submarine in towed and self-propelled configurations. *Comput. Fluids* **2018**, *165*, 116–126. [CrossRef]
33. Posa, A.; Balaras, E. A numerical investigation about the effects of Reynolds number on the flow around an appended axisymmetric body of revolution. *J. Fluid Mech.* **2020**, *884*, A41. [CrossRef]
34. Zhang, N.; Lv, S.J.; Xie, H.; Zhang, S.L. Numerical simulation of unsteady flow and flow induced sound of airfoil and wing/plate junction by LES and FW-H acoustic analogy. *Appl. Mech. Mater.* **2014**, *444–445*, 479–485. [CrossRef]
35. Zhang, N.; Xie, H.; Wang, X.; Wu, B.S. Computation of vortical flow and flow induced noise by large eddy simulation with FW-H acoustic analogy and Powell vortex sound theory. *J. Hydrodyn.* **2016**, *28*, 255–266. [CrossRef]
36. Zhang, N.; Li, Y.; Huang, M.M.; Chen, M. Numerical prediction approach for hydrodynamic force and noise of propeller in submarine propeller interaction condition. *J. Ship Mech.* **2021**, *25*, 1439–1451. (In Chinese)
37. Wang, X.M.; Zhang, N. Large eddy simulation of the flow field and hydrodynamic force of submarine in crashback. *J. Ship Mech.* **2022**, *26*, 1595–1610. (In Chinese)
38. Kroll, T.; Morse, N.; Horne, W.; Mahesh, K. Large eddy simulation of marine flows over complex geometries using a massively parallel unstructured overset method. In Proceedings of the 33th Symposium on Naval Hydrodynamics, Osaka, Japan, 31 May–5 June 2020.
39. Loid, H.P.; Bystrom, L. Hydrodynamic aspects of the design of the forward and aft bodies of the submarine. In Proceedings of the RINA International Symposium on Naval Submarines, London, UK, 17–20 May 1983.
40. Germano, M.; Piomelli, U.; Moin, P.; Cabot, W. A dynamic subgrid-scale eddy viscosity model. *Phys. Fluids A* **1991**, *3*, 1760–1765. [CrossRef]
41. Lilly, D.K. A proposed modification of the Germano subgrid–scale closure model. *Phys. Fluids A* **1992**, *4*, 633. [CrossRef]
42. Mahesh, K.; Kumar, P.; Gnanaskandan, A.; Nitzkorski, Z. LES applied to ship research. *J. Ship Res.* **2015**, *59*, 238–245. [CrossRef]
43. Kumar, P.; Mahesh, K. Large eddy simulation of propeller wake instabilities. *J. Fluid Mech.* **2017**, *814*, 361–396. [CrossRef]
44. Kumar, P.; Mahesh, K. Large-eddy simulation of flow over an axisymmetric body of revolution. *J. Fluid Mech.* **2018**, *853*, 537–563. [CrossRef]
45. ITTC. Recommended procedures and guidelines: Practical guidelines for ship CFD applications. In Proceedings of the 26th International Towing Tank Conference, Rio de Janeiro, Brasil, 2–5 March 2014; Volume 7.
46. Akkermans, R.A.D.; Cieslik, A.R.; Kamp, L.P.J.; Trieling, R.R.; Clercx, H.J.; van Heijst, G.J. The three-dimensional structure of an electromagnetically generated dipolar vortex in a shallow fluid layer. *Phys. Fluids* **2008**, *20*, 116601. [CrossRef]

Journal of
*Marine Science
and Engineering*

MDPI

Article

Numerical Study of the Force Characteristics and Flow Field Patterns of a Cylinder in the Internal Solitary Wave

Miao Zhang [1], Haibao Hu [1,*], Abdellatif Ouahsine [2], Peng Du [1,3,*], Xiao Huang [1] and Luo Xie [1]

1 School of Marine Science and Technology, Northwestern Polytechnical University, 127 Youyi Road, Beilin, Xi'an 710072, China; zhangmiao@mail.nwpu.edu.cn (M.Z.); huangxiao@nwpu.edu.cn (X.H.); xieluo@nwpu.edu.cn (L.X.)

2 Laboratoire Roberval, Sorbonne Université, Université de Technology de Compiègne, Centre de Recherches Royallieu, CS 60319, CEDEX, 60203 Compiègne, France; ouahsine@utc.fr

3 Xi'an Tianhe Defense Technology Co., Ltd., 158 Xibudadao Road, Gaoxin, Xi'an 710119, China

* Correspondence: huhaibao@nwpu.edu.cn (H.H.); dupeng@nwpu.edu.cn (P.D.)

Abstract: The density of the ocean is unevenly distributed along the depth direction, showing a stratified structure. When there is an external disturbance, large-scale internal solitary waves are easily generated. The internal solitary waves are bounded by the intermediate pycnocline, and the currents in the upper and lower layers will flow in opposite directions. This generates strong shear forces that threaten the safety of marine structures. In this paper, the flow field distribution characteristics of a cylinder under the action of internal solitary waves at different scales are analyzed as a research object. The whole cylinder is discretized into 40 regions, and the horizontal force applied to each section of the cylinder is extracted. The force characteristics of the cylinder are analyzed. It is concluded that the pressure is the main factor determining the magnitude of the total combined force. In addition, the paper extracts the main flow structures from the modal decomposition point of view and explains the reasons affecting the force behavior of the cylinder.

Keywords: internal solitary waves; force; cylinder; modal decomposition

1. Introduction

Differences in temperature and salinity cause the density of the ocean to be distributed differently among the depth layers. When there are external disturbances, such as currents and the motion of ships, the interfaces between the layers are easily disturbed, causing fluctuations. These fluctuations are called internal solitary waves (ISWs). The ISW has a pycnocline as its interface, above and below which the current flows in the opposite direction. This produces strong underwater shear effects, like a "big pair of scissors", that threaten the safety of marine structures [1]. In addition, ISWs are characterized by large amplitude, high energy, difficult to dissipate, and widely distributed throughout the world's oceans. This indicates that the encounter of marine structures with ISWs is frequent and dangerous. Therefore, it is of great engineering value to conduct research on the forces acting on marine structures under the action of ISWs [2].

Cylinders are the basic component form of offshore structures such as Spar, columns of tension-leg platform (TLP), and deep-sea oil drilling rigs [3]. Therefore, understanding the flow field characteristics of cylinders in ISWs is of great importance for the smooth operation of offshore oil and gas development projects. To this end, Wei et al. [4] fabricated a wavemaker capable of generating nonlinear large-amplitude ISWs in a large-section stratified liquid tank. The interaction characteristics between ISWs and cylinders were revealed by accurately measuring the small changes in the ISW force applied to the cylinders in the tank. Wang et al. [5] carried out experiments related to the action of ISWs on vertical cylinders in a stratified-flow tank. The effect of wave amplitude on the flow field characteristics of the cylinder was investigated. Wang et al. [6] studied the pressure distribution around the cylinder under the action of ISWs by means of a large eddy simulation method. It was revealed that vortex is an important factor affecting the pressure resistance of the cylinder in the ISW.

In addition to understanding the flow field distribution characteristics of the cylinder in ISWs, it is also important to be able to accurately calculate the loads exerted on the cylinder by ISWs to ensure the safety of marine structures. Cai et al. [7] introduced the Morison formula and gave the empirical coefficients in the Morison formula based on the surface wave theory. In their paper, they calculated the ISW on the vertical cylindrical horizontal force and its moment, which showed the non-negligible effect of ISW on the vertical cylinder. Thereafter, Song et al. [8] investigated the kinematic response of the cylinder under the action of ISWs based on the KDV theory in combination with the Morison formula. LÜ et al. [9] simulated the motion response of a vertical cylinder encountering different background parabolic currents. The forces on the vertical cylinder during the process were calculated using the Morison formula.

In all of the above studies, the overall forces on a whole cylinder through the depth of water or with a large draft were analyzed. However, due to the different flow characteristics at different depths, especially in different layers, there may be some differences in the force characteristics of the cylinder along the depth direction. These differences are crucial when we go to analyze the dynamic response of ISWs traversing the cylinder. Therefore, scholars have carried out studies on the force characteristics of columns located at different depths. Si et al. [10] used the generalized KdV (GKdV) model to obtain the vertical distribution of the horizontal force acting on the cylinder by a large amplitude ISW. They pointed out that the shear force reaches its maximum at the inflection point of the horizontal flow velocity. Wang et al. [11] measured the hydrodynamic coefficients of each cylinder section based on the experimental method. The paper suggested that different empirical coefficients should be used when calculating the ISW forces in different layers of the structural section. Yu et al. [12] calculated, compared, and analyzed the interactions between ISWs and a vertical cylinder using numerical simulations using the eKdV equations as the initial and boundary conditions. They divided the whole cylinder into six segments and analyzed the force variation characteristics of each segment when the cylinder spans the ISW.

Although many scholars have studied the force characteristics of the cylinder in ISWs, the number of segments of the whole cylinder is relatively small, and, in addition, the flow mechanism is rarely studied. In this study, the whole cylinder is divided into 40 equally spaced sections. The variation of the lift force and drag force of each section of the cylinder during the interaction with the ISW is studied separately. It is found that the drag force on the cylinder is 10 times higher than the lift force. It is proven that the shear phenomenon induced by the horizontal force is the main cause of the safety hazard of the structures. It is also found that the horizontal force is mainly caused by pressure, and the horizontal force induced by wall shear stress is only a small proportion. To explore the flow mechanism, we extract the main flow modes in the flow field based on the dynamic mode decomposition method. We analyze the main reasons that trigger the force distribution characteristics of the cylinder from the perspective of modes.

2. Numerical Methodology

2.1. Continuity and Momentum Equations

Based on the continuity assumption, the Navier–Stokes (N–S) equations include the conservation of mass and the conservation of momentum, which are expressed by Equations (1) and (2), respectively [13–15].

$$\frac{\partial \rho}{\partial t} + \frac{\partial(\rho u_i)}{\partial x_i} = 0 \tag{1}$$

$$\frac{\partial(\rho u_i)}{\partial t} + \frac{\partial(\rho u_i u_j)}{\partial x_j} = -\frac{\partial p}{\partial x_i} + \frac{\partial}{\partial x_j}\left(\mu \frac{\partial u_i}{\partial x_j}\right) + f_i \tag{2}$$

where ρ is the fluid density, t is the time, i (i = 1, 2, 3) is the three directions of the Cartesian coordinates, x_i is the spatial coordinate, u is the flow velocity, p is the pressure, μ is the kinematic viscosity, and f is the body force.

2.2. Turbulence Model

The SST k-ω model, while retaining its essence, was developed by building upon the standard k-ω model. By incorporating a blending function, it combines the strengths of both the k-ε and k-ω models. Specifically, the standard k-ω model is employed in close proximity to the wall boundaries, whereas the k-ε model is utilized in the boundary layer edge and fully developed flow regions. This unique approach allows the SST k-ω model to effectively handle near-wall flows. In addition, the SST k-ω model offers additional features such as the consideration of rotating reference frames and shear flows, accounting for the propagation of turbulent shear stresses. Consequently, it exhibits enhanced accuracy in simulating separation points and the sizes of separation zones induced by pressure gradients [16,17]. Its transport equations are as follows:

$$\frac{\partial}{\partial t}(\rho k) + \frac{\partial}{\partial x_i}(\rho k u_i) = \frac{\partial}{\partial x_j}\left(\Gamma_k \frac{\partial k}{\partial x_j}\right) + G_k - Y_k + S_k \tag{3}$$

$$\frac{\partial}{\partial t}(\rho \omega) + \frac{\partial}{\partial x_j}(\rho \omega u_j) = \frac{\partial}{\partial x_j}\left(\Gamma_\omega \frac{\partial \omega}{\partial x_j}\right) + G_\omega - Y_\omega + D_\omega + S_\omega \tag{4}$$

where G_k and G_ω represent the generation of the turbulent kinetic energy and dissipation, respectively. Γ_k and Γ_ω represent the effective diffusivity of k and ω, respectively. Y_k and Y_ω represent the dissipation of k and ω due to turbulence, respectively. D_ω represents the cross-diffusion term, calculated as described below. S_k and S_ω are user-defined source terms.

The numerical solution of the governing equations is accomplished through the utilization of the ANSYS Fluent 2019 commercial software. The SIMPLE algorithm is used to discretize the equations. For the gradient terms of convection and diffusion, a least-squares cell-based approach is utilized. The pressure interpolation at cell faces is carried out using the PRESTO! The second-order implicit scheme is used globally, which guarantees computational accuracy.

2.3. Modeling of Numerical Tank

To generate the ISW, we first need to build the numerical tank. In the 3D Cartesian coordinate system, the numerical tank is 12 m long, 0.8 m wide, and 0.8 m deep, as shown in Figure 1. The corresponding computational mesh structure is shown in Figure 2. In the real marine environment, along the vertical direction, it can be divided into three parts according to the density distribution. They are the upper layer, the lower layer, and the pycnocline in the middle. The density distribution of the upper and lower layers is more uniform while the density of the middle pycnocline is constantly changing along the water depth. In this paper, the density of the upper layer is 999 kg/m^3 and the depth is h_1. The density of the bottom layer is 1018.5 kg/m^3 and the depth is h_2. The density distribution in the pycnocline is specified using the hyperbolic tangent function as seen in Equation (5) [18].

In this study, we use the gravity collapse method to generate the ISWs. Specifically, a long rectangular gravitational potential well of length L and depth D was established on the left side of the tank. The gravitational collapse zone is separated from the rest of the zone by a gate. A low-density fluid is injected above the gravity collapse region, creating a height difference at the interface between the two sides of the gate. When the gate is opened, an initial disturbance in the form of a step is created. At this point, the denser fluid will sink and the lighter fluid will float under the action of gravity. Thus, the fluid below the right side moves to the left and the fluid above the left side moves to the right. This creates a restoring moment that forms an ISW. An ISW is usually unstable when it first forms. It takes some time for the interactions between the peaks and troughs of the wave to adjust and balance, resulting in a more stable wave structure. So, the cylinder is placed in the center of the tank, which allows the ISW to stabilize for a sufficiently long period of time. This also eliminates the effects of reflected waves. The diameter of the cylinder, d, is chosen to be 0.05 m, taking into account the relative size of the actual ISW to the marine structure. In addition, the interaction of the ISW with the cylinder is an unsteady process. Therefore, we need to specify the time step in the simulation. The time step should be chosen to satisfy the Courant–Friedrichs–Lewey (CFL) stability criterion [19]. According to the CFL criterion, the distance traveled in a time step must be less than the length of a grid. ($\Delta t < \min(\Delta x/|U|, \Delta z/|W|)$, where Δt is the time step, Δx is the unit grid size in the horizontal direction, U is the horizontal velocity, Δz is the unit grid size in the vertical direction, W is the vertical velocity). In this paper, $\Delta x_{\max} = 0.02$, $\Delta z_{\max} = 0.01$, $U_{\max} = 0.09$, and $W_{\max} = 0.05$. Therefore, we choose the time step $\Delta t = 0.01$ s, which satisfies the CFL criterion.

$$\begin{cases} \rho_1 & \left(z > z_c + \frac{d_c}{2}\right) \\ \frac{\rho_1+\rho_2}{2}\left\{1 - \frac{\rho_2-\rho_1}{\rho_1+\rho_2}\tanh\left(\frac{z-z_c}{d_c}\right)\right\} & \left(z_c - \frac{d_c}{2} < z < z_c + \frac{d_c}{2}\right) \\ \rho_2 & \left(z < z_c - \frac{d_c}{2}\right) \end{cases} \tag{5}$$

where z is the vertical position, and ρ_1 and ρ_2 represent the density of the upper and lower fluids, respectively. z_c is the vertical location of the pycnocline center. The pycnocline center is obtained by calculating the gradient of the density field. The maximum value of the gradient represents the position of the center of the pycnocline. d_c is the thickness of the pycnocline.

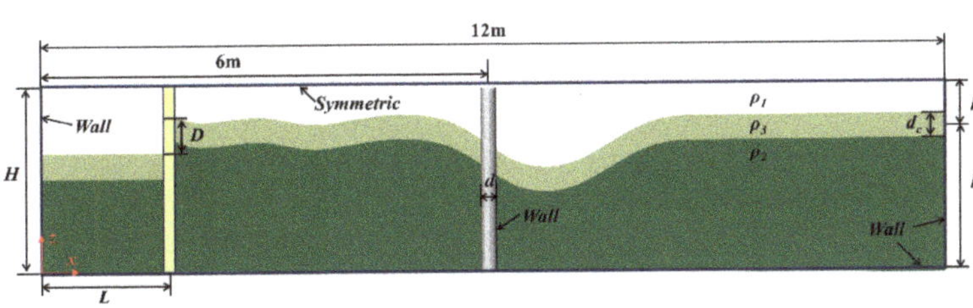

Figure 1. Schematic diagram of the numerical tank setup.

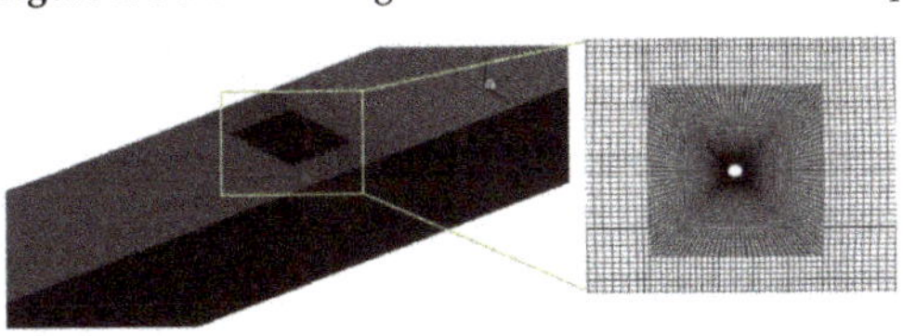

Figure 2. The computational mesh structure in the domain.

3. Model Verification

In order to avoid the interference of mesh size on the numerical simulation results, we first need to verify the mesh independence. Specifically, we used 3.08 million, 3.54 million, and 4.36 million mesh elements to represent coarse, medium, and fine configurations, respectively. Under the three mesh configurations, we computed and obtained the results of the ISW flow field generated at a collapse height of 0.1 m. We extracted the ISW troughs where the wave was located. We extract the horizontal velocity on the vertical profile where the trough of the ISW is located, and a comparison of the computational results under the three grid configurations is shown in Figure 3. We find that the horizontal velocities on the vertical line of the ISW trough computed by the three grids are essentially the same. Therefore, we use the mesh of the dense case as the mesh. This ensures the accuracy of the numerical simulation. The above mesh-independent analysis can show the accuracy of our calculations.

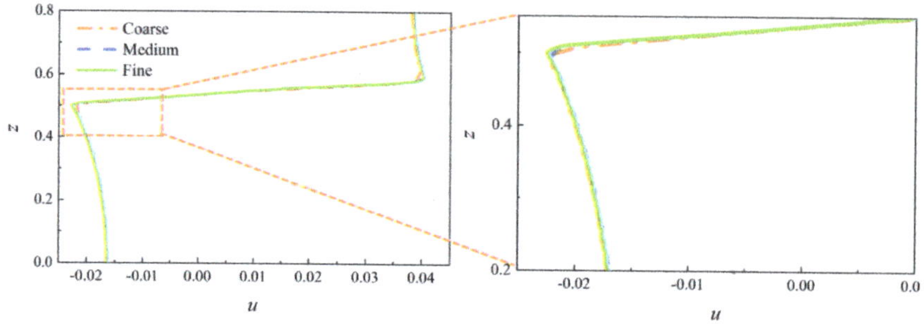

Figure 3. Horizontal velocity on the vertical line of the ISW trough for three numbers of mesh elements.

We also compared the computational and experimental results in addition to the mesh independence analysis. Du et al. [20–22] summarized the relationship between the gravity collapse coefficient S and the collapse length, height, and upper and lower layer heights in their experiments, seen in Equation (6). We compared our numerical results with their experimental results according to their settings of the relevant parameters in their experiments, seen in Figure 4. The results show that our numerical results are in general agreement with the experimental data. This shows the accuracy of our calculations.

$$S = \sqrt{\frac{1.5D(h_1 - h_2)}{h_1^2 h_2^2}L} \tag{6}$$

$$c_0 = \sqrt{gh_1h_2(\rho_2 - \rho_1)/(\rho_2h_1 + \rho_1h_2)} \tag{7}$$

where S represents the gravity collapse coefficient. h_1 and h_2 represent the height of the upper and lower fluids, respectively. D represents the gravity collapse height (the vertical distance between the interfaces on both sides of the gate). L represents the length of the collapse zone. c_0 is the linear long wave velocity. g is the gravitational acceleration. ρ_1 and ρ_2 represent the density of the upper and lower fluids, respectively.

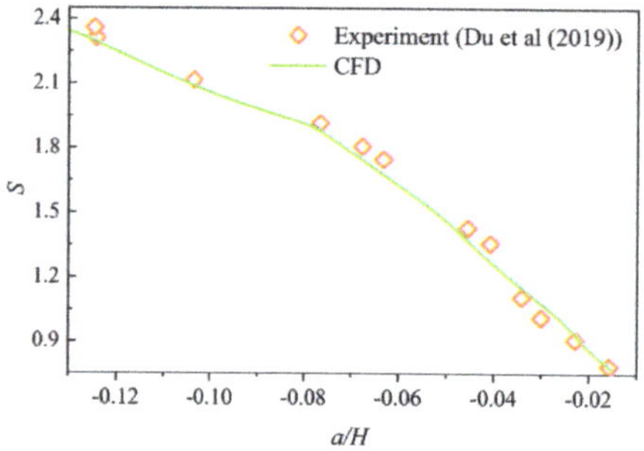

Figure 4. Comparison of generated ISW results between CFD and experimental results of Du et al. [20].

4. Results and Discussions

4.1. Evolution of the Flow Field around a Cylinder in the ISW

We created ISW flow fields with collapse heights of 0.1 m, 0.2 m, and 0.3 m, respectively, using the gravity collapse method. The velocity distributions in the horizontal direction are shown in Figure 5. Comparison reveals that the horizontal velocities induced by the ISW are bounded by the pycnocline, and the upper and lower layers of fluid move in opposite directions. The horizontal velocity on the vertical line at the wave trough position is the largest in the whole flow field. For this reason, the horizontal velocities on the vertical line at the wave trough position for the three cases are extracted and dimensionless. The correlation curves are shown in Figure 6. In order to perform the dimensionless treatment, we introduced the linear long wave velocity c_0 as shown in Equation (7). Due to the uniform thickness of the pycnocline, the horizontal velocity of the ISW gradually increases with the increase in the collapse height h. The horizontal velocity of the ISW increases gradually. This leads to a more drastic change in the value of the velocity in the pycnocline. The analysis of the flow field here provides the basis for the following analysis of the force law of the column in the internal wave at different scales.

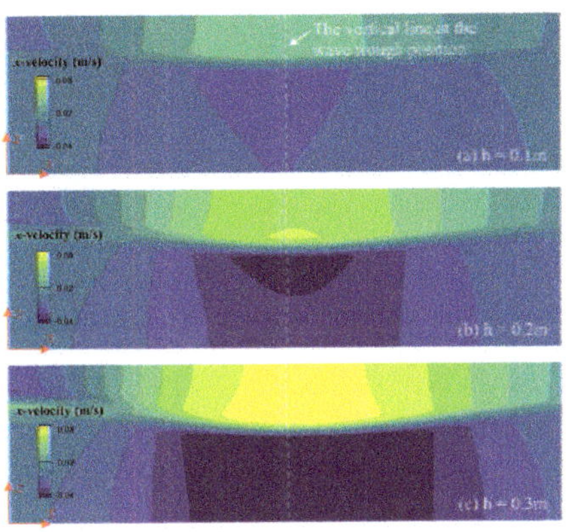

Figure 5. Velocity distribution of ISWs of different amplitudes.

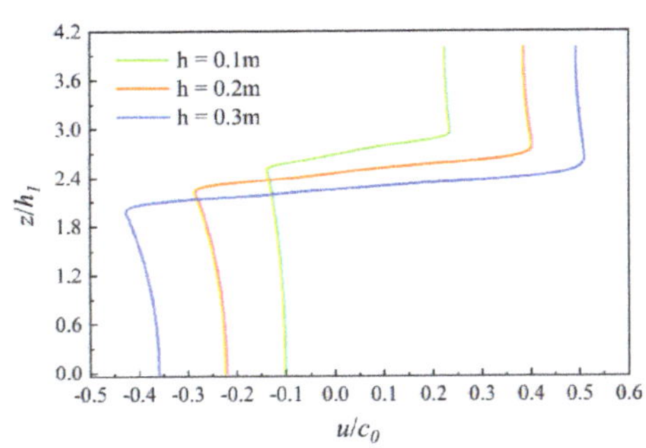

Figure 6. The horizontal velocities on the vertical line at the wave trough position.

In order to explore the evolution law of the flow field around the cylinder in the ISW, we extracted the results of the time-series flow field of the ISW interacting with the cylinder at three scales, as shown in Figure 7. We find that the ISW induces an elliptical flow field with its nucleus located at the wave trough. It rotates clockwise and moves forward. The tangential direction of the streamlines represents the direction of the velocity at the local location. Based on the orientation of the streamlines, we can see that the horizontal velocities of the upper and lower layers are opposite. At the initial stage of the contact between the ISW and the cylinder, there is almost no backflow on the backflow region of the cylinder. As time progresses, the backflow area in the backflow region gradually increases and the value of the wake velocity gradually increases. In addition, after the wave trough reaches the location of the cylinder, the cylindrical wake gradually becomes chaotic as the scale of the ISW increases. We can find from the distribution of the streamlines. In addition, we also analyzed the change in wavefront and the distribution of shear stress on the wall of the cylinder when the ISW flows through the cylinder, as shown in Figure 8. There is a lifting phenomenon of the wavefront at the back of the cylinder. As the scale of the ISW increases and the time advances, the area of the lifted wavefront behind the cylinder increases gradually and becomes chaotic. This is the effect of the cylinder on the wave surface, and in turn, the ISW generates shear stress on the surface of the cylinder. There will be symmetrical areas of high shear stress from the flow onset point to the separation point. For the cylinders in the upper and lower layers, the wall shear stresses applied are in opposite directions. In addition, the wall shear stress distribution in the wake region gradually becomes chaotic, a behavior caused by the shedding of the wake vortex of the cylinder.

To further quantitatively analyze the variation of the cylindrical wall shear stress, we extracted the maximum value of the forward direction and the backward direction of the cylindrical wall shear stress, which are shown in Figures 9 and 10. The changes in their values reflect the extreme stress state of the cylinder. We find that both of them have the trend of gradually increasing to the maximum value and then decreasing. However, the wall shear stress in the forward direction will be higher than that in the backward direction. For example, in the case of h = 0.3 m, the peak value of wall shear stress in the forward direction is 0.214 Pa, while the peak value of wall shear stress in the backward direction is 0.139 Pa. The peak value in the forward direction is 35% higher than in the reverse direction. This is due to the difference in horizontal velocity values between the upper and lower layers of the ISW. We also find that there is a fluctuation in the value of the wall shear stress as the trough of the ISW gradually moves away from the cylinder. The greater the collapse height h, the more pronounced the fluctuation.

4.2. Force Characteristics of Cylinders in the ISW

In order to investigate the force law of the cylinder in the ISW, we divide the whole cylinder into 40 segments at equal distances along the vertical direction, and the length of each segment is 0.02 m, since the horizontal force exerted on the cylinder is the dominant factor threatening the safety of the cylinder when it interacts with the ISW. Therefore, in this paper, the horizontal force ensemble is calculated by integrating the pressure and velocity shear on the surface of the cylinder. The force on the whole cylinder can be viewed as the sum of the contribution of the force on each segment of the cylinder at the time, i.e., as the sum of the normal pressure and tangential viscous components [23].

$$\vec{F} = \sum_i -s_{f,i} P_i \vec{n}_{\vec{i}} + \sum_i \mu s_{f,i} u_{s,i} \vec{\tau}_i \quad (8)$$

where s_f is the face area, P is the pressure, μ is the dynamic viscosity, u_s is the velocity shear at the cylinder surface, h is the distance between cylinder section and bottom boundary, $\vec{n}$ is the unit surface normal around the cylinder, and $\vec{\tau}$ is the unit tangential vector along the velocity direction.

The force on the cylinder is a time-varying process, as the ISW travels through the cylinder. We extracted the lift force F_l and drag force F_d of the cylinder, the variation of which is shown in Figures 11 and 12, where F_l is positive vertically upward and F_d is positive horizontally to the right. For F_l, it exhibits a gradual increase and then decrease in each of the up and down directions. It can be considered that the cylinder has a tendency to vibrate up and down. For F_d, the force is realized as an increase and then a decrease, and the force is always positive. It can be seen that F_d is 10 times greater than F_l. Therefore, the horizontal force on the cylinder should be more concerned for safety reasons. To analyze the force details of the cylinder, we further extracted the lift and drag forces applied to each segment of the cylinder, see Figures 13 and 14. For each segment of the cylinder, the drag force continues to be much greater than the lift force. In addition, we find that the direction of the force on each segment of the cylinder is almost the same at all times. Although some segments of the force are not in the same direction, the shear phenomenon induced by the lift force is much weaker compared to the drag force. Therefore, for the structural safety of the cylinder in the ISW, we should pay more attention to the variation of the drag force in its horizontal direction.

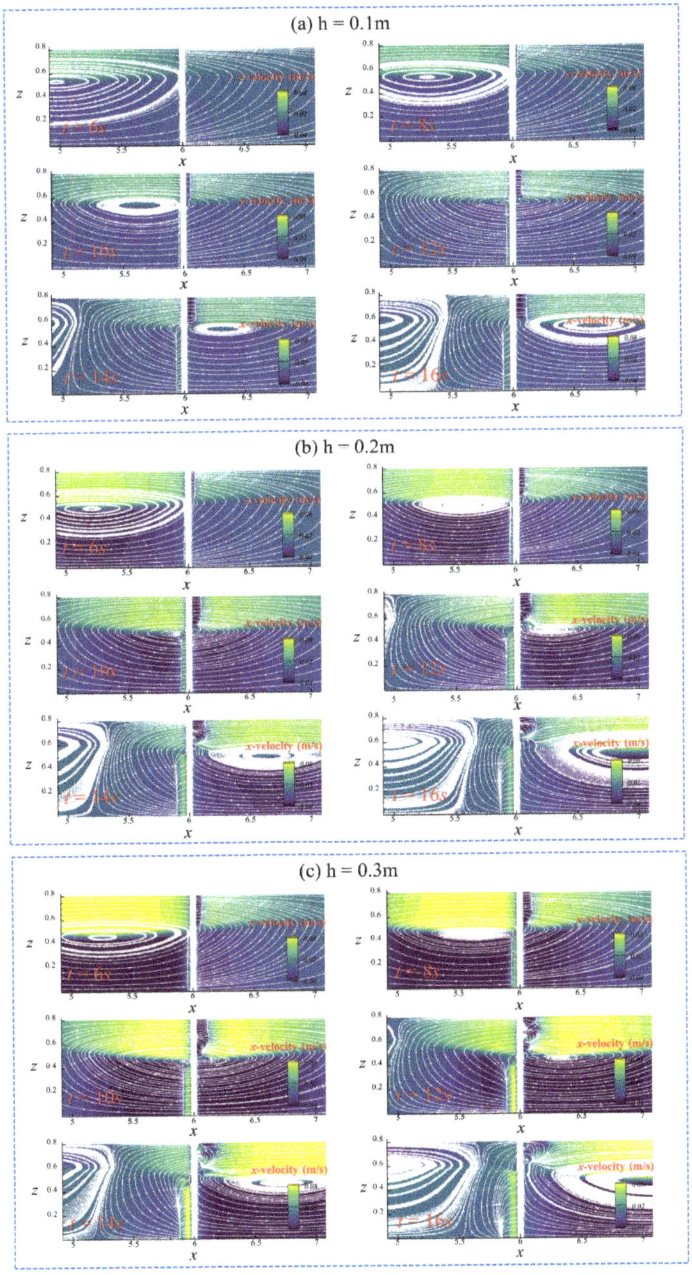

Figure 7. Horizontal velocity and streamline distribution around the cylinder at different times. (**a–c**) show the evolution of the flow field at different moments for gravity collapse heights of 0.1 m, 0.2 m and 0.3 m, respectively.

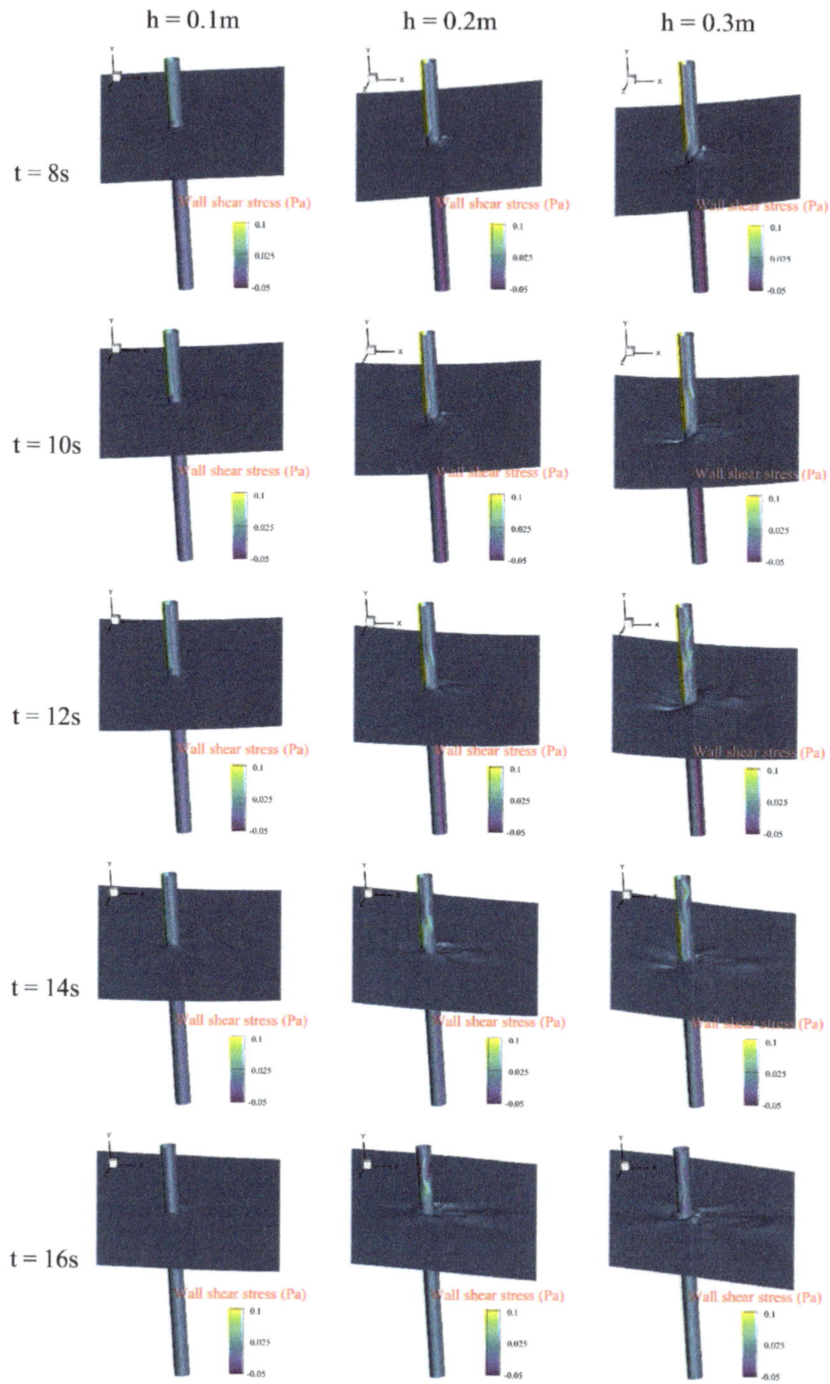

Figure 8. The change in wavefront and the distribution of shear stress on the wall of the cylinder when the ISW flows through the cylinder.

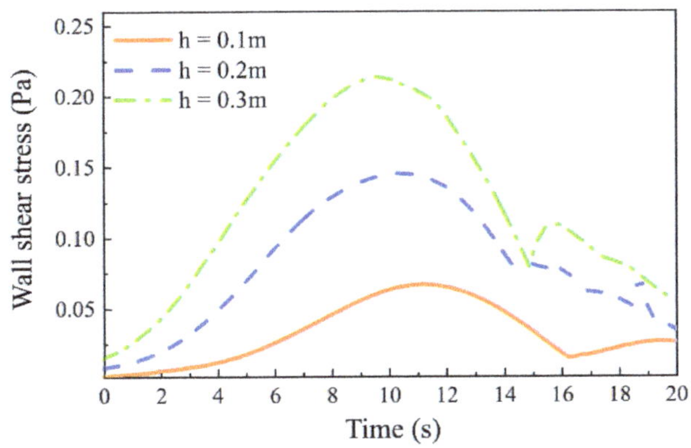

Figure 9. Maximum value of wall shear stress in the forward direction at different moments of time.

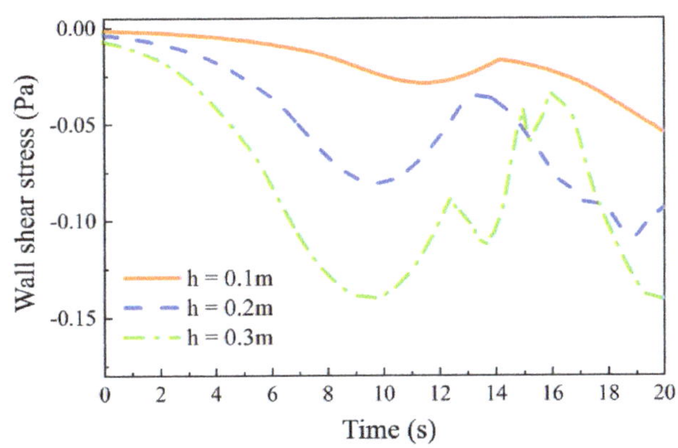

Figure 10. Maximum value of wall shear stress in the backward direction at different moments of time.

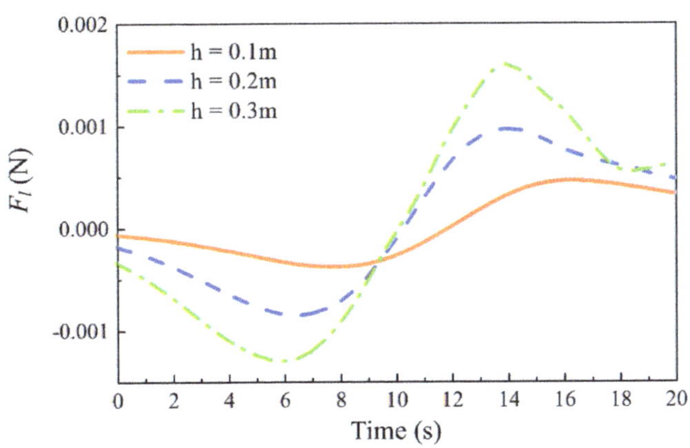

Figure 11. Lift force of the whole cylinder at different moments of time.

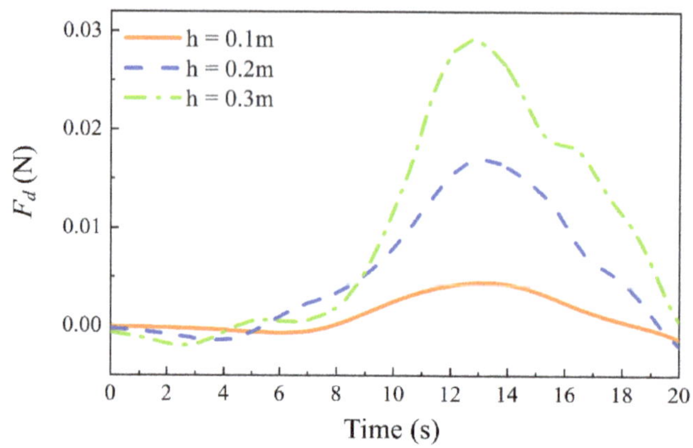

Figure 12. Drag force of the whole cylinder at different moments of time.

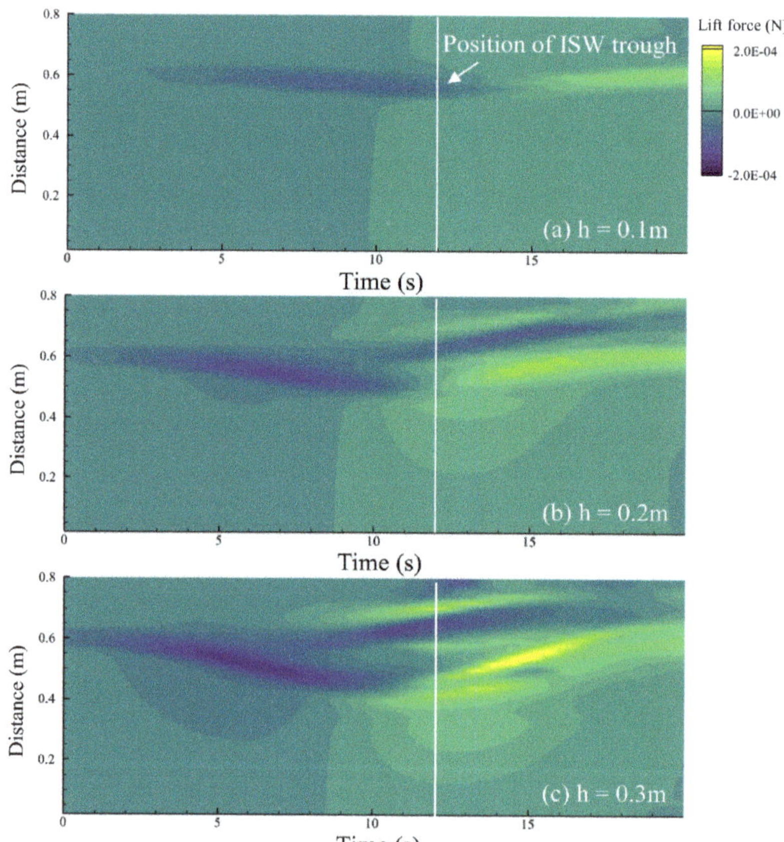

Figure 13. Lift force of each segment of the cylinder in the ISWs at different moments of time. (**a**) Lift force of each segment of the cylinder in an ISW with h = 0.1 m. (**b**) Lift force of each segment of the cylinder in an ISW with h = 0.2 m. (**c**) Lift force of each segment of the cylinder in an ISW with h = 0.3 m.

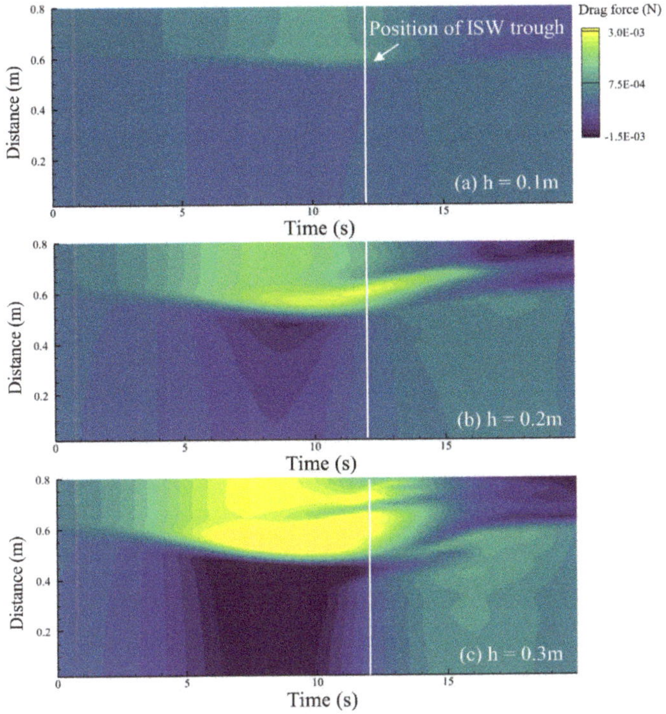

Figure 14. Drag force of each segment of the cylinder in the ISWs at different moments of time. (**a**) Drag force of each segment of the cylinder in an ISW with h = 0.1 m. (**b**) Drag force of each segment of the cylinder in an ISW with h = 0.2 m. (**c**) Drag force of each segment of the cylinder in an ISW with h = 0.3 m.

We can find that the drag forces on the cylinder are in opposite directions between the upper and lower layers as it traverses the ISW, which is bounded by the intermediate pycnocline. As the scale of the ISW becomes larger, the strength of the force on the cylinder increases. In addition, we can find that the trough of the ISW arrives at the position where the cylinder is located at t = 12 s. Therefore, the process of the cylinder travelling through the ISW can be divided into two stages: the cylinder is gradually approaching the trough from 0 to 12 s; and the cylinder is gradually moving away from the trough from 12 s to 20 s. In the first stage, the distribution of the force characteristics of the cylinder is similar to the horizontal velocity of the ISW flow field. In the second stage the horizontal force distribution becomes chaotic, and the larger the ISW scale is, the more chaotic it is.

In order to further analyze the force commonality of the cylinders under different scales of ISW, we carried out the dimensionless quantification of the horizontal force. The distribution of the horizontal force after dimensionless sizing is shown in Figure 15. In the first stage, when the scale of the ISW increases, the force on the cylinders also increases. In addition, the cylindrical forces in the upper fluid become progressively more complex with time evolution. In the second stage, the force on the cylinder in the upper fluid becomes smaller instead as the ISW scale increases. Figure 16 reflects the force behavior of the whole cylinder in the ISW. We find that the force at h = 0.2 m is consistently larger than the case of h = 0.1 m. When h = 0.3 m, the force profile of the cylinder changes more complicatedly. Sometimes it is lower than the previous two cases, but its peak point is higher than the two, and its trend is consistent with the previous two.

$$CF^* = \frac{2F_x}{\rho_1 U_{max}^2 DH} \tag{9}$$

where F_x is the horizontal force, ρ_1 represents the density of the upper layer, U_{max} is the maximum value of transverse velocity, D is the diameter of the cylinder, and H is the total water depth in the tank.

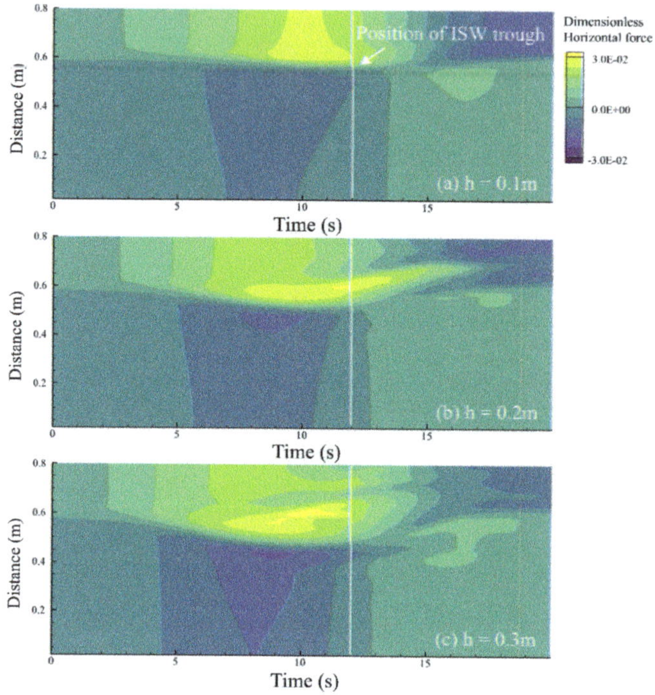

Figure 15. Temporal variation of dimensionless horizontal forces on the cylinder subjected to ISWs. (**a**) Dimensionless horizontal forces on the cylinder in an ISW with h = 0.1 m. (**b**) Dimensionless horizontal forces on the cylinder in an ISW with h = 0.2 m. (**c**) Dimensionless horizontal forces on the cylinder in an ISW with h = 0.3 m.

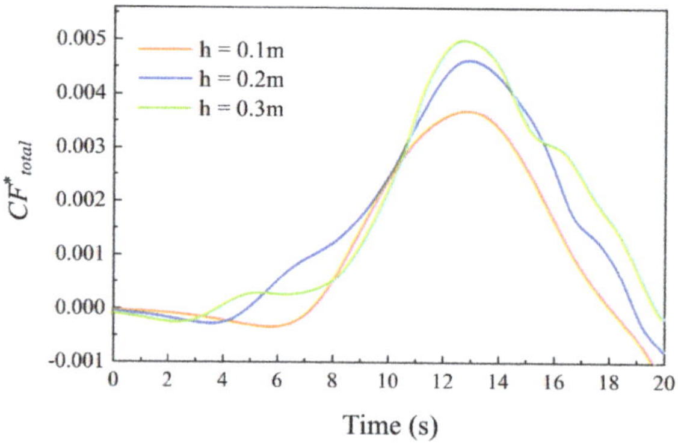

Figure 16. Temporal variation of the total combined horizontal forces of the cylinder.

In addition, we calculated the force on the cylinder under the influence of pressure to find out the specific reasons that affect the force evolution of the cylinder, as shown in Figure 17a. Figure 17b represents the total combined force, and Figure 17c represents the difference between the pressure-induced force on the cylinder and the total combined force. We use the concept of relative error as shown in Equation (10). After comparison, the share of viscous shear stress is larger when the combined force is close to zero. However, most of the time, the pressure-induced horizontal force dominates the total horizontal force. Therefore, we extracted the pressure distributions at the positions of h = 0.1 m and cylinder heights of 0.5 m and 0.7 m. The height of 0.5 m corresponds to the lower part of the pycnocline, and 0.7 m is located in the upper part of the pycnocline. In this way, we can analyze the pressure variations of the cylinder as it traverses the ISW. For the first stage, we show the pressure distribution at four moments, t = 4 s, 6 s, 8 s, and 10 s, as shown in Figure 18. The value of the pressure on the walls of the cylinders gradually increases from the beginning of the contact of the cylinder with the ISW to the trough until the arrival of the cylinder at the trough position. The pressure on the upper cylinder is greater than that on the lower one. This is because the horizontal velocity field induced by the ISW is different in magnitude between the upper and lower layers. In addition, we can also find that the pressure extremes appear near the flow separation point on the headward side. The pressure values in the separation zone on the backflow side of the cylinder also increase gradually with time. When the cylinder reaches the trough, the evolution of the pressure distribution on its wall is shown in Figure 19. We can see that the pressure on the head side is gradually weakened, and the pressure on the back side is gradually strengthened. When t = 16 s, the pressure on the backflow surface has exceeded the pressure on the onflow surface. At this time, the direction of the horizontal force on the cylinder is reversed. Thereafter, the pressure on the backflow surface continues to increase, which leads to a consequent increase in the horizontal force on the cylinder.

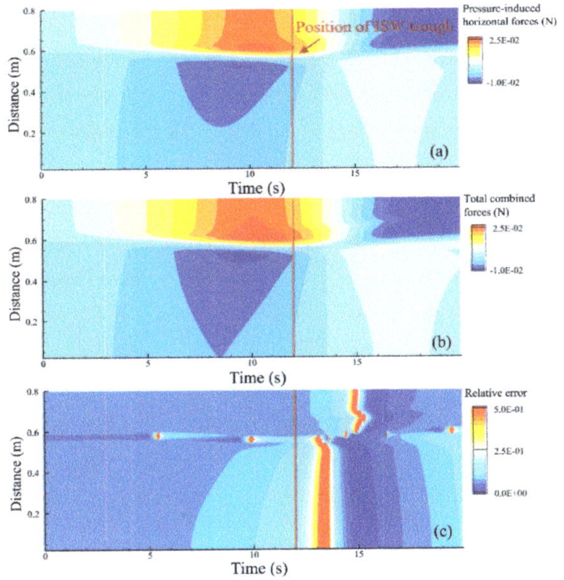

Figure 17. Pressure-induced horizontal forces versus total combined forces. (**a**) Pressure-induced horizontal forces. (**b**) Total combined forces. (**c**) The difference between the pressure-induced force and the total combined forces.

$$RE = \left| \frac{F_{xP} - F_x}{F_x} \right| \quad (10)$$

where RE represents relative error, F_x is the horizontal force, and F_{xp} is the pressure-induced horizontal force.

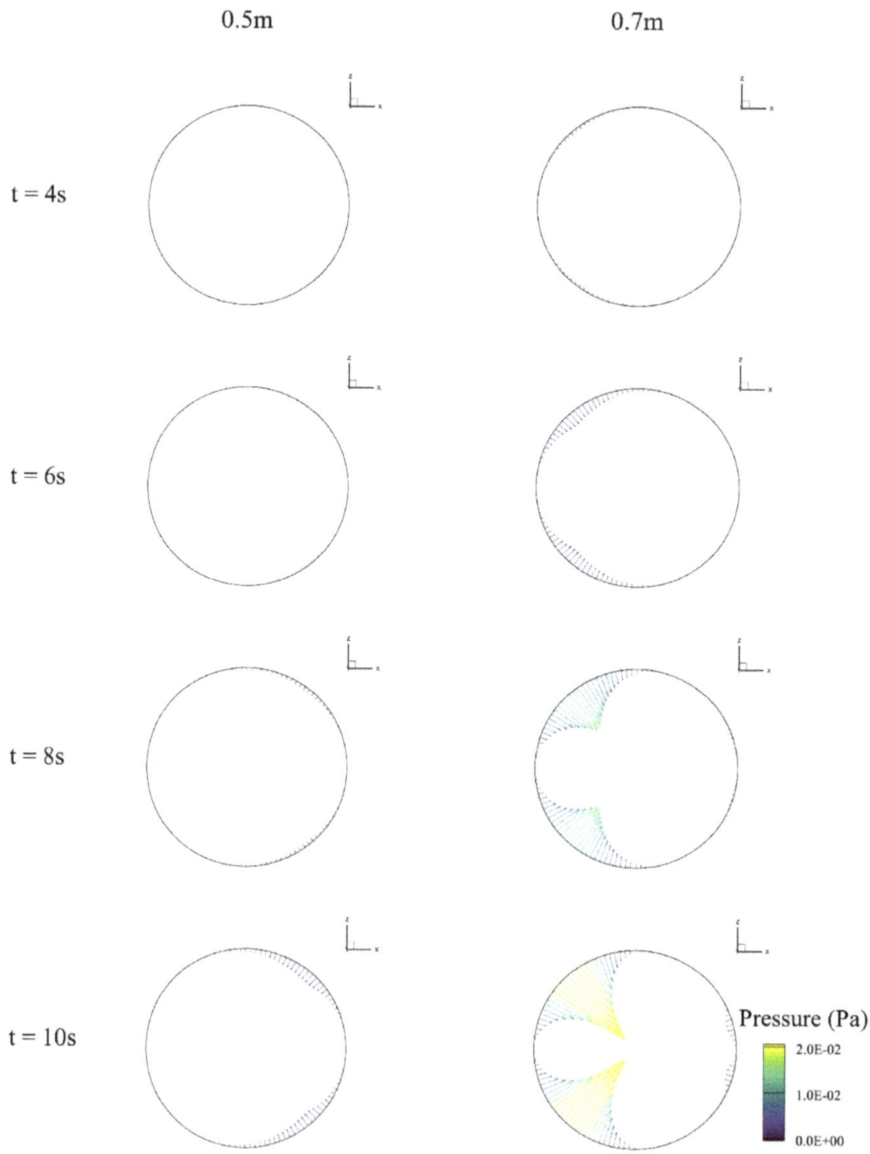

Figure 18. Dynamic pressure changes on a cylinder as the trough of the ISW gradually approaches.

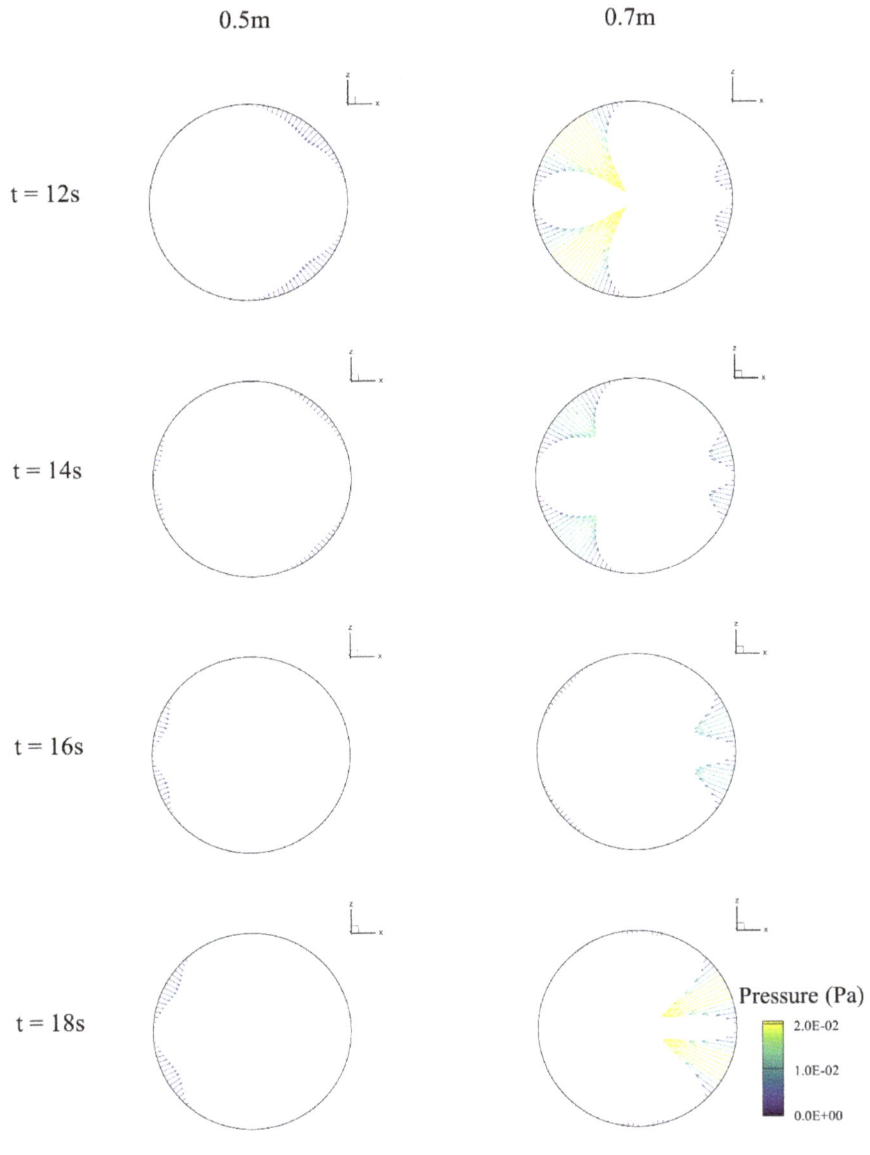

Figure 19. Dynamic pressure changes on the cylinder as the trough of the ISW gradually moves away from it.

4.3. Spatiotemporal Evolution Patterns of Cylindrical Trailing Vortex

In the above article, we found that the horizontal force on the cylinder will show a certain pattern of change when it crosses the ISW. Previous studies have also shown that the force on the cylinder is closely related to the shedding phenomenon of its wake [24]. Therefore, we try to analyze the force characteristics of the cylinder from the evolution characteristics of the cylindrical trailing vortex. We adopt the dynamic mode decomposition (DMD) method to extract the main modes. These main modes represent the dominant flow structure in the whole time period [25]. The intrinsic flow of information can be mined by analyzing its spatial and temporal evolution.

Before using the DMD method, multiple snapshots of time over a period of time must be integrated into a two-dimensional matrix [26]. The complete time snapshot matrix is then divided into two matrices, **X X**$'$,.

$$\begin{aligned} \mathbf{X} = & \begin{bmatrix} \mathbf{x}_1 & \mathbf{x}_2 & \ldots & \mathbf{x}_{m-1} \end{bmatrix} \\ \mathbf{X}' = & \begin{bmatrix} \mathbf{x}_2 & \mathbf{x}_3 & \ldots & \mathbf{x}_m \end{bmatrix} \end{aligned} \tag{11}$$

Assuming that $\mathbf{x}$ is linearly correlated with $\mathbf{X}'$, we can obtain the matrix A. The A matrix can be calculated by eigendecomposition $\mathrm{A}\Phi = \Phi\Lambda$. This step is obtained based on the low-rank approximation of the data matrix.

$$\mathbf{X}' = \mathbf{AX} \tag{12}$$

Eigenvalues and eigenvectors are obtained utilizing rank truncation. These eigenvalues move forward in time. The state at each time in the original data can be calculated by the linear combination of DMD modes (columns ϕ_k of $\mathbf{\Phi}$), eigenvalues ($\lambda_k = \Lambda_{kk}$), and corresponding modal amplitudes b:

$$\mathbf{x}(t) = \sum_{k=1}^{r} b_k \phi_k(\xi) \exp(i\omega_k t) \tag{13}$$

In this study, the interaction of ISWs with the cylinder is an unsteady process. Therefore, we selected 2000 snapshots. The time interval between two snapshots is 0.01 s. The total duration is 20 s. In addition, we found that the trough of the ISW arrives at the location of the cylinder at t = 12 s. Therefore, the whole flow field is divided into two segments with t = 12 s as discontinuity. For the first segment, we extracted the cylindrical trailing vortex modes in three scales of ISWs. The lower the corresponding frequency of the mode, the higher the energy contributed. That is, the order of frequencies determines the order of modes. For the trailing vortex field around the cylinder under the influence of the three scaled ISWs, we extract the first three orders of the modes.

In Figure 20, we have extracted two vortex equivalent surfaces for $\omega_y = -0.008$ and $\omega_y = 0.008$, respectively, to demonstrate their tailing vortex patterns. The corresponding time evolution curves for each mode are shown in Figure 21. By comparing the mode 1 distributions for the three cases, it can be seen that the tailing vortex of the cylinder gradually grows as the ISW scale increases. However, as the ISW scale continues to increase, the distribution of the tailing vortex in the upper layer of the cylinder becomes irregular. The time evolution curves also show that the intensity of the tailing vortex gradually increases when the cylinder is close to the wave trough. The larger the ISW scale, the stronger and faster the intensity of the tailing vortex grows. This can also be compared to the force behavior of the cylinder, see Figure 14. In the first stage, as the ISW scale gradually increases, the force of the cylinder also increases and becomes chaotic. We can also see that the gradient of the change in force increases. For modes 2 and 3, they represent the higher frequency features of the flow field compared to mode 1, and the distribution of mode 2 is similar to that of mode 1. Mode 3 represents the smaller vortexes in the flow field, which are more complex than the first two modes. However, its total energy content is much smaller than that of the first two modes. In addition, according to the time evolution curves, it is found that the amplitude change in the corresponding modes is proportional to the scale of the internal wave. Moreover, the frequencies of the corresponding modes are almost the same under the three working conditions.

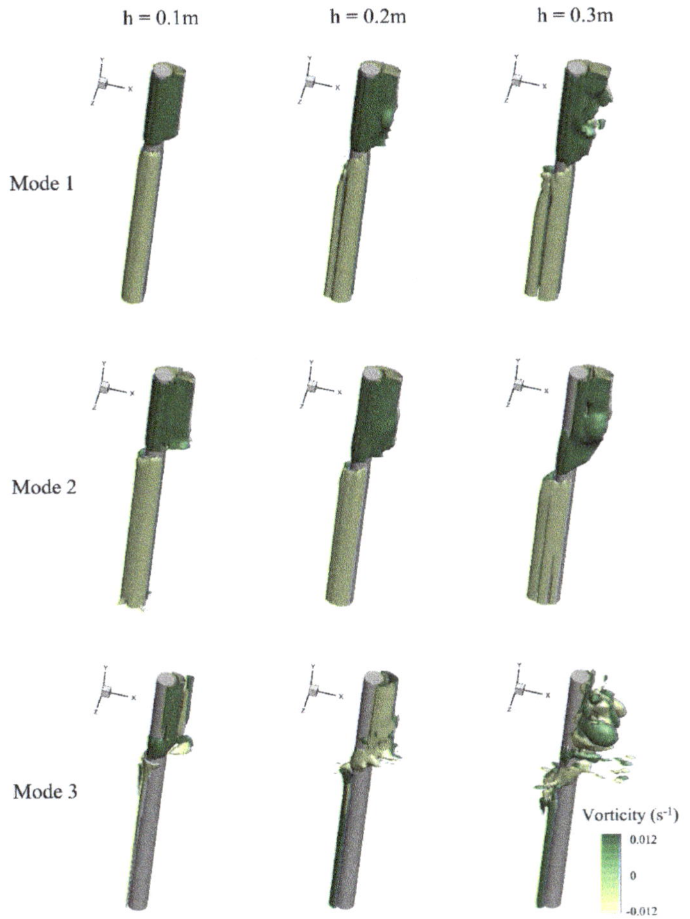

Figure 20. Mode distribution of vorticity around the cylinder in the first stage.

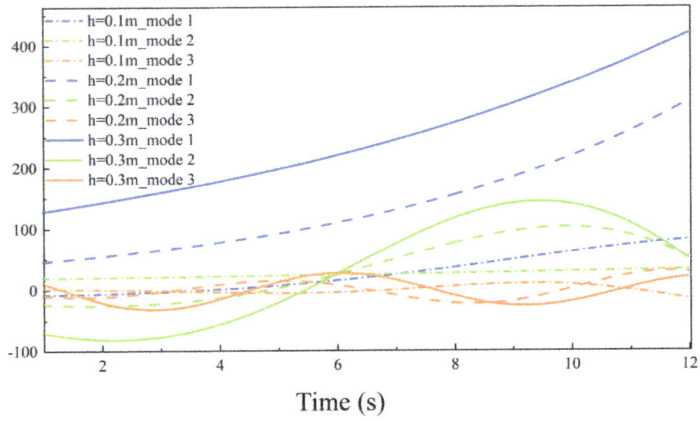

Figure 21. Time evolution of different modes in the first stage.

For the second stage in which the trough gradually moves away from the cylinder, we similarly extracted the first three orders of modes of the cylindrical trailing vortex during this process, which are shown in Figure 22. The time evolution curves corresponding to each mode are shown in Figure 23. Upon comparison, it is found that the trend of the evolution of the corresponding modes in this phase is exactly opposite to that of the first phase. In addition, the modal distribution of the vorticity in this stage is very different from that of the first stage. Taking mode 1 as an example, the wake of the cylinder in the upper layer gradually becomes chaotic when the trough gradually moves away from the cylinder. In terms of time evolution, the change in mode 1 is exactly opposite to the first stage, which is gradually weakened, and the final value comes to the same. And the larger the ISW is, the stronger the corresponding mode strength is at the same moment, the more it decreases. The above changes in the cylindrical tailing vortex can exactly confirm the force manifestation of the cylinder in the second stage in Figure 17. The force distribution of the part of the cylinder in the upper layer also becomes cluttered with the increase in the ISW scale. Comparing the performances of the first three stages of modes, we find that the cylinder can still maintain a stable vortex shedding in the ISW generated at h = 0.1 m. By the time h = 0.2 m, the vortex shedding of the cylinder becomes extremely unstable. Until h = 0.3 m, almost no stable vortex shedding appears. This is something we can analyze against from the dimensionless horizontal force distribution, as shown in Figure 15.

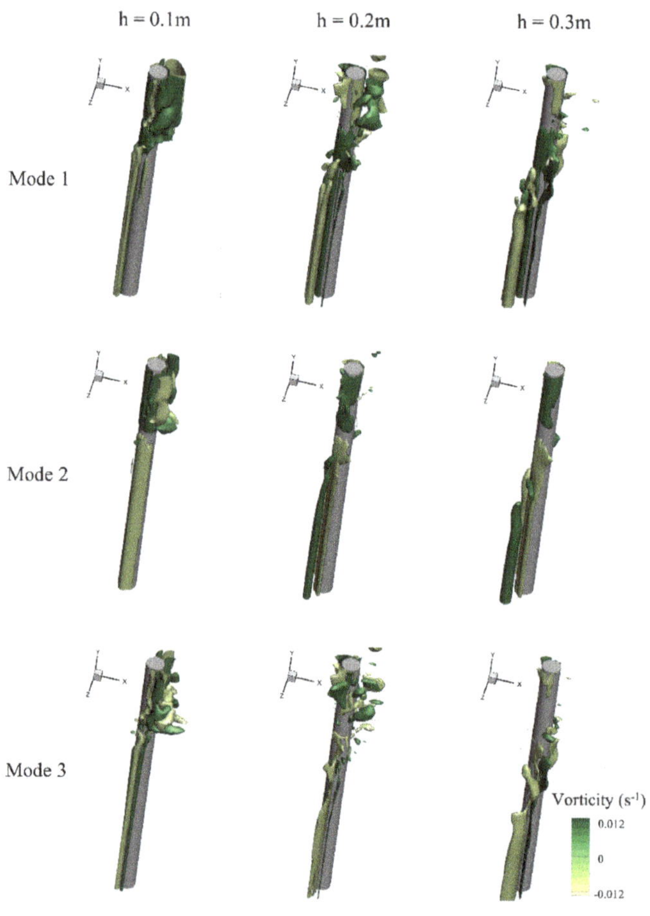

Figure 22. Mode distribution of vorticity around the cylinder in the second stage.

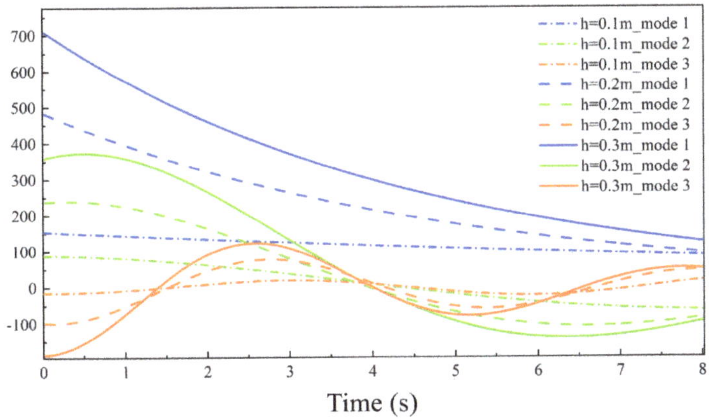

Figure 23. Time evolution of different modes in the second stage.

5. Conclusions

In this paper, we generate different scales of ISWs based on the gravity collapse method and analyze the evolution of the flow field around a cylinder. It was found that the flow field induced by the ISW is an elliptical, clockwise-rotating flow field. Its flow core is located in the intermediate pycnocline. The ISW flow field is bounded by the pycnocline, and the upper and lower layers of fluid move in opposite directions. As a result, there will be a symmetric region of high shear stress from the flow meeting point to the separation point, and the forces in the upper and lower layers are in opposite directions. With time, the wall shear stresses in the wake region of the cylinder gradually become chaotic.

In order to investigate the force pattern of the cylinder in the ISW, the whole cylinder is divided into 40 segments at equal distances along the vertical direction. The horizontal force ensemble is calculated by integrating the pressure and velocity shear on the surface of the cylinder. Through our analysis, we found that the force distribution pattern of the cylinder is consistent with the horizontal velocity of the ISW at the stage when the wave trough is gradually approaching the cylinder. In the latter half of the stage, the force gradually becomes chaotic as the cylindrical tailing vortex grows and breaks up. The larger the scale of the ISW, the more complicated the forces behave. In addition, we demonstrated that the normal pressure dominates in the total ensemble force, and analyzed the pressure changes on the cylindrical wall in different stages.

Finally, we extracted the main flow modes in each of the two stages based on the dynamic mode decomposition. When the ISW scale increased, we analyzed the change in the trailing vortex flow field of the cylinder in terms of the mode distribution as well as its time evolution. In addition, we analyzed the flow characteristics in the two phases to further reveal the change in the flow state as the cylinder crosses the wave trough. More importantly, we used the flow modes to explain the reasons affecting the force performance of the cylinder.

Author Contributions: Conceptualization, M.Z. and H.H.; methodology, M.Z.; software, P.D.; validation, M.Z. and H.H.; formal analysis, M.Z.; investigation, M.Z.; resources, M.Z.; data curation, M.Z.; writing—original draft preparation, M.Z.; writing—review and editing, P.D. and A.O.; visualization, P.D.; supervision, P.D.; project administration, X.H.; funding acquisition, L.X. All authors have read and agreed to the published version of the manuscript.

Funding: This research was supported by the National Natural Science Foundation of China (No. 52201380), Fundamental Research Funds for the Central Universities (No. D5000230080), Capability Support Program of Shaanxi (No. 2024ZC-KJXX-081) and Innovation Foundation for Doctor Dissertation of Northwestern Polytechnical University (No. CX2024049).

Institutional Review Board Statement: Not applicable.

Informed Consent Statement: Not applicable.

Data Availability Statement: Data are contained within the article.

Conflicts of Interest: Author Peng Du was employed by the company Xi'an Tianhe Defense Technology Co., Ltd. The remaining authors declare that the research was conducted in the absence of any commercial or financial relationships that could be construed as a potential conflict of interest.

References

1. Bourgault, D.; Galbraith, P.S.; Chavanne, C. Generation of internal solitary waves by frontally forced intrusions in geophysical flows. *Nat. Commun.* **2016**, *7*, 13606. [CrossRef] [PubMed]
2. Zhang, M.; Hu, H.; Du, P.; Chen, X.; Li, Z.; Wang, C.; Cheng, L.; Tang, Z. Detection of an internal solitary wave by the underwater vehicle based on machine learning. *Phys. Fluids* **2022**, *34*, 115137. [CrossRef]
3. Cheng, M.H.; Hwang, R.R.; Hsieh, C.M. Numerical study on the transformation of an internal solitary wave propagating across a vertical cylinder. *Appl. Ocean Res.* **2020**, *95*, 102016.
4. Wei, G.; Du, H.; Xu, X.H.; Zhang, Y.M.; Qu, Z.Y.; Hu, T.Q.; You, Y.X. Experimental investigation of the generation of large amplitude internal solitary wave and its interaction with a submerged slender body. *Sci. China Phys. Mech. Astron.* **2014**, *57*, 301–310. [CrossRef]
5. Wang, F.; Guo, H.Y.; Li, M.L.; Meng, F.S. Experimental study on ocean internal wave force on vertical cylinders in different depths. *China Ocean Eng.* **2016**, *30*, 459–468. [CrossRef]
6. Wang, Y.; Wang, L.L.; Ji, Y.; Zhang, J.; Xu, M.; Xia, X.; Wang, C.L. Research on the force mechanism of two tandem cylinders in a stratified strong shear environment. *Phys. Fluids* **2022**, *34*, 053308. [CrossRef]
7. Cai, S.Q.; Long, X.M.; Wang, S.G. Forces and torques exerted by internal solitons in shear flows on cylindrical piles. *Appl. Ocean Res.* **2008**, *30*, 72–77. [CrossRef]
8. Song, Z.J.; Teng, B.; Gou, Y.; Lu, L.; Shi, Z.M.; Xiao, Y.; Qu, Y. Comparisons of internal solitary wave and surface wave actions on marine structures and their responses. *Appl. Ocean Res.* **2011**, *33*, 120–129. [CrossRef]
9. Lü, H.B.; Xie, J.S.; Xu, J.X.; Chen, Z.W.; Liu, T.Y.; Cai, S.Q. Force and torque exerted by internal solitary waves in background parabolic current on cylindrical tendon leg by numerical simulation. *Ocean Eng.* **2016**, *114*, 250–258. [CrossRef]
10. Si, Z.S.; Zhang, Y.L.; Fan, Z.S. A numerical simulation of shear forces and torques exerted by large-amplitude internal solitary waves on a rigid pile in South China Sea. *Appl. Ocean Res.* **2012**, *37*, 127–132. [CrossRef]
11. Wang, F.; Sun, R.; Wang, C.X.; Fu, Q.; Li, P.; Guo, H.Y. Experimental study on flow field induced by internal solitary wave and load characteristics on pile sections at different depth. *Ocean Eng.* **2019**, *188*, 106292. [CrossRef]
12. Yu, W.; Wang, F.G.; Lin, J.G.; Li, D. Numerical Simulation of the Force Acting on the Riser by Two Internal Solitary Waves. *Appl. Sci.* **2022**, *12*, 4873. [CrossRef]
13. Xie, Z.L.; Jiao, J.; Zhao, B.; Xu, F.C. Theoretical and experimental research on the effect of bi-directional misalignment on the static and dynamic characteristics of a novel bearing. *Mech. Syst. Signal Process.* **2024**, *208*, 111041. [CrossRef]
14. Jiao, J.L.; Chen, Z.W.; Xu, S. CFD-FEM simulation of water entry of aluminium flat stiffened plate structure considering the effects of hydroelasticity. *Brodogradnja* **2024**, *75*, 75108. [CrossRef]
15. Gong, S.K.; Gao, J.L.; Mao, H.F. Investigations on fluid resonance within a narrow gap formed by two fixed bodies with varying breadth ratios. *China Ocean Eng.* **2023**, *37*, 962–974.
16. Qi, C.; Lyu, X.J.; Wang, X.; Ye, H.; Shi, H.J.; Wan, Z.H. Experimental and numerical studies on vertical water entry of a cylinder under the influence of current. *Phys. Fluids* **2024**, *36*, 033322. [CrossRef]
17. Liu, S.; Ong, M.C.; Obhrai, C.; Gatin, I.; Vukčević, V. Influences of free surface jump conditions and different $k-\omega$ SST turbulence models on breaking wave modelling. *Ocean Eng.* **2020**, *217*, 107746. [CrossRef]
18. Zhang, M.; Hu, H.B.; Guo, B.B.; Liang, Q.Y.; Zhang, F.; Chen, X.P.; Xie, Z.L.; Du, P. Predicting shear stress distribution on structural surfaces under internal solitary wave loading: A deep learning perspective. *Phys. Fluids* **2024**, *36*, 035153.
19. Shiau, Y.H.; Peng, Y.F.; Hwang, R.R.; Hu, C.K. Multistability and symmetry breaking in the two-dimensional flow around a square cylinder. *Phys. Rev. E* **1999**, *60*, 6188–6191. [CrossRef]
20. Du, H.; Wei, G.; Wang, S.D.; Wang, X.L. Experimental study of elevation-and depression-type internal solitary waves generated by gravity collapse. *Phys. Fluids* **2019**, *31*, 102104. [CrossRef]
21. Peng, P.; Du, H.; Wei, G.; Wang, S.D.; Xuan, P.; Cai, S.Q.; Xie, J.S. Experimental investigation on the vertical structure characteristics of internal solitary waves. *J. Mar. Sci. Eng.* **2022**, *10*, 1045. [CrossRef]
22. Wang, S.D.; Du, H.; Wei, G.; Peng, P.; Xuan, P. Experimental modification of the internal solitary wave force exerted on a horizontal transverse cylinder due to wave-flow and vortex shedding. *Ocean Eng.* **2023**, *269*, 113513. [CrossRef]
23. Lin, Z. Numerical study on the forces and moments on a bottom-mounted cylinder by internal solitary wave. *J. Fluids Struct.* **2023**, *121*, 103952. [CrossRef]
24. Janocha, M.J.; Yin, G.; Ong, M.C. Modal analysis of wake behind stationary and vibrating cylinders. *J. Offshore Mech. Arct. Eng.* **2021**, *143*, 041902. [CrossRef]

25. Min, G.; Jiang, N. Data-driven identification and pressure fields prediction for parallel twin cylinders based on POD and DMD method. *Phys. Fluids* **2024**, *36*, 02361. [CrossRef]
26. Zhang, J.M.; Zou, L.; Sun, T.Z.; Wen, Z.H.; Yu, Z.B. Experimental investigation on the propagation characteristics of internal solitary waves based on a developed piecewise dynamic mode decomposition method. *Phys. Fluids* **2020**, *32*, 082102. [CrossRef]

Journal of
Marine Science and Engineering

MDPI

Article

Numerical Simulation and Analysis of Added Mass for the Underwater Variable Speed Motion of Small Objects

Xuanquan Wang [1,2], Suwei Xiao [3,*], Xinchun Wang [4] and Debo Qi [5]

1 School of Optics and Photonics, Beijing Institute of Technology, Beijing 100081, China
2 Yangtze Delta Region Academy, Beijing Institute of Technology, Jiaxing 314003, China
3 Rubber Research Institute, Chinese Academy of Tropical Agricultural Sciences, Haikou 570216, China
4 Beijing Aerospace Automatic Control Institute, Beijing 100854, China
5 College of Intelligent Manufacturing, Yangzhou Polytechnic Institute, Yangzhou 225127, China
* Correspondence: xiaosuwei123@sina.com

Abstract: Unlike uniform motion, when an object moves underwater with variable speed, it experiences additional resistance from the water, commonly referred to as added mass force. At present, several methods exist to solve this force, including theoretical, experimental, and simulation approaches. This paper addresses the challenge of determining the added mass force for irregularly shaped small objects undergoing variable speed motion underwater, proposing a method to obtain the added mass force through numerical simulation. It employs regression analysis and parameter separation analysis to solve the added mass force, added mass, viscous drag coefficient, and pressure drag coefficient. The results indicate that an added mass force exists during both the acceleration and deceleration of the object, with little difference between them. Under the same velocity conditions, significant differences exist in pressure drag forces, while differences in viscous drag forces are not significant. This suggests that the primary source of added mass force is pressure drag, with viscous drag having little effect on it. During acceleration, the surrounding fluid accelerates with the object, increasing the pressure drag with a high-pressure area concentrating at the object's front, forming an added mass force that is directed backward. By contrast, during deceleration, the fluid at the object's front tends to detach, and the fluid at the rear rushes forward, leading to a smaller high-pressure area at the front and a larger one at the rear, reducing the pressure drag and forming an added mass force that is directed forward. By comparing the added mass of a standard ellipsoid obtained from numerical simulation with theoretical values, the regression analysis method is proven to be highly accurate and entirely applicable for solving the added mass of underwater vehicles.

Keywords: added mass; added mass force; pressure drag; viscous drag; computational fluid mechanics

1. Introduction

When a rigid body accelerates relative to the surrounding fluid, it experiences additional inertial forces. That is, if two rigid bodies are moving at the same speed, the one that is accelerating will be subjected to a greater force from the water. The underlying reason for this is that when the rigid body accelerates, the surrounding fluid will also accelerate, as if additional mass was attached to the rigid body, resulting in an increase in the body's inertia [1,2]. Added mass, a classical concept in the field of fluids, was proposed by researchers and has been widely studied for many years. The added mass effect was discovered by Du Buat in 1786 and Friedrich Bessel in 1828, independently. Added mass is of fundamental importance in hydrodynamic research [1].

Currently, researchers employ various methods to determine the added mass of objects, including Analytical and Semi-Empirical (ASE), Experimental Fluid Dynamics (EFD), and Computational Fluid Dynamics (CFD) approaches [3]. Analytical methods mainly consist of Strip Theory (ST) and the Equivalent Ellipsoid Method (EMM), which calculate the added mass based on the object's geometry. These methods solve for velocity potential using boundary conditions and determine the pressure distribution around the object using Bernoulli's equation, thus solving for the hydrodynamic forces and moments acting on the object. ST divides the object into a finite number of transverse two-dimensional slices, solving for the added mass by integrating slice values along the length of the object [4]. In EMM, different parts of the object are assumed to be equivalent ellipsoids, and the velocity potential is solved using the Laplace equation, thereby obtaining the added mass of each equivalent ellipsoid. By integrating the added mass of different parts in a specific direction, the added mass in that direction can be solved [5]. Overall, ASE methods are only suitable for objects with simple shapes, as increased complexity can reduce accuracy. These methods require sufficient experimental data and must consider the complexity of the object's shape [6,7].

In EFD methods, researchers have developed numerous experimental techniques to study added mass, such as free oscillation [8–11] and forced oscillation [12–16]; these inertia methods, with extensive experimental setups like the Rotating Arm (RA) and Planar Motion Mechanism (PMM), are being developed [17,18]. Most added masses are currently analyzed from the hydrodynamic characteristics of oscillation, but new methods for solving added mass, such as PIV testing and trajectory tracking, have emerged [19–21], which can solve the added mass of deformable objects. EFD methods are the most direct way to solve added mass, with relatively mature related technologies. However, they are not suitable for very large or small objects. On one hand, large sizes can affect the implementation of related experiments; on the other hand, small sizes can lead to significant errors, affecting the resulting accuracy. Moreover, due to the need for special facilities and equipment (e.g., PMM) to generate an object's acceleration, these methods are not only costly but also time-consuming.

CFD includes two methods for estimating added mass: the first is based on potential flow assumptions, including the Panel Method (PM), Boundary Element Method (BEM), and Finite Element Method (FEM). Many researchers have used PM [22–25], BEM [26–30], and FEM [31,32] to predict the added mass coefficients of objects. In practice, the viscosity of the fluid around the object can reduce the accuracy of the results obtained from these assumptions. The second approach uses commercial CFD software (https://link.springer.com/article/10.1007/s11804-010-9009-9, accessed on 28 March 2024) to simulate dynamic testing processes like PMM and RA. In these simulation methods, the RANS equations are solved in the computational domain, considering the effect of fluid viscosity. With the continuous development of commercial CFD software, numerous related studies have emerged [33–39], making this method a supplementary approach to experimental testing.

In recent years, the added mass effect has received increasing attention and research [40–44]. With the continuous development of marine technology, underwater vehicles are evolving toward miniaturization, possessing advantages such as compact size, agile movement, low resistance, good concealment, and low noise. They play an important role in intelligence gathering, target tracking and striking, monitoring enemy actions, and cluster warfare, with a very broad prospect for military applications [45–51]. Due to the flexible and variable movement of miniature underwater robots, with constantly changing directions and speeds, the added mass effect becomes more significant and is an important factor that cannot be ignored. On the other hand, with the continuous exploitation of marine resources and advancements in underwater visual capabilities, underwater robots equipped with multi-degree-of-freedom manipulators will continue to emerge. They can perform targeted collection and exploration of marine resources, and the irregular motion of the manipulators and underwater robots will also generate significant added mass forces. Therefore, there is an urgent need to find a simple, quick, and convenient method to solve for added mass. With the continuous advancement of numerical simulation technology, several studies have applied this technique to complex fluid dynamics analysis [52–54]. This paper employs numerical simulation as a means, combined with effective data analysis methods, to analyze and study the added mass of objects in variable speed motion underwater. Compared to existing methods, the approach proposed in this paper does not require complex theoretical derivations and assumptions. It relies on the dependency relationship between fluid resistance and motion variables, utilizing a large number of data samples to solve mechanical parameters. This method is not only simple, understandable, and easy to operate, but it also maintains good analytical precision under the premise of ensuring the accuracy of simulation results. All of these indicate that this method is applicable to the actual mechanical analysis of underwater vehicles, simplifying the analysis process and significantly shortening the analysis cycle.

2. Methods

As previously mentioned, when a rigid body moves underwater linearly with variable speed, it experiences an additional force known as added mass force. In laminar flow conditions, an object moving underwater at a constant speed is subject only to viscous drag and pressure drag. Therefore, the water resistance experienced by an object in variable speed motion can be decomposed into the viscous drag and pressure drag corresponding to the current speed of motion, in addition to the extra added mass force. Pressure drag is caused by the difference in fluid pressure between the front and back surfaces of the rigid body. Viscous drag, also referred to as wall shear stress, is induced by the viscosity of the fluid. When a rigid body moves underwater, the viscous fluid rubs against the surface of the body, hindering its movement and forming viscous drag. Generally, for small objects moving at low speeds, the viscous drag is dominant, while for large objects moving at high speeds, the pressure drag predominates. The viscous drag is approximately proportional to the first power of the relative speed between the object and the fluid, whereas the pressure drag is approximately proportional to the square of their relative speed. These two types of drag exist during the motion of the rigid body, regardless of whether the motion is uniform.

The third type of force, the added mass force, is an additional force that arises due to the variable speed motion of the object in the fluid. When two identical objects move through a fluid at the same speed, the one accelerating experiences greater fluid resistance. The difference between the two is referred to as the added mass force. This phenomenon occurs because, when an object accelerates in a fluid, the surrounding fluid will also accelerate due to frictional forces. The inertial force of this portion of the fluid acts back on the object, making it more difficult for the object to accelerate. This effect appears as if the mass of the object has increased, and this additional mass is referred to as added mass. Figure 1 demonstrates the drags acting on a rigid body under three distinct motion conditions. From the figure, it is evident that pressure drag and viscous drag consistently act in opposition to the direction of motion. Conversely, the added mass force opposes the direction of the rigid body's acceleration. The computational formula for fluid resistance is provided in Equation (1), while Equation (2) delineates the relationship between the added mass force and the added mass itself.

$$F_{dw} = F_{dp} + F_{dv} + F_{am} \tag{1}$$

where F_{dw} represents the total resistance experienced by an object moving underwater (N); F_{dp} denotes the pressure drag on the object during uniform motion (N); F_{dv} is the viscous drag on the object during uniform motion (N); F_{am} indicates the added mass force acting on the object (N).

$$F_{am} = m^* a \tag{2}$$

where m^* represents the added mass (kg); a is the acceleration of the object ($m{\cdot}s^{-2}$), with the constraint that the fluid is either stationary or moving at a constant speed.

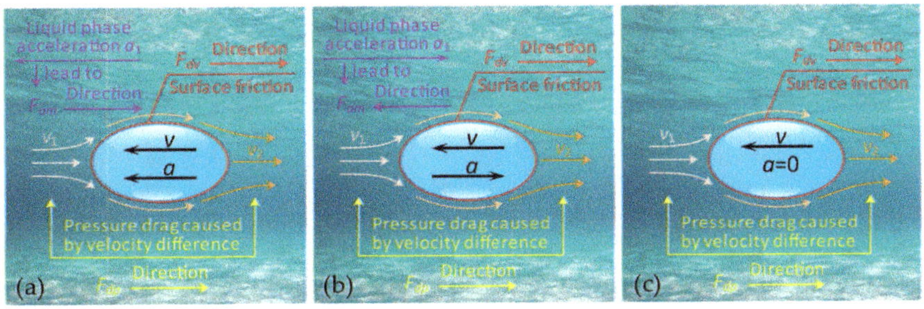

Figure 1. Resistance encountered by an object during underwater motion. (**a**) Accelerating motion; (**b**) decelerating motion; (**c**) uniform motion.

Currently, numerous papers have introduced methods for determining added mass from both experimental and simulation perspectives. For regular objects, added mass can be obtained through theoretical derivation. However, for smaller, irregularly shaped objects, which also tend to have less mass, it is challenging to ascertain their added mass during underwater motion through experimental and theoretical derivation methods. The numerical simulation methods available at present are also somewhat cumbersome.

To address the challenges encountered in the solution process mentioned above, this paper proposes a solution based on CFD simulation methods for determining the added mass of irregularly shaped small objects. The methodological schematic is shown in Figure 2, with the specific steps outlined as follows:

(1) Based on three-dimensional models of a standard ellipsoid and an irregular object, the fluid domain mesh is partitioned. The Fluent 19.2 software is utilized to simulate the process of a standard ellipsoid and an irregular object undergoing variable acceleration movement underwater.

(2) The fluid resistance curve is fitted using the least squares method, with the model's speed and acceleration (v, a) as independent variables, and the total water resistance (F_{dw}) as the dependent variable. The specific fitting function is shown in Equation (3). This method is primarily based on the components of F_{dw}, by utilizing a large set of sample data to solve the three parameters $\hat{a}$, $\hat{b}$, and $\hat{c}$.

$$F_{dw} = F_{db} + F_{am} = F_{dp} + F_{dv} + F_{am} = \hat{a}v^2 + \hat{b}v + \hat{c}a \tag{3}$$

where F_{dw} represents the total water resistance experienced by the model during its motion (N); F_{db} denotes the water resistance acting on the model during motion, excluding the added mass force (N); F_{am} is the added mass force (N); $\hat{a}$, $\hat{b}$, and $\hat{c}$ are constants to be determined through fitting; v is the speed of the model (m·s^{-1}); a represents the acceleration of the model (m·s^{-2}).

(3) Based on the components of F_{db} and the relationship between F_{am} and m^*, the values of C_1, C_2, and m^* can be determined, as shown in Equations (4)–(6).

$$C_1 = \frac{\hat{b}}{\mu L} \tag{4}$$

$$C_2 = \frac{2\hat{a}}{\rho S_W} \tag{5}$$

$$m^* = |\hat{c}| \tag{6}$$

where C_1 and C_2 are the coefficients for the viscous drag and pressure drag, respectively, and can be considered constants within a certain range of Reynolds numbers; μ is the dynamic viscosity of water (Pa·s); L is the characteristic length scale; ρ is the density of water (kg·m^{-3}); S_W is the model's maximum cross-sectional area facing the flow (m^2).

(4) As a method of verification in contrast to process (2), this step employs the parameter separation method to solve for the three parameters $\hat{a}$, $\hat{b}$, and $\hat{c}$. Seven sets of uniform motion simulations are added. According to the cause of the added mass force, by comparing the water resistance of each group with that of the same speed during the acceleration process, the added mass force under the accelerated state can be obtained. From the relationship graph between the seven sets of added mass forces and acceleration, the parameter $|\hat{c}|$, which represents the added mass of both the standard ellipsoid and irregular object, can be obtained. Similarly, during the simulation process, viscous drag and pressure drag data at different speeds are extracted. By understanding the relationship between the viscous drag or pressure drag and the speed, parameters $\hat{a}$ and $\hat{b}$ can be solved.

(5) By comparing the differences between the added mass of the standard ellipsoid and its theoretical values, the accuracy of the methods used in processes (2) and (3) can be assessed, thereby obtaining a reasonable approach to solving for added mass.

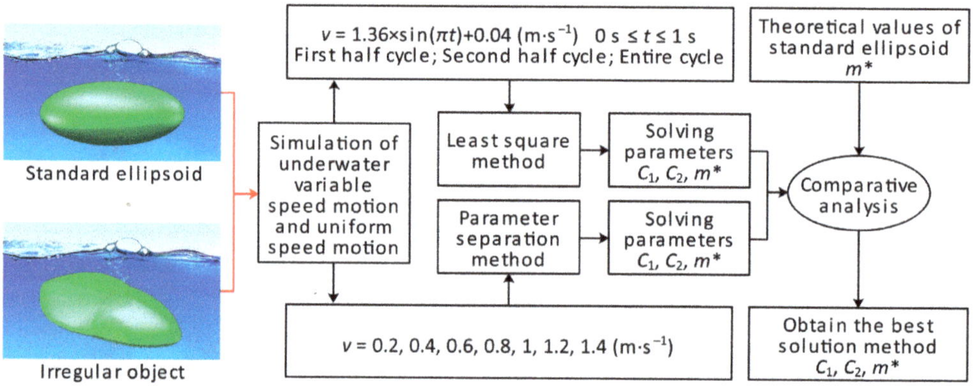

Figure 2. Schematic diagram of the method for solving added mass.

3. Numerical Simulation Methodology

3.1. Governing Equation

This numerical simulation, targeting the process of the underwater variable speed motion of objects, is classified as viscous incompressible single-phase flow, adhering to the laws of mass conservation and momentum conservation. The fluid control equations can be provided by the mass conservation of Equation (7) and the momentum conservation of Equation (8).

$$\frac{\partial\rho}{\partial t}+\frac{\partial(\rho v_x)}{\partial x}+\frac{\partial(\rho v_y)}{\partial y}+\frac{\partial(\rho v_z)}{\partial z}=0 \tag{7}$$

$$\begin{aligned}
\rho(\frac{\partial v_x}{\partial t}+v_x\frac{\partial v_x}{\partial x}+v_y\frac{\partial v_x}{\partial y}+v_z\frac{\partial v_x}{\partial z})&=\rho f_x-\frac{\partial p}{\partial x}+\mu(\frac{\partial^2 v_x}{\partial x^2}+\frac{\partial^2 v_x}{\partial y^2}+\frac{\partial^2 v_x}{\partial z^2})\\
\rho(\frac{\partial v_y}{\partial t}+v_x\frac{\partial v_y}{\partial x}+v_y\frac{\partial v_y}{\partial y}+v_z\frac{\partial v_y}{\partial z})&=\rho f_y-\frac{\partial p}{\partial y}+\mu(\frac{\partial^2 v_y}{\partial x^2}+\frac{\partial^2 v_y}{\partial y^2}+\frac{\partial^2 v_y}{\partial z^2})\\
\rho(\frac{\partial v_z}{\partial t}+v_x\frac{\partial v_z}{\partial x}+v_y\frac{\partial v_z}{\partial y}+v_z\frac{\partial v_z}{\partial z})&=\rho f_z-\frac{\partial p}{\partial z}+\mu(\frac{\partial^2 v_z}{\partial x^2}+\frac{\partial^2 v_z}{\partial y^2}+\frac{\partial^2 v_z}{\partial z^2})
\end{aligned} \tag{8}$$

where ρ represents the fluid density (kg·m^{-3}); μ is the dynamic viscosity of the fluid (Pa·s); p denotes pressure (pa); v_x, v_y, v_z are the components of the velocity vector (m·s^{-1}); f_x, f_y, f_z are the components of the volume force's unit vector.

In this numerical simulation, the fluid medium water is considered incompressible, with ρ being constant. Thus, the continuity equation can be obtained as shown in Equation (9).

$$\frac{\partial v_x}{\partial x} + \frac{\partial v_y}{\partial y} + \frac{\partial v_z}{\partial z} = 0 \tag{9}$$

3.2. Solution Model Selection

When an object moves underwater, the resistance acting on it typically depends on numerous factors, including the density and dynamic viscosity of the fluid, as well as the object's own geometric shape, surface characteristics, and velocity. The Reynolds number (Re) plays a crucial role in determining fluid flow characteristics and resistance [55], as shown in Equation (10). The values of density and dynamic viscosity of liquid water at different temperatures are presented [56], as shown in Table 1.

$$\text{Re} = \frac{\rho L v}{\mu} \tag{10}$$

where ρ represents the density of the fluid (kg·m^{-3}); v is the relative velocity of the object with respect to the fluid (m·s^{-1}); μ is the dynamic viscosity of the fluid (Pa·s); L denotes the characteristic length of the object (m).

Table 1. Values of the density and dynamic viscosity of liquid water at different temperatures.

Temperature/°C	Density ρ/kg·m^{-3}	Dynamic Viscosity μ/10^{-6} Pa·s
20	998.2336	1002.0
25	997.0751	890.3
30	995.6783	797.5
35	994.0635	719.5

In this numerical simulation, the minimum and maximum velocities of the model moving underwater are 0.04 m/s and 1.4 m/s, respectively, with the characteristic length of the model being 36 mm. Assuming the temperature of the water is at room temperature, 25 °C, the range of Reynolds numbers (Re) during the model's motion is from 1613 to 56,445. This indicates that both laminar and turbulent flow states exist during the model's movement.

In Fluent software, five main categories of turbulence models are provided, including the Spalart–Allmaras model, k-ε models (further divided into the standard k-ε model, RNG k-ε model, and k-ε model with turbulence correction), k-ω models (further divided into the standard k-ω model and pressure-corrected k-ω model), the Reynolds Stress Model (RSM), and Large Eddy Simulation (LES). The first four methods obtain flow field data by solving the Reynolds-averaged Navier–Stokes (RANS) equations, a method that averages over time to eliminate the effects of vortices, known as the RANS method. LES is an intermediate method based on Direct Numerical Simulation (DNS) and RANS. In LES, coarse large-scale vortices are simulated using DNS, while smaller-scale vortices are modeled using RANS, requiring a relatively high mesh quality. DNS is a numerical solution method based on the Navier–Stokes equations that fully resolves every vortex in the fluid. This method is the most accurate in terms of computation but also the most resource-intensive.

In this numerical simulation, due to the relatively small size of the model and the moderate Reynolds number, the RANS method was chosen. Within the RANS methodologies, the Spalart–Allmaras model is designed for aviation applications, primarily focusing on wall-bounded flows. Among the k-ε models, the standard k-ε model is intended for high Reynolds number flows. The RNG k-ε model offers an analytical formula that accounts for low Reynolds number flow viscosity. Both the turbulence-corrected k-ε model and the RNG k-ε model exhibit better performance than the standard k-ε model in conditions of strong streamline curvature, vortices, and rotation. In the k-ω models, the standard k-ω model is based on the Wilcox k-ω model, modified for low Reynolds numbers, compressibility, and shear flow propagation. The SST k-ω model integrates cross-diffusion from the ω equation and takes into account the turbulent shear stress propagation, offering higher accuracy and reliability than the standard k-ω model. The RSM model rigorously accounts for streamline bend, vortices, rotation, and rapid changes in tension, providing higher accuracy for complex flows. However, pressure tension and dissipation rates are the main factors that can reduce the prediction accuracy of the RSM model. It is necessary to use the RSM model when considering the anisotropy of Reynolds stresses, such as in hurricane flows, secondary flows in pipes, etc. Consequently, for this numerical simulation, the RNG k-ε model was selected.

3.3. Mesh Division of Fluid Computational Domain

The smaller the mesh size, the finer the calculation domain will be divided, and the more accurate the simulation results will be. However, an excessive number of meshes can significantly extend computation time, resulting in the wastage of computational resources. By densifying the mesh in areas where the flow changes abruptly and appropriately and increasing the mesh size in areas with less significant changes, accurate simulation results will also be achieved. The computational domain in this numerical simulation is divided into four fluid regions: Fluid 1, Fluid 2, Fluid 3, and Fluid 4, as illustrated in Figure 3.

The entire fluid domain is cylindrical, and the hybrid mesh is employed, with unstructured mesh used to densify the surrounding area of the model, while structured mesh is used in other areas. Among these, the cylindrical fluid domain around the model is Fluid 4, which is designated as a dynamic mesh area using unstructured tetrahedral meshes. The fluid domain behind Fluid 4 is Fluid 1, and the fluid domain in front of it is Fluid 2. Dynamic layering is used to update the mesh in these areas. During the model's motion, the mesh in the Fluid 1 area will stretch, and the mesh in the Fluid 2 area will compress. When stretching or compression reaches a set value, the mesh in that area will be updated and reorganized. The peripheral fluid domain of Fluid 1, Fluid 2, and Fluid 4 is Fluid 3, which serves to connect the wall and the inner fluid areas. Fluid 1, Fluid 2, and Fluid 3 all utilize hexahedral meshes. To ensure the accuracy of the numerical simulation, the mesh of the cylindrical fluid domain Fluid 4 is densified around the model, while the meshes in other fluid domains are relatively sparse, which can somewhat shorten the computation time. The mesh parameters for each fluid domain are shown in Table 2. The data in the table indicate that the mesh quality exceeds 0.2 for all cases, meeting the computational requirements.

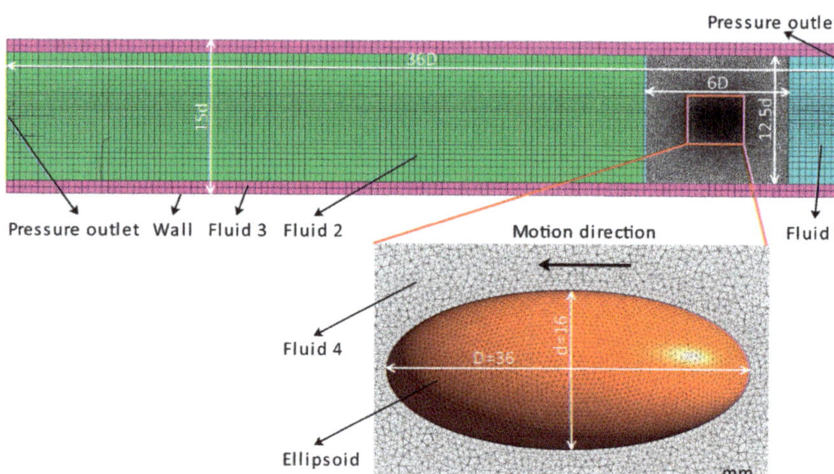

Figure 3. Division of mesh in the fluid computational domain.

Table 2. Detailed mesh parameters.

Region	Mesh Size/mm	Number of Elements	Mesh Quality
Ellipsoid Surface	0.3–0.5	10,866	>0.58
Fluid 4	0.5–4	2,783,156	>0.65
Fluid 1	4–10	5152	>0.75
Fluid 2	4–10	64,400	>0.75
Fluid 3	10–14	14,560	>0.9

3.4. Solution Method

In this simulation, the numerical discrete method adopted a pressure-based incompressible PISO algorithm, the transient time dispersion method adopted the Euler implicit format, the gradient term dispersion method adopted the Green–Gauss node-based format, and the pressure dispersion method adopted the body force weighted format. The fixed time step length is set according to the maximum speed of the object. A User-Defined Function (UDF) is employed to capture instantaneous data on the resistance and coordinates of the object moving underwater at each time step. The viscous drag and pressure drag are extracted every 10 time steps through defined Commands and Macros. To achieve the underwater variable speed motion of the model, the simulation process uses UDFs to assign velocities to the model, with the movement speeds of the standard ellipsoid and irregular model presented in Table 3. Additionally, the fluid surrounding the model is stationary.

Table 3. Movement speeds of the standard ellipsoid and irregular model.

Model	Variable Acceleration Process Velocity/m·s^{-1}	Uniform Motion Process Velocity/m·s^{-1}
standard ellipsoid	$v = 1.36\sin(\pi t) + 0.04(0\text{ s} \le t \le 1\text{ s})$	v = 0.2, 0.4, 0.6, 0.8, 1, 1.2, 1.4
irregular model	$v = 1.36\sin(\pi t) + 0.04(0\text{ s} \le t \le 1\text{ s})$	v = 0.2, 0.4, 0.6, 0.8, 1, 1.2, 1.4

4. Result Analysis and Discussion

4.1. Fluid Resistance Change Characteristics during Variable Speed Process

As depicted in Figure 4a, within the acceleration phase, the object starts with a velocity of nearly zero, and the initial acceleration is at its maximum positive value. As the acceleration continues, the velocity of the object reaches its peak, while the acceleration gradually decreases to zero. Upon entering the deceleration phase, the object's velocity gradually reduces, eventually nearing zero, as the negative value of the acceleration increases, reaching its maximum negative value by the end of the deceleration phase. Overall, the curve of the velocity is symmetrical about the t = 0.5 s axis, and the acceleration change curve is symmetrical around the coordinate point (0.5, 0).

As shown in Figure 4b, the relationship between acceleration and velocity can be approximately described as follows: the greater the absolute value of the acceleration, the lower the object's velocity, and conversely, the smaller the absolute value of the acceleration, the higher the object's velocity. This facilitates the distinction between pressure drag, viscous drag, and added mass force during the object's variable speed movement.

From Figure 4c, it is evident that during the acceleration phase, the standard ellipsoid's pressure drag is significantly higher than its viscous drag, and the maximum difference approaches 6 mN, which is approximately 50% of the viscous drag. However, upon entering the deceleration phase, the pressure drag decreases rapidly, whereas the reduction in the viscous drag is relatively slow; at around 0.75 s, the pressure drag equals the viscous drag, after which the pressure drag further decreases, resulting in the pressure drag being 3 mN less than the viscous drag during the later stages of the deceleration phase. By the end of the deceleration phase, the pressure drag turns positive, indicating that it has transitioned from resistance to propulsion. During the deceleration phase, the pressure drag changes by approximately 22 mN, while the viscous drag changes by about 15 mN. This reveals that pressure drag is more sensitive to changes in the direction of the acceleration. Similarly, during the initial phase of the acceleration phase, the difference between the pressure drag and viscous drag for an irregular object is minimal. Subsequently, the difference between the two continues to increase; during the later phase of the deceleration phase, the pressure drag is notably less than viscous drag, and the difference reaches its maximum value of 3 mN, further illustrating the pressure drag's greater sensitivity to acceleration direction changes. Additionally, there are significant differences in the proportion of the pressure drag and viscous drag between different objects, primarily due to the pressure drag being largely influenced by the object's shape.

As Figure 4d shows, the difference in the viscous drag between the acceleration and deceleration phases is minimal, with only slight variations at lower velocities, whereas pressure drag exhibits more substantial differences. This indicates that under identical velocity conditions, the viscous drag remains essentially unchanged, while differences in acceleration significantly impact the pressure drag.

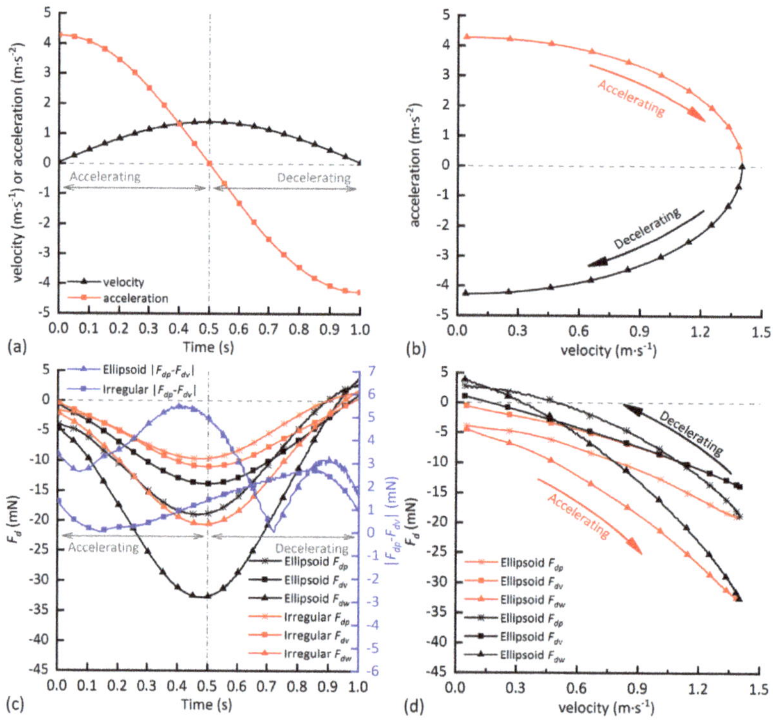

Figure 4. Numerical simulation results of the variable speed process for standard ellipsoid and irregular model. (**a**) Velocity and acceleration change curves over time; (**b**) relationship diagram between acceleration and velocity; (**c**) change curves of various resistances over time; (**d**) change curves of various resistances with velocity for the standard ellipsoid.

As shown in Figure 5a, the difference in the pressure drag at the same speed points during the acceleration and deceleration phases for a standard ellipsoid is approximately twice that of the irregular model, where the velocities are identical, but the directions of the acceleration are opposite. Since acceleration is the source of added mass force, this difference is actually caused by the added mass force, indicating that the added mass force of the standard ellipsoid is approximately twice that of the irregular object. At the end of the deceleration process, the pressure drag for both the irregular object and the standard ellipsoid is greater than zero, indicating that the pressure drag is now acting as a propulsive force. Despite the forward motion of the object, the direction of pressure drag has shifted from pointing backward to forward. The differences in the pressure fields during the motion process will be compared and analyzed in later sections.

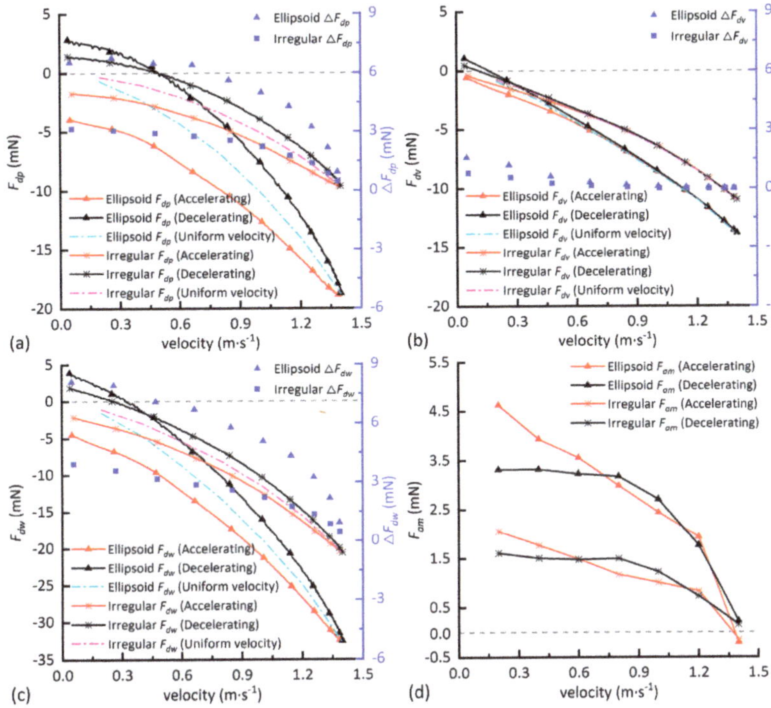

Figure 5. Comparison of various resistance forces for standard ellipsoid and irregular model. (**a**) Change curve of pressure drag with velocity during uniform and variable speed motion; (**b**) change curve of viscous drag with velocity during uniform and variable speed motion; (**c**) change curve of total resistance with velocity during uniform and variable speed motion; (**d**) change curve of added mass force with velocity during variable speed motion process.

According to Figure 5b, during the acceleration and deceleration phases, the maximum difference in viscous drag for the irregular object and the standard ellipsoid does not exceed 1.5 mN. After the speed exceeds 0.6 m/s, this difference essentially approaches 0. This indicates that changes in the direction of the acceleration do not significantly affect viscous drag, given the same velocity. Since the change in the acceleration direction is a source of added mass force, it can be inferred that the added mass force is primarily caused by changes in pressure drag.

Figure 5c reveals that the total resistance curves of both the irregular object and the standard ellipsoid under uniform motion are positioned approximately midway between the total resistance curves for the acceleration and deceleration phases. This suggests that the added mass forces for an irregular object and a standard ellipsoid are basically equal in value but in opposite directions during acceleration and deceleration processes. The total resistance during acceleration is greater than that during uniform motion, whereas the total resistance during deceleration is less than that during uniform motion. This indicates that the added mass force during acceleration points backward relative to the motion, while the added mass force during deceleration aligns with the direction of motion.

As shown in Figure 5d, the added mass force for the standard ellipsoid and the irregular object during the acceleration and deceleration processes demonstrates excellent consistency, indicating that the added mass remains essentially constant between the two variable speed processes. However, at the initial stage of the acceleration, the added mass force for both the standard ellipsoid and the irregular object is higher. The reason may be that, when the acceleration begins, the surrounding fluid is stationary, and the object must overcome the fluid's inertia to accelerate forward together. This results in greater resistance faced by the object, which is then added to the added mass force, leading to a higher added mass force at this stage. Upon entering a stable period during the acceleration phase, the added mass force exhibits good consistency.

4.2. Comparison of Pressure and Velocity Fields during Acceleration and Deceleration Phases

As shown in Figure 6, a1 to a6 represent the pressure contours of a standard ellipsoid during the acceleration phase, while d1 to d6 are those during the deceleration phase. Both the ellipsoids in Figure 6(a1,d1) have the same velocity, with the directions of the velocity pointing to the right. The directions of the acceleration are opposite, with rightward pointing considered positive and leftward negative. This condition also applies to other figures with the same numbering. As the sequence number increases, the velocity of the ellipsoid continuously increases, while the absolute value of the acceleration decreases.

From Figure 6(a1,d1), it can be observed that under conditions of a lower velocity and a higher absolute acceleration value, the accelerating ellipsoid exhibits a large high-pressure area at the front, with no significant positive pressure area at the rear. Conversely, the decelerating ellipsoid shows no prominent high-pressure area at the front but forms negative pressure areas on both shoulders. This indicates that due to the ellipsoid's deceleration, the fluid acting on the front maintains a certain velocity due to inertia, leading to fluid detachment from the ellipsoid's head, thus reducing the pressure there. A significant high-pressure area forms at the rear of the ellipsoid, suggesting that fluid continuously flows toward the ellipsoid from behind, creating a high-pressure area at the rear and generating vortices on both sides of the tail.

With an increasing velocity and a decreasing absolute acceleration, the difference in the high-pressure area at the front of the ellipsoid becomes less noticeable, indicating that the increase in velocity significantly reduces the impact of the acceleration on the front high-pressure area. For the ellipsoid in deceleration, the high-pressure area at the rear is larger, highlighting a more pronounced effect of the acceleration on the rear high-pressure area.

As shown in Figure 7, a1 to a6 are the velocity contours of a standard ellipsoid during the acceleration phase, and d1 to d6 are for the deceleration phase. In Figure 7(a1,d1), the ellipsoids have the same velocity, with the directions of the velocity pointing to the right. The directions of the acceleration are opposite, with rightward pointing considered a positive value and leftward a negative value. This condition applies similarly to other figures with matching sequence numbers. As the sequence number increases, the velocity of the ellipsoid continuously rises, and the absolute value of acceleration decreases.

Figure 7(a1) reveals that at lower speeds, for the ellipsoid in acceleration motion, the fluid at its front is squeezed away and flows toward the rear, forming a subtle velocity vortex in the middle and reconverging at the tail. Figure 7(d1) shows that at lower speeds, for the ellipsoid in deceleration motion, there is a significant difference in the velocity field compared to during the acceleration, with a large area of high-speed fluid at the tail continuously converging on both sides of the tail and merging with the fluid flowing from the front to form a velocity vortex. The high-speed fluid rushing forward extends to the middle of the ellipsoid. This rushing fluid is precisely the primary source of added mass force during deceleration, while the difference in the velocity field at the front is less noticeable, indicating that the change in the acceleration direction has a more pronounced impact on the flow field at the rear of the object.

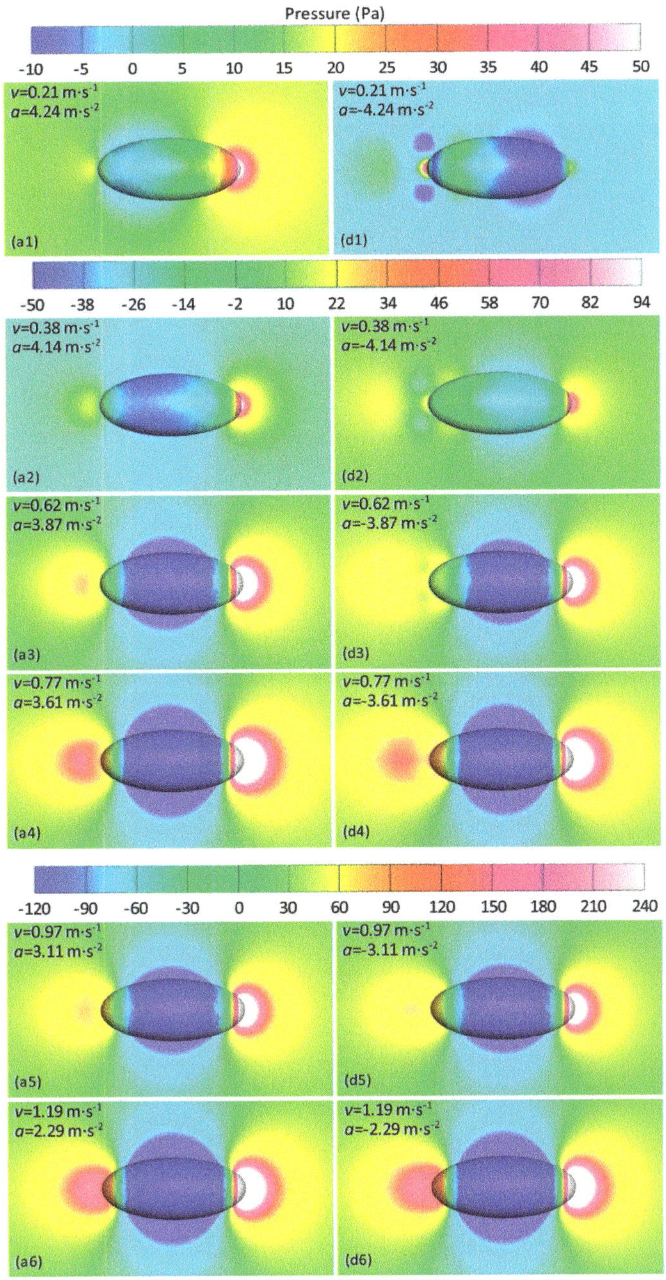

Figure 6. Pressure contours for the standard ellipsoid during acceleration and deceleration movements. (a1–a6) are the pressure contours for acceleration; (d1–d6) are the pressure contours for deceleration.

As the velocity increases and the absolute value of the acceleration decreases, the decelerating ellipsoid no longer forms a velocity circulation at the tail but forms a small-scale velocity circulation at the waist, and the velocity circulation becomes increasingly smaller. The difference in the velocity field at the tail of the ellipsoids between acceleration and deceleration motions becomes less noticeable, suggesting that the increase in velocity and the decrease in the absolute value of the acceleration reduce the impact of the acceleration direction change on the rear velocity field.

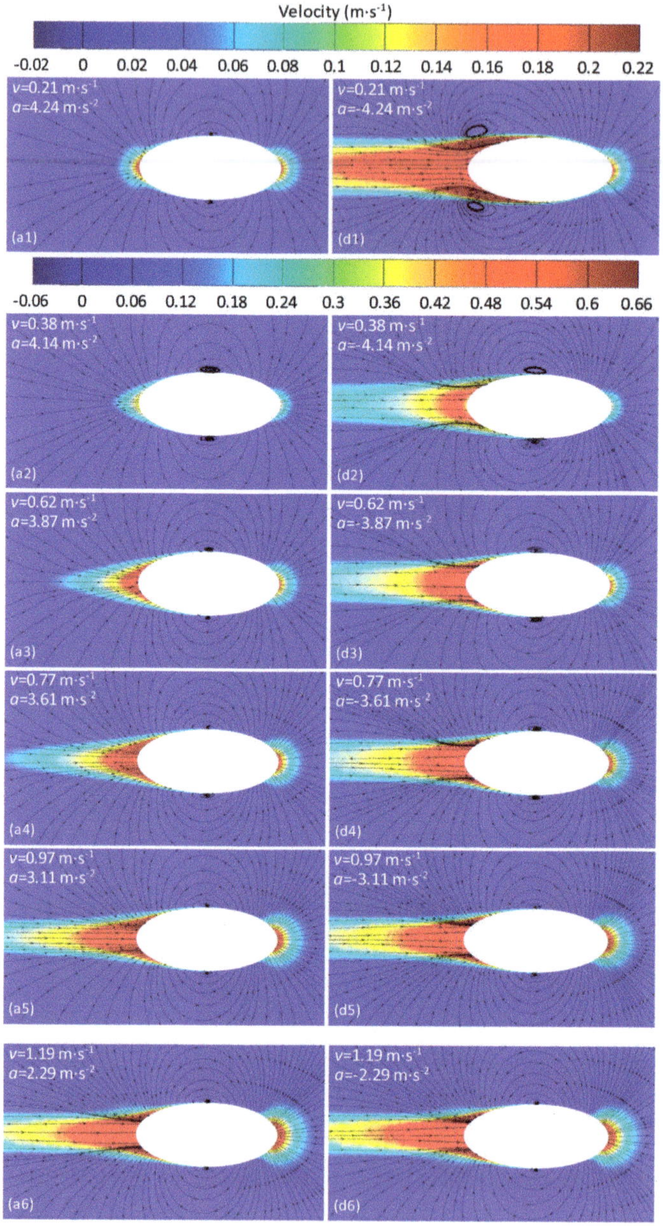

Figure 7. Velocity contours for the standard ellipsoid during acceleration and deceleration movements. (a1–a6) are the velocity contours for acceleration; (d1–d6) are the velocity contours for deceleration.

As depicted in Figure 8, the characteristics of the pressure distribution at the head and tail of an irregular object are similar to those of ellipsoids. At lower velocities and greater absolute values of acceleration, the irregular object undergoing acceleration has a larger high-pressure area at its head and a smaller one at its tail, which is the opposite of the pressure distribution during the deceleration, and the reason is as previously described. With the increase in the velocity and the decrease in the absolute value of the acceleration, this difference in the flow field gradually becomes less pronounced. A distinction from the pressure distribution of ellipsoids is the asymmetrical pressure distribution on both shoulders caused by a geometric asymmetry.

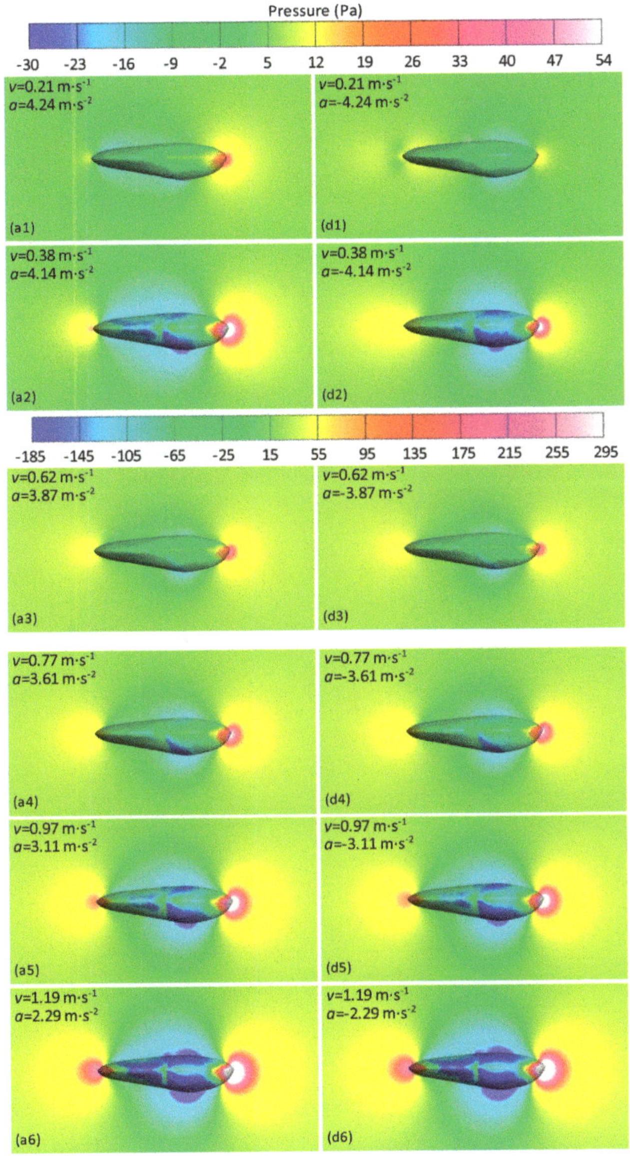

Figure 8. Pressure contours for irregular object during acceleration and deceleration movements. (a1–a6) are the pressure contours during acceleration; (d1–d6) are the pressure contours during deceleration.

As illustrated in Figure 9, due to the geometric asymmetry of an irregular object, when the velocity is low and the absolute value of the acceleration is high, the velocity distribution is noticeably asymmetrical. Fluid on both sides of the irregular object's tail tends to converge more easily on the side with a gentler slope, forming a velocity recirculation with the outwardly retreating fluid, while no velocity recirculation is formed on the side with a steeper slope, indicating that there is less forward-moving fluid on that side. As the velocity increases and the absolute value of acceleration decreases, the forward-moving fluid continuously diminishes, causing this asymmetrical distribution of the velocity field to gradually become more similar.

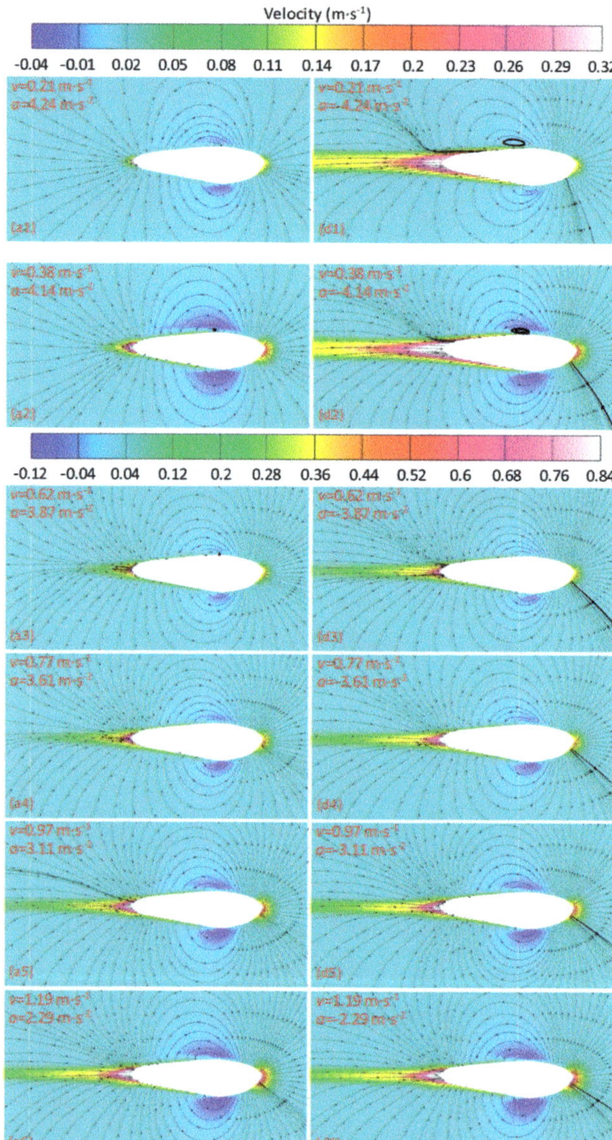

Figure 9. Velocity contours for irregular object during acceleration and deceleration movements. (a1–a6) are the velocity contours during acceleration; (d1–d6) are the velocity contours during deceleration.

4.3. Least Squares Method for Fluid Resistance Regression Analysis

From the previous text, it is known that viscous drag is approximately proportional to the first power of velocity, pressure drag is approximately proportional to the square of velocity, and added mass force is approximately proportional to acceleration. Based on the proportional relationships between these three forces and their respective motion variables, the total resistance curve shown in Figure 4c was analyzed using the least squares regression analysis, as demonstrated by the calculation formula in Equation (3). The regression analysis was conducted for the total resistance during the acceleration cycle (Acc. cycle), the deceleration cycle (Dec. cycle), and the entire variable speed cycle (En. cycle). The fitting results are shown in Figure 10, and the diagram indicates that there are significant differences between the fitting curves at point A for the acceleration cycle and the entire variable speed cycle. Additionally, there are connectivity issues between the fitting curves for the acceleration and deceleration cycles at points B and C. The corresponding coefficients $\hat{a}$, $\hat{b}$, and $\hat{c}$ are shown in Table 4. The data in the table reveal some discrepancies in the total resistance regression analysis results for the acceleration phase compared to the other two groups, characterized by larger values for $\hat{a}$ and $\hat{c}$ and a smaller value for $\hat{b}$. This is mainly due to the fact that at the initial stage of acceleration, the surrounding fluid is stationary, and the object experiences greater pressure drag, while the viscous drag remains relatively low, leading to differences in the parameters. Subsequently, the coefficient regression analysis results in the table will be compared and analyzed with the results obtained from the parameter separation method to verify the accuracy of solving parameters using the least squares regression analysis.

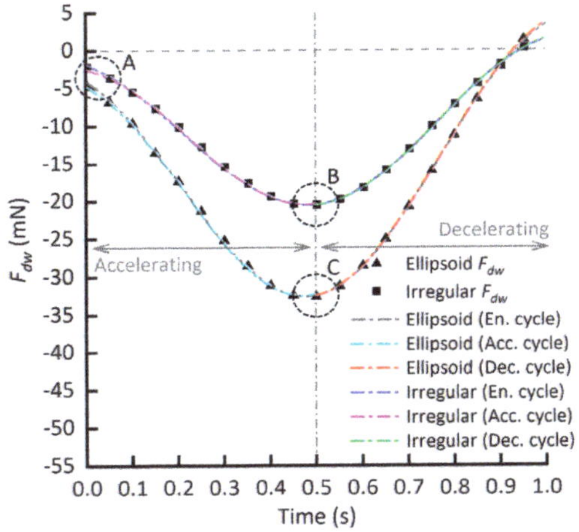

Figure 10. Fitting results of the curve of water drag.

Table 4. Fitting results of the fluid resistance curve during variable speed motion.

Number i	Model	Time Range (s)	$F_{dw} = \hat{a}v^2 + \hat{b}v + \hat{c}a$ (mN)			R^2	RMSE 10^{-4}
			$\hat{a}_i$	$\hat{b}_i$	$\hat{c}_i$		
1	Standard ellipsoid	[0.004, 0.5]	−12.40	−5.713	−1.0440	0.9997	1.608
2		[0.5, 1]	−10.29	−8.808	−0.9502	0.9999	1.226
3		[0.004, 1]	−10.94	−7.835	−0.8759	0.9997	1.991
4	Irregular model	[0.004, 0.5]	−8.749	−2.270	−0.5463	0.9994	1.478
5		[0.5, 1]	−7.229	−4.531	−0.4084	0.9997	1.275
6		[0.004, 1]	−7.397	−4.240	−0.3860	0.9993	1.820

4.4. Parameter Separation Method for Fluid Resistance Regression Analysis

To verify the accuracy of the parameters solved by least squares regression analysis, this study utilizes data on viscous drag and pressure drag from a standard ellipsoid and an irregular object during uniform motion, along with total resistance data from both uniform and variable speed motions. The parameter separation method is then applied to conduct regression analysis on coefficients $\hat{a}$, $\hat{b}$, and $\hat{c}$.

As shown in Figure 11a, the change curves of the pressure drag F_{dp} with respect to v^2 for both the standard ellipsoid and the irregular object exhibit good linearity, indicating that the pressure drag is proportional to the square of the velocity, i.e., $F_{dp} = k_{dp}v^2$, where k_{dp} is the slope of the curve. From the components of the total resistance F_{dw}, it is known that $\hat{a} = k_{dp}$. The linearity of the viscous drag F_{dv} with respect to v for the standard ellipsoid and the irregular object is slightly poorer, containing nonlinear components. However, since the pressure drag of the standard ellipsoid is significantly greater than its viscous drag, the nonlinear components of F_{dv} with respect to v can be disregarded. In contrast, for the irregular object, the difference between the viscous drag and pressure drag is not significant, so the nonlinear components of F_{dv} with respect to v cannot be ignored. Therefore, when conducting the regression analysis using the parameter separation method, the nonlinear components were retained, as shown in Table 5.

Since the added mass force is generated by the variable speed motion of an object, a body in uniform motion is not affected by an added mass force. Therefore, to separate the added mass force, we just subtract the total resistance during uniform motion from the total resistance during variable speed motion. As shown in Figure 11b, the change curves of the added mass force F_{am} with respect to the acceleration a for both the standard ellipsoid and the irregular object exhibit good linearity, with larger deviations only in areas of higher acceleration. As indicated by Figure 4b, this corresponds to the initial stages of the acceleration phase, and the reasons for the increased added mass force have been explained previously. Thus, it can be considered that the added mass force is proportional to the acceleration, i.e., $|F_{am}| = k_{am}|a|$, where k_{am} is the slope of the curve. From the components of the total resistance F_{dw} and the direction of the added mass force, it is known that $\hat{c} = -k_{am}$. This allows for the separation and solving of the three parameters $\hat{a}$, $\hat{b}$, and $\hat{c}$, and the results are presented in Table 5.

By comparing the values of parameters $\hat{a}$, $\hat{b}$, and $\hat{c}$ in Tables 4 and 5, it can be observed that the values of $\hat{a}$, $\hat{b}$, and $\hat{c}$ obtained through the parameter separation method are closer to the data in groups 3 and 6 of Table 4, especially with very small errors in $\hat{c}$. This indicates that performing regression analysis on the total resistance data for the entire variable speed cycle using the least squares method can yield fairly accurate values for the added mass, with certain errors present in $\hat{a}$ and $\hat{b}$. Specifically, the discrepancy between $\hat{a}_{dp}$ for the irregular object in Table 5 and $\hat{a}_6$ in Table 4 is significant, mainly due to the nonlinear term of the velocity v in the viscous drag F_{dv} and the closeness of the viscous drag to the pressure drag. If the nonlinear term is taken into account, then the regression parameter $\hat{a}$ for the pressure drag of the irregular object in Table 5 should be equal to $\hat{a}_{dv} + \hat{a}_{dp}$, and finally, the values are no longer significantly different. This shows that performing regression analysis on the resistance data for the entire variable speed cycle using the least squares method can achieve a certain degree of accuracy for the coefficients of pressure drag and viscous drag. In practical engineering applications, the least squares method can be used for rapid preliminary regression analysis of the total resistance data throughout the entire variable speed cycle. Subsequent validation with several sets of resistance data under uniform motion conditions can then provide accurate coefficients for pressure drag, viscous drag, and added mass.

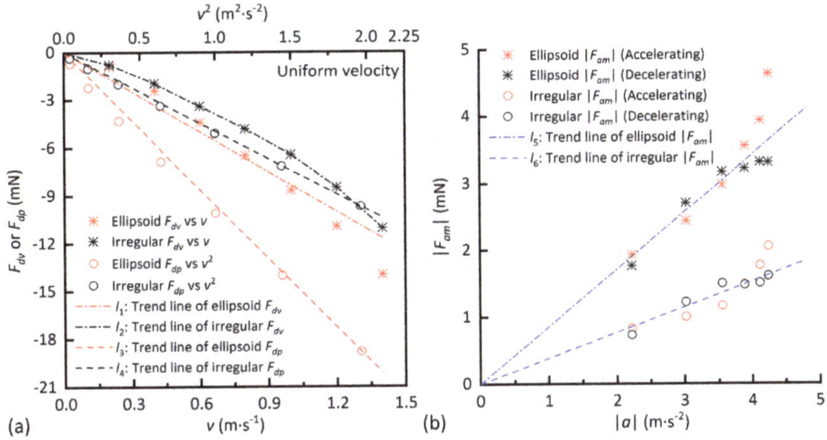

Figure 11. Relationships between various resistances and motion variables for standard ellipsoid and irregular model. (**a**) Relationships between viscous drag or pressure drag and velocity; (**b**) relationship between added mass force and acceleration.

Table 5. Functional equations of trend lines l_1~l_6 in Figure 11.

Model	Curve Number	Mechanical Variable (mN)	Functional Equation	Parameter Value
Standard ellipsoid	l_1	F_{dv}	$F_{dv} = \hat{b}v$	$\hat{b}_1 = -8.3571$
	l_3	F_{dp}	$F_{dp} = \hat{a}v^2$	$\hat{a}_1 = -9.6062$
	l_5	F_{am}	$\lvert F_{am}\rvert = \hat{k}_{am}\lvert a\rvert$ $F_{am} = \hat{c}a$	$\hat{k}_{am1} = 0.8556$ $\hat{c}_1 = -\hat{k}_{am1} = -0.8556$
Irregular model	l_2	F_{dv}	$F_{dv} = \hat{b}v + \hat{a}_{dv}v^2$	$\hat{b}_2 = -3.5306$ $\hat{a}_{dv} = -3.0396$
	l_4	F_{dp}	$F_{dp} = \hat{a}_{dp}v^2$	$\hat{a}_{dp} = -4.9459$
	l_6	F_{am}	$\lvert F_{am}\rvert = \hat{k}_{am}\lvert a\rvert$ $F_{am} = \hat{c}a$	$\hat{k}_{am2} = 0.3822$ $\hat{c}_2 = -\hat{k}_{am2} = -0.3822$

4.5. Theoretical Comparison

To verify the accuracy of the two analytical methods, the theoretical added mass of the ellipsoid was calculated. According to reference [57], when the ellipsoid accelerates along its major axis, its added mass can be calculated using Equation (11).

$$m^* = \frac{\alpha_0}{2-\alpha_0} \times \frac{4}{3}\rho\pi ab^2 \tag{11}$$

$$\alpha_0 = \frac{1-e^2}{e^3}\left(\ln\frac{1+e}{1-e} - 2e\right) \tag{12}$$

where m^* represents the added mass of the ellipsoid; ρ is the density of water; a and b are the major and minor axes of the ellipsoid, respectively; e is the eccentricity of the ellipsoid cross-section through the axis of symmetry.

The results of regression analysis using the least squares method and the parameter separation method indicate that the added mass of the ellipsoid is 0.8759 g and 0.8556 g, respectively. Meanwhile, the theoretical value of the ellipsoid's added mass calculated through the formula is 0.8679 g. The relative errors of the two calculated added masses are 0.92% and −1.42%, respectively. This demonstrates that the added mass obtained through both methods exhibits good accuracy.

5. Conclusions

This paper conducted numerical simulations on both standard ellipsoid and irregular objects undergoing variable and uniform motion underwater. The changing characteristics of the various drags were analyzed, and the least squares method was proposed for calculating the added mass. The accuracy of this method was verified using parameter separation regression analysis and theoretical solutions.

The analysis of fluid resistance change characteristics during the variable speed process reveals that changes in the direction of acceleration significantly affect the pressure drag but have a negligible impact on the viscous drag. This indicates that the source of added mass force is primarily the changes in the pressure drag. During the deceleration process, the direction of the added mass force aligns with the direction of motion, acting as a propulsive force.

The analysis of differences in the pressure fields during acceleration and deceleration processes shows that the distribution of high pressure at the head and tail of the ellipsoid presents opposite phenomena. This is because, during the deceleration process, the surrounding fluid continues to flow forward due to inertia, leading to a tendency for the fluid at the head to detach, while the fluid at the tail has a tendency to surge forward. As the velocity increases and the absolute value of acceleration decreases, this difference in the pressure field becomes less pronounced.

The analysis of differences in velocity fields during acceleration and deceleration processes indicates that, at lower speeds, for the ellipsoid undergoing deceleration, there is a significant presence of high-speed fluid at the tail, which continuously converges toward both sides of the tail and merges with the fluid flowing from the front to form a velocity vortex. These forward-moving fluids are precisely the primary sources of the added mass force during deceleration, while the difference in the velocity field at the head is less pronounced. This demonstrates that changes in the direction of acceleration have a more noticeable impact on the flow field at the rear of the object.

The regression analysis using both the least squares method and the parameter separation method indicates that conducting a least squares regression analysis on the total resistance throughout the entire variable speed cycle can yield relatively accurate added mass values, consistent with the results of the parameter separation method. By comparing these values with the theoretical value of the added mass for the ellipsoid, the relative errors of the added mass solved by these two methods are only 0.92% and -1.42%. This demonstrates that utilizing the least squares method for parameter analysis can achieve relatively accurate results.

Author Contributions: X.W. (Xuanquan Wang): methodology, writing—original draft. S.X.: resources, funding acquisition, supervision. X.W. (Xinchun Wang): software, investigation. D.Q.: writing—review and editing, data curation. All authors have read and agreed to the published version of the manuscript.

Funding: This research was supported by the Hainan Provincial Natural Science Foundation of China, grant number No: 323MS074.

Institutional Review Board Statement: Not applicable.

Informed Consent Statement: Not applicable.

Data Availability Statement: The data that support the findings of this study are available from the corresponding author upon reasonable request.

Conflicts of Interest: The authors declare no conflicts of interest.

References

1. Li, J.; Lu, C.J.; Huang, X. Calculation of added mass of a vehicle running with cavity. *J. Hydrodyn.* **2010**, *3*, 312–318. [CrossRef]
2. Raza, N.; Mehmood, I.; Rafiuddin, H.; Rafique, M. Numerical simulation of added mass determination of standard ellipsoids. In Proceedings of the 2012 9th International Bhurban Conference on Applied Sciences and Technology (IBCAST), Islamabad, Pakistan, 9–12 January 2012; pp. 270–274.

3. Javanmard, E.; Mansoorzadeh, S.; Mehr, J.A. A new CFD method for determination of translational added mass coefficients of an underwater vehicle. *Ocean Eng.* **2020**, *215*, 107857. [CrossRef]
4. Sen, D.T.; Vinh, T.C. Determination of added mass and inertia moment of marine ships moving in 6 degrees of freedom. *Int. J. Transp. Eng. Technol.* **2016**, *2*, 8–14.
5. Korotkin, A.I. *Added Masses of Ship Structures*; Krylov Shipbuilding Research Institute: St. Petersburg, Russia; Springer: St. Petersburg, Russia, 2009.
6. Pan, Y.-C.; Zhang, H.-X.; Zhou, Q.-D. Numerical prediction of submarine hydrodynamic coefficients using CFD simulation. *J. Hydrodyn.* **2012**, *24*, 840–847. [CrossRef]
7. Go, G.; Ahn, H.T. Hydrodynamic derivative determination based on CFD and motion simulation for a tow-fish. *Appl. Ocean Res.* **2018**, *82*, 191–209. [CrossRef]
8. Kuwabara, S.S.; Okamoto, K. Experimental investigation of added mass coefficient with a free oscillating circular cylinder. *JPN. Soc. Mech. Eng.* **2008**, *74*, 1396–1401. [CrossRef]
9. Hsieh, W.M. An experimental method for determining the frequency-dependent added mass and added mass moment of inertia for a floating body in heave and pitch motions. *Ocean Eng.* **2001**, *28*, 417–438.
10. Rodriguez, C.; Egusquiza, E.; Escaler, X.; Liang, Q.; Avellan, F. Experimental investigation of added mass effects on a Francis turbine runner in still water. *J. Fluids Struct.* **2006**, *22*, 699–712. [CrossRef]
11. Sedlar, D.; Lozina, Z.; Vucina, D. Experimental investigation of the added mass of the cantilever beam partially submerged in water. *Tech. Gaz.* **2011**, *4*, 589–594.
12. Pan, G.; Wei, G.; Du, X.X. Whole scheme of added mass's forced oscillatory mechanism. *Meas. Control Technol.* **2007**, *26*, 35–37.
13. De La Torre, O.; Escaler, X.; Egusquiza, E.; Farhat, M. Experimental investigation of added mass effects on a hydrofoil under cavitation conditions. *J. Fluids Struct.* **2013**, *39*, 173–187. [CrossRef]
14. Fanning, P.J.; Carden, E.P. Experimentally validated added mass identification algorithm based on frequency response functions. *J. Eng. Mech.* **2004**, *130*, 1045–1051. [CrossRef]
15. Caspersen, C.; A Berthelsen, P.; Eik, M.; Pâkozdi, C.; Kjendlie, P.-L. Added mass in human swimmers: Age and gender differences. *J. Biomech.* **2010**, *43*, 2369–2373. [CrossRef] [PubMed]
16. Wang, Z.; Fan, D.; Triantafyllou, M.S. Illuminating the complex role of the added mass during vortex induced vibration. *Phys. Fluids* **2021**, *33*, 085120. [CrossRef]
17. Lee, S.K.; Cheon, S.J.; Jang, T.S.; Lee, J.H.; Joung, T.H. Evaluation of the Added Mass for a Spheroid-type UUV by VPMM Test. In Proceedings of the International Offshore and Polar Engineering Conference, Osaka, Japan, 21–26 June 2009.
18. Ueno, M.; Yukawa, K.; Nimura, T.; Asanuma, T.; Ohmatsu, S.; Takai, R. PMM test analysis for added mass coefficients of asymmetric bodies. *IFAC Proc. Vol.* **2003**, *36*, 31–36. [CrossRef]
19. Corkery, S.J.; Babinsky, H.; Graham, W.R. Quantification of added-mass effects using particle image velocimetry data for a translating and rotating flat plate. *J. Fluid Mech.* **2019**, *870*, 492–518. [CrossRef]
20. Salibindla, A.K.; Masuk, A.U.M.; Ni, R. Experimental investigation of the acceleration statistics and added-mass force of deformable bubbles in intense turbulence. *J. Fluid Mech.* **2021**, *912*, A5. [CrossRef]
21. Pantaleone, J.; Messer, J. The added mass of a spherical projectile. *Am. J. Phys.* **2011**, *79*, 1202–1210. [CrossRef]
22. Sahin, I.; Crane, J.; Watson, K. Added Mass Coefficients for Submerged Bodies by a Low-Order Panel Method. *J. Fluids Eng.* **1993**, *115*, 452–456. [CrossRef]
23. Sahin, I.; Crane, J.W.; Watson, K.P. Application of a panel method to hydrodynamics of underwater vehicles. *Ocean Eng.* **1997**, *24*, 501–512. [CrossRef]
24. Zhang, M.; Wang, X.; Duan, D. Panel method predictions of added mass for flexible airship. *Aeronaut. J.* **2013**, *117*, 519–531. [CrossRef]
25. Liu, L.; Sun, M. The added mass forces in insect flapping wings. *J. Theor. Biol.* **2018**, *437*, 45–50. [CrossRef] [PubMed]
26. Lin, Z.; Liao, S. Calculation of added mass coefficients of 3D complicated underwater bodies by FMBEM. *Commun. Nonlinear Sci. Numer. Simul.* **2011**, *16*, 187–194. [CrossRef]
27. Qi, Z.; Jiang, M.; Jia, L.; Zou, B.; Zhai, J. The effect of mass ratio and damping coefficient on the propulsion performance of the semi-active flapping foil of the wave glider. *J. Mar. Sci. Eng.* **2020**, *8*, 303. [CrossRef]
28. Gardano, P.; Dabnichki, P. Application of Boundary Element Method to modelling of added mass and its effect on hydrodynamic forces. *Comput. Model. Eng. Sci.* **2006**, *15*, 87–98.
29. Lin, C.Y.; Zhu, J. Numerical computation of added mass of submarine maneuvering with small clearance to sea-bottom. *Ship Eng.* **2003**, *5*, 26–29.
30. Ghassemi, H.; Yari, E. The added mass coefficient computation of sphere, ellipsoid and marine propellers using boundary element method. *Pol. Marit. Res.* **2011**, *18*, 17–26. [CrossRef]
31. Leung, A.Y.T.; Fok, A.S.; Dai, H.; Su, R.K.L. The fractal finite element method for added-mass-type problems. *Int. J. Numer. Meth. Eng.* **2008**, *75*, 1194–1213. [CrossRef]
32. Levy, S.; Wilkinson, J.P.D. Calculation of added water mass effects for reactor system components. In *Structural Mechanics in Reactor Technology*; IASMiRT: London, UK, 1975; Volume F2/5, pp. 1–14.

33. Phillips, A.; Furlong, M.; Turnock, S.R. The use of computational fluid dynamics to determine the dynamic stability of an autonomous underwater vehicle. In Proceedings of the 10th Numerical Towing Tank Symposium (NuTTS'07), Hamburg, Germany, 23–25 September 2007; pp. 6–11.
34. Sakamoto, N. URANS, DES Simulations of Static and Dynamic Manoeuvring for Surface Combatant. Ph.D. Thesis, University of Iowa, Iowa City, IA, USA, 2009.
35. Tang, S.; Ura, T.; Nakatani, T.; Thornton, B.; Jiang, T. Estimation of the hydrodynamic coefficients of the complex-shaped autonomous underwater vehicle TUNA-SAND. *J. Mar. Sci. Technol.* **2009**, *14*, 373–386. [CrossRef]
36. Zhang, H.; Xu, Y.-R.; Cai, H.-P. Using CFD software to calculate hydrodynamic coefficients. *J. Mar. Sci. Appl.* **2010**, *9*, 149–155. [CrossRef]
37. Kim, H.; Akimoto, H.; Islam, H. Estimation of the hydrodynamic derivatives by RaNS simulation of planar motion mechanism test. *Ocean Eng.* **2015**, *108*, 129–139. [CrossRef]
38. Song, L. Calculation of added mass for underwater vehicles based on FVM. In Proceedings of the International Conference on Advances in Mechanical Engineering and Industrial Informatics, Hangzhou, China, 9–10 April 2016.
39. Zakaria, M.S.; Abdullah, H.; Ahmad, K.A. Fluid structure interaction simulation of large deformation and added-mass effect using OpenFOAM. In Proceedings of the Mechanical Engineering Research Day, Malang, Indonesia, 23–25 October 2018; pp. 79–80.
40. Perrault, D.; Bose, N.; O'young, S.; Williams, C.D. Sensitivity of AUV added mass coefficients to variations in hull and control plane geometry. *Ocean Eng.* **2003**, *30*, 645–671. [CrossRef]
41. Vu, V.H.; Thomas, M.; Lakis, A.A.; Marcouiller, L. Effect of added mass on submerged vibrated plates. In Proceedings of the 25th Seminar on Machinery Vibration, Canadian Machinery Vibration Association CMVA, St. John, NL, Canada, 25 June 2021.
42. Paniccia, D.; Graziani, G.; Lugni, C.; Piva, R. On the role of added mass and vorticity release for self-propelled aquatic locomotion. *J. Fluid Mech.* **2021**, *918*, 645–671. [CrossRef]
43. Zhou, Y.; Li, Y.; Yoshida, A. Effect of added mass on wind-induced vibration of a circular flat membrane by wind tunnel tests. *Int. J. Struct. Stab. Dyn.* **2018**, *18*, 1850156. [CrossRef]
44. Mohaghegh, F.; Udaykumar, H. Comparison of sharp and smoothed interface methods for simulation of particulate flows II: Inertial and added mass effects. *Comput. Fluids* **2017**, *143*, 103–119. [CrossRef]
45. Chu, Z.; Zhu, D.; Sun, B.; Nie, J.; Xue, D. Design of a dead reckoning based motion control system for small autonomous underwater vehicle. In Proceedings of the IEEE 28th Canadian Conference on Electrical and Computer Engineering (CCECE), Halifax, NS, Canada, 3–6 May 2015; pp. 728–733.
46. Zhang, T.; Zhou, H.; Wang, J.; Liu, Z.; Xin, J.; Pang, Y. Optimum design of a small intelligent ocean exploration underwater vehicle. *Ocean Eng.* **2019**, *184*, 40–58. [CrossRef]
47. Zhang, H. Research on key technologies of small underwater robots. *Int. J. Simul. Syst. Sci. Technol.* **2016**, *17*, 14.1–14.7.
48. Ba, X.; Luo, X.H.; Shi, Z.C.; Zhu, Y.Q. A vectored water jet propulsion method for autonomous underwater vehicles. *Ocean Eng.* **2013**, *74*, 133–140.
49. Jiao, W.; Cheng, L.; Zhang, D.; Zhang, B.; Su, Y.; Wang, C. Optimal design of inlet passage for waterjet propulsion system based on flow and geometric parameters. *Adv. Mater. Sci. Eng.* **2019**, *2019*, 2320981. [CrossRef]
50. Fish, F.E. Advantages of aquatic animals as models for bio-inspired drones over present AUV technology. *Bioinspiration Biomim.* **2020**, *15*, 025001. [CrossRef]
51. Raj, A.; Thakur, A. Fish-inspired robots: Design, sensing, actuation, and autonomy—A review of research. *Bioinspiration Biomim.* **2016**, *11*, 031001. [CrossRef] [PubMed]
52. Ding, D.; Ouahsine, A.; Xiao, W.; Du, P. CFD/DEM coupled approach for the stability of caisson-type breakwater subjected to violent wave impact. *Ocean Eng.* **2021**, *223*, 108651. [CrossRef]
53. Huang, Z.; Ouahsine, A.; Du, P.; Ding, D. Computational fluid dynamics predictions of critical hydrodynamics thresholds in the erodibility of inland waterway bank by ship-induced waves. *Ocean Eng.* **2022**, *266*, 112825. [CrossRef]
54. Du, P.; Ouahsine, A.; Sergent, P.; Hoarau, Y.; Hu, H. Investigation on resistance, squat and ship-generated waves of inland convoy passing bridge piers in a confined waterway. *J. Mar. Sci. Eng.* **2021**, *9*, 1125. [CrossRef]
55. Long, L.N.; Weiss, H. The velocity dependence of aerodynamic drag: A primer for mathematicians. *Am. Math. Mon.* **1999**, *106*, 127–135. [CrossRef]
56. Kestin, J.; Sokolov, M.; Wakeham, W.A. Viscosity of liquid water in the range −8 °C to 150 °C. *J. Phys. Chem. Ref. Data.* **1978**, *7*, 941–948. [CrossRef]
57. Brennen, C.E. *A Review of Added Mass and Fluid Inertial Forces*; Technical Report–CR82.010; Naval Civil Engineering Laboratory: Port Hueneme, CA, USA, 1982.

Journal of
Marine Science and Engineering

MDPI

Article

A Numerical Simulation Study and Effectiveness Evaluation on the Flow Field Effect of Trapezoidal Artificial Reefs in Different Layouts

Xiaolong Chen, Xuan Che *, Yin Zhou, Changfeng Tian and Xinfeng Li

Fishery Machinery and Instrument Research Institute, Chinese Academy of Fishery Sciences, Shanghai 200092, China; chenxiaolong@fmiri.ac.cn (X.C.); zhouyin@fmiri.ac.cn (Y.Z.); tianchangfeng@fmiri.ac.cn (C.T.); lixinfeng@fmiri.ac.cn (X.L.)
* Correspondence: chexuan@fmiri.ac.cn

Abstract: The combined release of artificial reefs in different quantities and arrangements leads to different flow field effects. This study designs a small trapezoidal artificial reef. To optimize the quantity and layout of these reefs, trapezoidal reefs in three different layouts were selected for analysis at five different velocity gradients (0.1, 0.5, 1.0, 1.5, and 2.0 m/s). The effects of disposal spacing and layout on the flow field effect of trapezoidal artificial reefs at different flow velocities were simulated using Ansys Fluent. According to the findings: after simulation, flow velocity could indirectly reflect the distribution of upwelling and back eddy, the scale and strength of upwelling increased as flow velocity increased, and the back eddy showed no obvious variation with flow velocity. In transverse combination mode, both the scale and strength of the upwelling and back eddy were maximized when the reef spacing was 1.0 L; in longitudinal combination mode, upwelling and back eddy reached maximum scale and strength when the disposal spacing of the reefs was 1.5 L of a single reef. In 2020, flow mapping and fishery surveys were carried out in the engineering pilot area. The results showed that the number and species of fish populations with 1.5 L spacing in the vertical combination method were significantly higher than those in other forms, and the structure of the fish reef was stable without any flipping or sliding phenomenon. This study can provide a theoretical reference for the design and the actual deployment of artificial reefs to improve the ecological restoration of the water.

Keywords: ocean engineering; artificial reef; effect of flow field; numerical simulation; ecological restoration

1. Introduction

With the rapid development of urbanization in China, a lot of industrial and domestic wastewater is being discharged into the ocean, leading to serious offshore environmental pollution. At present, frequent red tides are strong signals of the impact this pollution is having. Artificial reefs can form a protective biological circle, creating a harmonious and mutually beneficial ecological environment. This will help promote biodiversity, both by circulating nutrients and by creating symbiotic relationships between different marine creatures and the environment. Artificial reefs can provide various microorganisms with substrates, forming a natural ecosystem. Upwelling, streamline flow, and eddies formed by the reef area also favor the water cycle, meaning the regional ocean environment will improve as time passes.

Artificial reefs as foundational structures have been widely used in marine ranches, which have been vigorously developed in China in recent years [1,2]. Artificial reefs are artificial constructs deployed in the corresponding sea areas, and are used to create marine ranches. This helps improve the habitat for aquatic organisms, regulates and optimizes the ocean environment, and provides marine organisms with places to forage, shelter, grow, develop, and propagate. All of this plays a role in increasing and enhancing fishery resources [3–5]. As an effective method for protecting marine biodiversity and promoting the restoration of coastal habitats, the deployment of artificial reefs plays an important role in changing ocean flow fields. The use of artificial reefs will change the particle size distribution of marine sediments, which in turn favors the accumulation of marine organic matter [6–8].

Research has shown that the proportion of fish species (as a portion of all marine organisms) improves in sea areas where artificial reefs have been deployed. This indicates that the reefs exert a significant positive effect on the density of fishery resources [9,10]. By a process of investigation, it has been found that trapezoidal artificial reefs have a good ecological effect on shallow sea areas. These types of reef are able to change the concentration of bacterial species present in the deployment area, thus significantly reducing the abundance of sulfur-oxidizing bacteria while increasing that of *Woeseia oceani* and desulfurizing bacteria [11]. The deployment of artificial reefs will also influence the flow field of the surrounding waters and significantly change the local water flow distribution; this includes flow velocity, turbulence models, and sediment conditions, providing fish with a suitable habitat [12]. Changing the flow velocity near the reefs is an important factor in increasing the presence of fish, and is related to reef shape and layout [13]. Waters where artificial reefs are deployed are relatively shallow and located in the mixed layer of the upper ocean. The momentum, heat, and flux of material exchange are all controlled by turbulent vertical mixing at this boundary layer [14]. Therefore, investigating the flow field effect of artificial reefs is a key factor when evaluating their performance, and is one that has a direct impact on the scale of construction of artificial reefs and on their subsequent evaluation.

Many studies on the flow field effect of artificial reefs have been carried out in China and overseas [15–18], with the flow fields displaying very unsteady characteristics [19–21]. The flume experiment was first used to investigate the flow field effect of artificial reefs, and as computer technology has developed, numerical simulation has also been widely used [22,23]. Various studies have verified these results, concluding that numerical simulation can accurately reflect the flow field distribution around artificial reefs [24,25]. The flow field effect of artificial reefs is influenced by many factors, such as the structure of a single reef, the reef combination model used, and the disposal spacing [26]. The structural design of the artificial reef has a significant effect on the flow field effect of a single artificial reef, which can be effectively improved by optimizing the reef structure [27]. At present, porous trapezoidal or rectangular artificial reefs are suitable for shallow sea areas due to their good stability, high water permeability, and large effective side air volume; they also favor the survival of aquatic animals, plants, and microorganisms. The direct deployment of reefs in shallow waters will maintain a natural link to the water–sediment interface on the seabed and favor a good link to the ecological chain throughout [28]. The structure and flow field of large artificial reefs have been studied throughout the world, but the flow velocity, upwelling, and back eddy of small trapezoidal artificial reefs with different layouts and different flow velocities have not been systematically analyzed. Therefore, numerical simulations of small trapezoidal artificial reefs with different layouts are needed to reflect the effects of disposal spacing on the flow field around them. This will provide a theoretical basis for the design, actual deployment, and ecological restoration of small trapezoidal artificial reefs.

According to the characteristics of shallow waters, this paper designed a small trapezoidal artificial reef with a porous structure, suitable for use in shallow waters with small flow velocity. This reef can also generate upwelling on the approaching flow surface of the reef to transport the nutrient salts from the bottom upwards, thus promoting the mass propagation and growth of all kinds of food organisms and providing good habitats, feeding and breeding sites for fish and other organisms. Fish will also be able to cluster together to form a fishery and achieve a similar ecological effect to that provided by large reefs. This study designs a small trapezoidal artificial reef, while Ansys Fluent 2021 R1 is used as the flow field analysis software. The flow field around reefs with disposal spacing using the ratio method is then investigated to analyze their flow field distribution characteristics. This can provide a theoretical reference for the design and the future deployment of trapezoidal artificial reefs.

2. Materials and Methods

2.1. Governing Equation

The CFD (computational fluid dynamics) method is used in this study. The compressibility of the fluid and changes in temperature are not taken into account. The governing equations include the continuity and momentum equations.

The continuity equation of fluid follows the law of conservation of mass. The equation can be expressed as:

$$\frac{\partial U_i}{\partial x_i} = 0 \tag{1}$$

where U_i is the average velocity component and x_i is the Cartesian coordinate.

The governing equation obtained according to Newton's second law is used in the flow field as the momentum equation. The momentum equation can be expressed as:

$$\frac{\partial U_i}{\partial t} + U_j \frac{\partial U_i}{\partial x_j} = -\frac{1}{\rho}\frac{\partial P}{\partial x_i} + \frac{\partial}{\partial x_j}\left[\nu\left(\frac{\partial U_i}{\partial x_j} + \frac{\partial U_j}{\partial x_i}\right)\right] - \frac{\partial \overline{u_i' u_j'}}{\partial x_j} + f_i \tag{2}$$

where t is the time, ρ is the density, P is the pressure, u_i' is the velocity fluctuations due to turbulence, ν is the kinematic viscosity, $\overline{u_i' u_j'}$ is the Reynolds stress term, f is the body force such as gravity acceleration. The Reynolds stress term is introduced in the Reynolds averaging process and is expressed as:

$$-\rho\overline{u_i' u_j'} = \mu_t\left(\frac{\partial U_i}{\partial x_j} + \frac{\partial U_j}{\partial x_i} - \frac{2}{3}\frac{\partial U_k}{\partial x_k}\delta_{ij}\right) - \frac{2}{3}\rho k \delta_{ij} \tag{3}$$

where $\mu_t = \rho C_\mu \frac{k^2}{\varepsilon}$ is the eddy viscosity; C_μ is a constant. δ_{ij} is the Kronecker delta; k is the turbulent kinetic energy.

2.2. Turbulence Model

The ocean waters are large, the flow pattern is complex, and there are more eddies, so the RNG $k-\varepsilon$ turbulence model, which is more widely used in engineering, is adopted to close the Reynolds averaged equation, and this model can effectively simulate the turbulent flow with a more homogeneous distribution and smaller turbulence structure, and it is suitable for the study of the flow field effect of the artificial fish reef body. In this model, the effects of small scales can be reflected by reflecting the small-scale effects in the large-scale motions and the modified viscosity term, so that these small-scale motions are systematically removed from the control equations. At the same time, the model takes into account the effects of eddies and low Reynolds number on turbulence, which improves the accuracy of calculations in the presence of swirling flows, and is particularly suitable for describing the environment of complex flows in the ocean with large strain rate shear flows, with swirling and separation. It satisfies the following governing equations [29]:

The transport equation for turbulent kinetic energy k is:

$$\frac{\partial(\rho k)}{\partial t} + \frac{\partial(\rho U_i k)}{\partial x_i} = \frac{\partial}{\partial x_j}\left[\left(\mu + \frac{\mu_t}{\sigma_k}\right)\frac{\partial k}{\partial x_j}\right] + P_k + P_b - \rho\varepsilon + S_k \tag{4}$$

where σ_k is the turbulent Prandtl number, P_k is the production of turbulent kinetic energy (TKE) due to average velocity gradient, P_b is the production of TKE due to buoyancy, is the dissipation rate, and S_k is the source.

The transport equation for turbulent dissipation rate ε is:

$$\frac{\partial(\rho\varepsilon)}{\partial t} + \frac{\partial(\rho U_i \varepsilon)}{\partial x_i} = \frac{\partial}{\partial x_j}\left[\left(\mu + \frac{\mu_t}{\sigma_\varepsilon}\right)\frac{\partial \varepsilon}{\partial x_j}\right] + C_{1\varepsilon}\frac{\varepsilon}{k}(P_k + C_{3\varepsilon}P_b) - C_{2\varepsilon}\rho\frac{\varepsilon^2}{k} + R_\varepsilon + S_\varepsilon \quad (5)$$

where σ_ε is the turbulent Prandtl number; S_ε is the source. $C_{1\varepsilon}$, $C_{2\varepsilon}$ are model coefficients that vary within $k-\varepsilon$ turbulence models. In Equation (5), $R_\varepsilon = \frac{C_\mu \rho \eta^3 (1-\eta/\eta_0)}{1+\beta\eta^3}\frac{\varepsilon^2}{k}$, where $\eta = Sk/\varepsilon$, $S = (2S_{ij}S_{ij})^{1/2}$, $S_{ij} = \frac{1}{2}(\frac{\partial U_i}{\partial x_j} + \frac{\partial U_j}{\partial x_i})$. In the RNG $k-\varepsilon$ turbulence model, $\sigma_k = 1.0$, $\sigma_\varepsilon = 1.2$, $C_\mu = 0.0845$, $C_{1\varepsilon} = 1.42$, $C_{2\varepsilon} = 1.68$, $C_{3\varepsilon} = 0$, $\eta_0 = 4.337$, $\beta = 0.012$.

In this paper, the finite volume method is used to discretize the control equations, SIMPLEC algorithm is used for pressure–velocity coupling, standard difference format is used for pressure term processing, second-order upward format is used for spatial discretization of equations, the residual value of calculation is taken to be 10^{-5}, and an unstructured mesh is used to simulate the hydrodynamic characteristics of the square reef numerically, to study the hydrodynamic characteristics of trapezoidal artificial reefs of different layout forms and its changing law under the action of different flow velocities.

2.3. Reef Model

The flow field experiment for trapezoidal reefs featured three different layouts, each of which was composed of 1 or 5 trapezoidal reefs. They refer to single reef, transverse combination reefs, and longitudinal combination reef. The reef structures are shown in Figure 1. Five actual flow velocities (0.1, 0.5, 1.0, 1.5, and 2.0 m/s) in the water were selected according to gradient designs of maximum flow velocity and experimental flow velocity in the reef-deployed waters. The specifications of the simulated reef were consistent with those of the actual model. The specific size was 0.7 × 0.5 m (length × height), with a bilateral hole diameter of 0.2 m, while the wall thickness was 0.05 m.

2.4. Computational Domain, Boundary Conditions, and Meshing

The length of the computational domain was 30 times the length of the reef, its width was 10 times the reef length, and its height was 6 times the reef length, i.e., 21 × 7 × 4.2 m (Figure 2), where the fore-reef was 10 times the reef length and the bottom of the reef overlapped with the computational domain. Firstly, it was supposed that the bottom is transverse and has not been washed over, with no need to consider the existence of creatures in other waters; secondly, it was supposed the initial flow velocity and flow direction of the incoming flow are constant; finally, it was supposed that the reef remains stable and does not roll or slide [30–32]. Velocity inlet boundary conditions were used to define the inlet boundary on the left side of the computational domain. In consideration of the average velocity conditions of the actual artificial reef-deployed waters, five water flow velocities were selected for simulation. The right side outlet boundary was set as the pressure outlet condition, while the bottom surface of the computational domain and the reef surface featured no-slip wall boundary conditions. Both side and top surfaces of the computational domain adopted symmetrical boundary conditions.

Meshing was performed on the computational domain using the Fluent Mesh module. All meshes were unstructured tetrahedral meshes. In order to increase computational accuracy, the center of the reef bottom was taken as the center; local mesh encryption was performed in the range of a circle, while the minimum mesh precision around the reef was 0.001 m, namely 1/500 of the height of the trapezoidal artificial reef. There were a total of 3.12×10^5 partitioned meshes. Meshing was shown in Figure 3. According to the calculation, the height of the first layer of the reef boundary layer is 0.002, the growth rate of the boundary layer is 1.1, and the number of layers is 12.

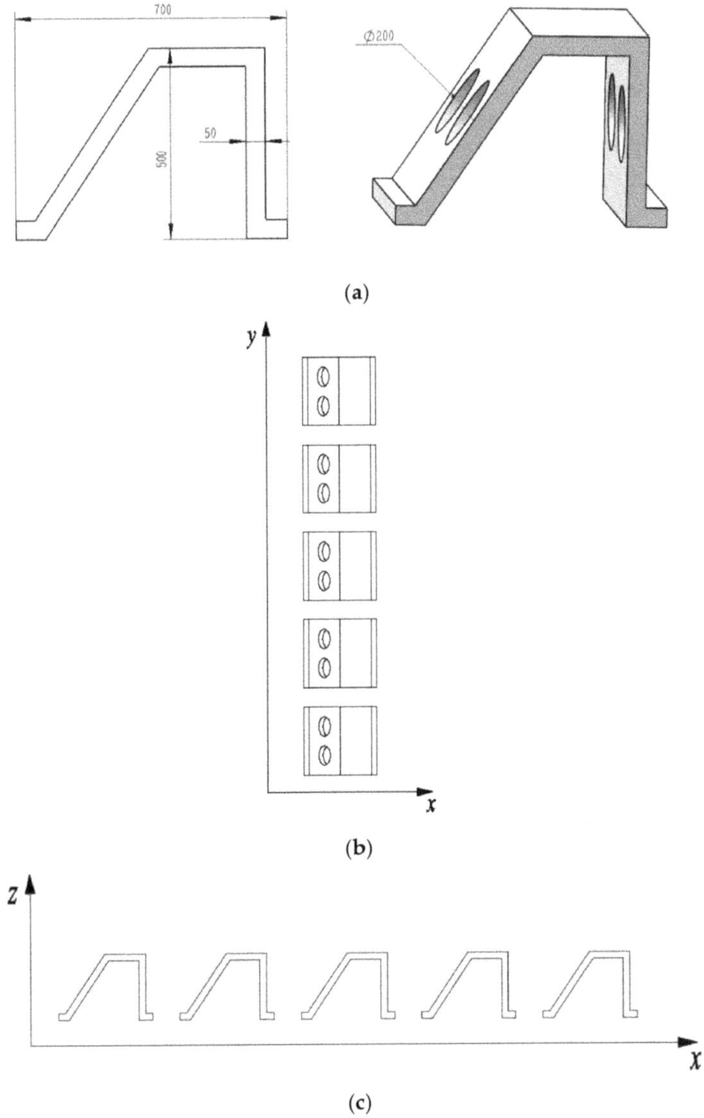

Figure 1. Models of trapezoidal reefs. (**a**) Structure of trapezoidal artificial reef. (**b**) Transverse combination. (**c**) Longitudinal combination.

2.5. Evaluation Indices and Calculation Method for Flow Field Effect

The flow field efficiency for artificial reefs was evaluated by measuring the scale of upwelling and back eddy. Researchers have found from comparative analysis of the flow field effect of commonly used artificial reefs that the difference in concrete conditions has a great effect on experimental results under the same reef height conditions. They believe that this could be used to evaluate the upwelling and back eddy effects of reefs in a more scientific and accurate manner than when using the reef volume method. In this study, upwelling and back eddy effects were evaluated by comparing the relative upwelling and back eddy volume obtained from their ratios to approaching flow volume. Therefore, the inflow direction was taken to be along the X-axis, while the Z-axis was the longitudinal direction. X = 0 m was a mid-vertical plane; The volume on both sides of the vertical plane directly behind the artificial reef, where the ratio of the vertical upward component velocity of the Z-axis to the inflow velocity, being greater than 10% of the volume, was used as a measurement index for the upwelling scale. The ratio of flow velocity in the X-axis direction to the inflow velocity was smaller than 75% of the volume, and was used as a measurement index for the back eddy scale.

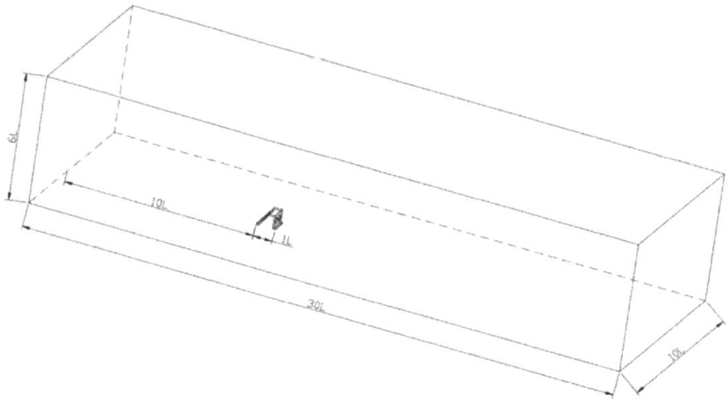

Figure 2. Computational domain of the artificial reef. Note: L is the length of reef model, 0.7 m.

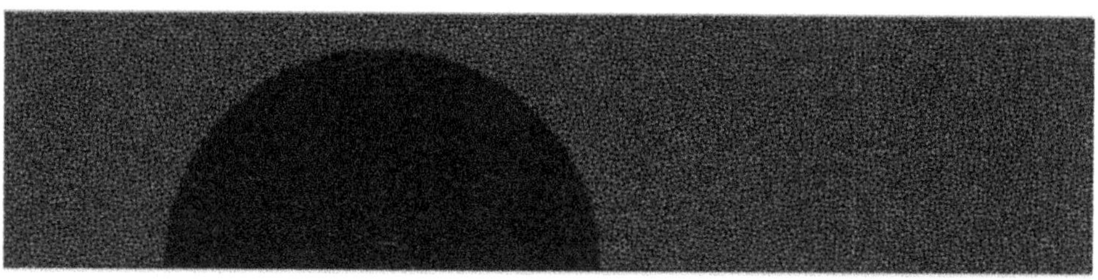

Figure 3. Meshing.

In this paper, four indexes, namely the ratio of the maximum upwelling velocity to the approaching velocity of the reef, the ratio of the upwelling height to the reef height, the ratio of the upwelling volume to the approaching volume behind the artificial reef, and the ratio of the back eddy volume to the approaching volume behind the artificial reef, are used to evaluate the strength of the upwelling and back eddy currents, which also provide a reliable method for the study of numerical simulation of artificial reefs for evaluating the effect of the flow field around the reefs.

At present, there is no agreed standard for the selection of calculation data for numerical simulation, and the selection range of this data has not been definitively described. In this study, the flow field data of three types of reefs (single reef, transverse combination reefs with a spacing of 1 L, and longitudinal combination reefs with a spacing of 1 L) at a flow approach angle of 0° and flow rate of 1.5 m/s were selected for analysis, with one datum saved every second. Based on this, a maximum upwelling velocity could be plotted (Figure 4). From Figure 4, after 63 s, it can be seen that the maximum upwelling velocity of the three types of reefs are generally stable, with no significant fluctuations. Therefore, the data selected for this study were those from the seven consecutive seconds after 63 s, which were imported into CFD-Post for averaging, after which they could undergo analysis and comparison.

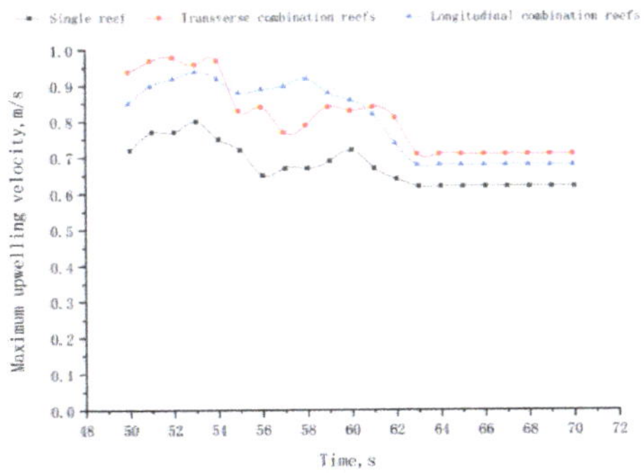

Figure 4. Upwelling volume–time curve.

3. Results and Analysis

In this study, the flow field effects of a single reef, a transverse reef combination with spacing of 0.5, 1.0, 1.5, and 2.0 L, and a longitudinal reef combination with spacing of 0.5, 1.0, 1.5, 2.0, 2.5, and 3.0 L were analyzed using the numerical simulation method, where L = 0.7 m.

3.1. Flow Field Distribution of a Single Trapezoidal Reef

In order to verify the effectiveness of the numerical simulation and testing of artificial reefs, a flume was used to test a single artificial reef, and the test results and numerical simulation results of the reef flow velocity of 1.0 m/s are shown in Figure 5. According to the above comparison results, it can be seen that the numerical model established in this paper obtained by the measurement point flow velocity and the physical model of the artificial reef experiments coincide with the results, which strongly proves that the method of this paper is selected to establish a numerical computational model to carry out the study of the dynamic characteristics of artificial reefs under the action of the water flow is feasible.

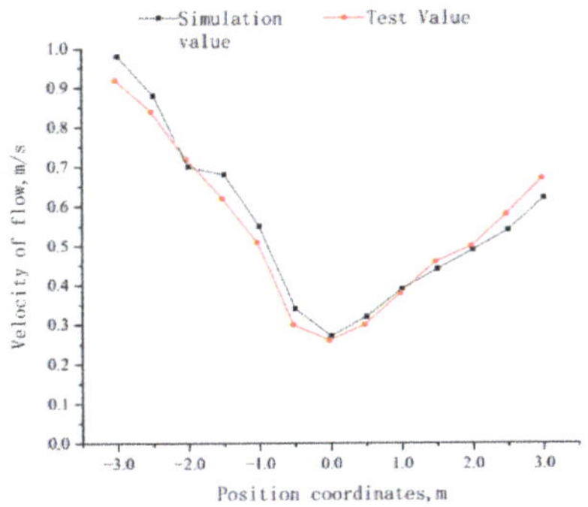

Figure 5. Comparison of simulated and experimental values of flow velocity at the measuring point.

(1) Analysis of upwelling characteristics. Figure 6a–c show the scale and strength of the upwelling of a single trapezoidal reef. It can be seen from the figures that the maximum upwelling velocity, the ratio of upwelling height to reef height, and the ratio of upwelling volume to approaching flow volume all increase as the flow velocity increases. The maximum upwelling velocity increases most rapidly; the upwelling height and volume increase slightly, while the ratio of the upwelling volume to the approaching flow volume ranges from 0.60 to 0.85; the upwelling height increases more rapidly as the flow velocity increases from 1.0 to 1.5 m/s.

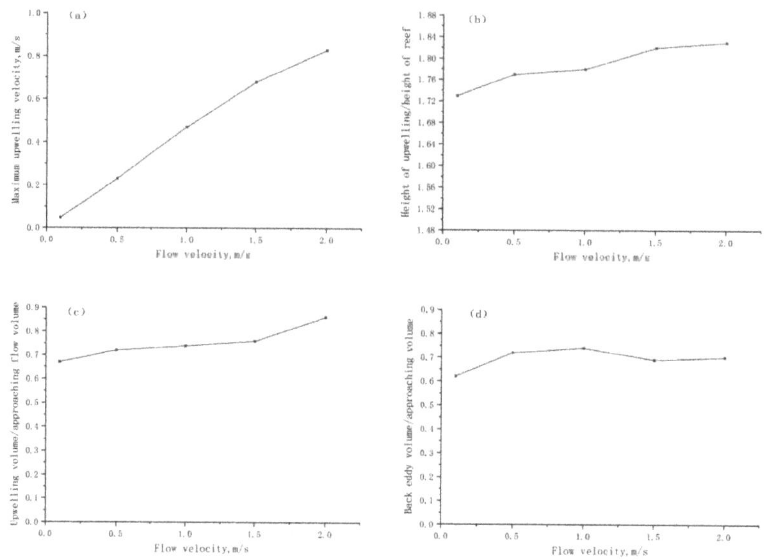

Figure 6. Scale and strength of upwelling and back eddy of a single trapezoidal reef at different flow velocities. (**a**) The variation curve of maximum upwelling velocity with flow velocity. (**b**) The variation curve of height of upwelling with flow velocity. (**c**) The variation curve of upwelling volume with flow velocity. (**d**) The variation curve of back eddy volume with flow velocity. Note: approaching volume refers to the volume of flow velocity in the *X*-axis direction.

(2) Analysis of back eddy characteristics. Figure 6d shows the scale and strength of the back eddy of a single three-round-pipe reef. It can be seen from the figure that the back eddy volume shows a tendency to increase, decrease, and then increase again as the inlet velocity increases; this reaches a maximum when the flow velocity is 1.0 m/s, before it decreases again. Generally speaking, though, it shows an increasing trend.

3.2. Effect of Transverse Combination Spacing on Flow Field around the Trapezoidal Reef

Figure 7 shows the numerical simulation results of flow field effect in a transverse combination of reefs with different disposal spacing at five different flow velocities. Each index changes as follows: As the disposal spacing increases, the maximum upwelling velocity, upwelling height, and upwelling volume show a tendency to first increase and then decrease, reaching their maximum values at a spacing of 1.0 L. The maximum upwelling velocity and upwelling height also show an insignificant tendency to increase at first and then decrease under low-velocity constraint conditions. The back eddy volume also reaches a maximum at a spacing of 1.0 L, but does not change significantly with increased disposal spacing, approximating the same back eddy volume as that of a single reef. At a given disposal spacing, and as flow velocity increases, the maximum upwelling velocity, upwelling height, and upwelling volume increase gradually, with the upwelling volume showing a consistent change trend with the single reef; the back eddy volume does not display any obvious change trend. In the transverse combination mode, the scale and strength of the upwelling and back eddy were greatest at a reef spacing of 1.0 L, meaning the maximum scale and strength of the upwelling and back eddy were obtained when the disposal spacing between two unit reefs was 1.0 L.

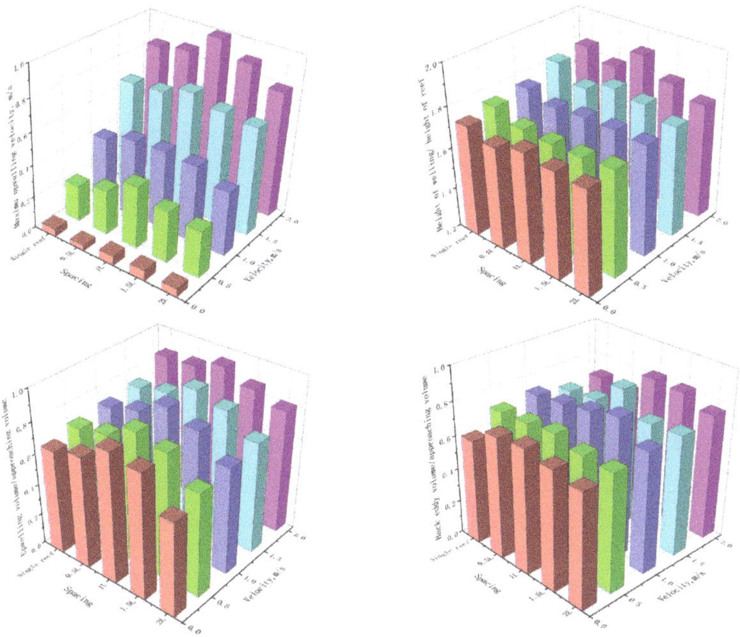

Figure 7. Scale and strength of upwelling and back eddy on trapezoidal reefs with transverse combination spacing at different flow velocities.

3.3. *Effect of Longitudinal Combination Spacing on Flow Fields around the Trapezoidal Reef*

Figure 8 shows the numerical simulation results for the flow field effect in a longitudinal combination of reefs with different disposal spacing at five different flow velocities. Each index changes as follows: At a given flow velocity, and as disposal spacing increases, the maximum upwelling velocity, upwelling height, and upwelling volume all show a tendency to increase at first and then decrease, reaching their maximum values at a spacing of 1.5 L. The upwelling volume shows a relatively flat tendency, with a consistent change trend for a single reef that reaches a maximum at a spacing of 1.5 L. The back eddy volume behind the reef is smaller than that of a single reef, because the sheltering effect of the front reef leads to a significant reduction in the function of the one behind. At a given disposal spacing, and as flow velocity increases the maximum upwelling velocity, the ratio of upwelling height to reef height and the ratio of upwelling volume to approaching flow volume also increase gradually, showing a consistent change trend with the single reef; the back eddy volume does not show any obvious change trend. In longitudinal combination mode, the scale and strength of the upwelling reached a maximum at 1.5 L; the scale of the back eddy on the second unit reef was significantly smaller than that for the first reef. This is possibly because the first unit reef weakens the back eddy on the second one; this means that a longitudinal layout has a large influence on the back eddy.

3.4. *Stability Analysis of Trapezoidal Reefs*

In order to stabilize the reef in the water, current force conditions on the different combinations of trapezoidal reef models were tested by means of model simulation. Pressure distribution of a trapezoidal reef under the impact of water flow. According to the results obtained, the maximum pressure of the trapezoidal reef is 985 Pa, and the overall flow pressure is calculated as F by the computer. The current forces for the reef model at five inflow velocities, specific flow velocities, and force conditions under various working conditions are shown in Table 1.

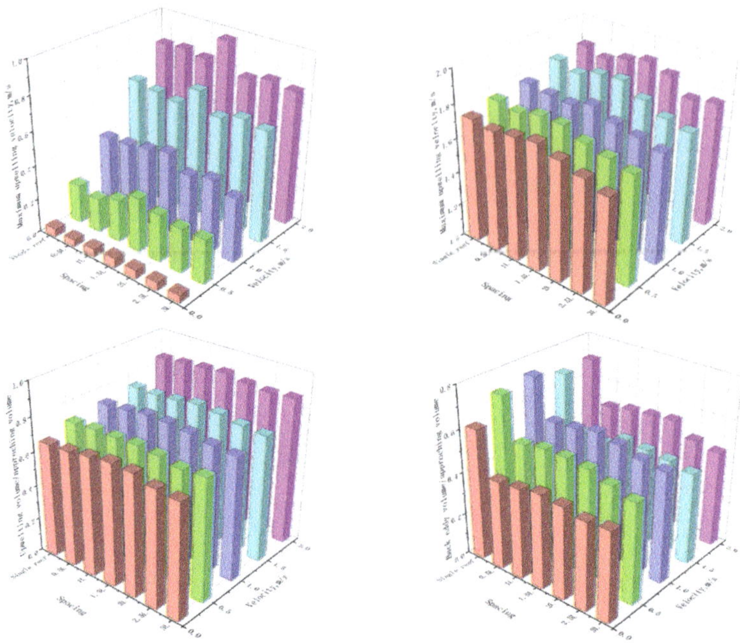

Figure 8. Scale and strength of upwelling and back eddy of trapezoidal reefs with longitudinal combination spacing at different flow velocities.

Table 1. Current forces exerted on the model at five flow velocities.

	1	2	3	4	5
Flow velocity, m/s	0.1	0.5	1.0	1.5	2.0
Current force, KN	0.0368	0.0435	0.0546	0.0767	0.0979

3.4.1. Analysis of Anti-Slide Safety

For the reef not to slide, the maximum static frictional force between reef and seabed must be greater than the fluid force exerted on the reef itself, and the anti-slide coefficient $S_1 > 1$. The calculation is as follows:

$$S_1 = \frac{W_\mu(1-\rho/\sigma)}{F_{max} > 1} \tag{6}$$

where W is the weight of the trapezoidal reef model, 1.0916 KN; μ is the maximum static friction coefficient using the particle size for medium sand (0.60); ρ is the water density, 1025 kg/m^3; σ is the weight of the reef per unit volume, 2500 kg/m^3; and F is the current force exerted on the reef. The analytical calculation results are shown in Table 2.

Table 2. Stability analysis for the reef model at five different flow velocities.

	1	2	3	4	5
Flow velocity, m/s	0.1	0.5	1.0	1.5	2.0
Current force, KN	0.0368	0.0435	0.0546	0.0767	0.0979
Anti-slide coefficient S_1	10.50	8.88	7.08	5.04	3.95
Anti-rolling coefficient S_2	15.31	12.95	10.32	7.35	5.76

3.4.2. Analysis of Anti-Roll Safety

For the reef not to roll, the resulting moment M_1 of the gravity and buoyancy of the reef must be greater than the maximum moment M_2 of fluid force exerted upon it, and the anti-rolling coefficient $S_2 > 1$. The calculation is as follows [27]:

$$S_2 = \frac{M_1}{M_2} = \frac{W(1-\rho/\sigma)l_w}{F_{\max}h_0} \quad (7)$$

where W is weight of the trapezoidal reef model, 1.0916 KN; ρ is the water density, 1025 kg/m^3; σ is the weight of the reef per unit volume, 2500 kg/m^3; l_w is the horizontal distance between the center of gyration and the center of gravity, 0.035 m; h_0 is the height of the fluid force, 0.0400 m; and $F_{\max}$ is the current force exerted on the reef. Analytical calculation results are shown in Table 2.

According to the findings of the force analysis of the reef model under different working conditions, the current force exerted on the reef increases as the inflow velocity increases. Its stability also decreases accordingly, but as a whole, the anti-slide coefficient and anti-rolling coefficient of the reef at different flow velocities are all greater than 1. That is to say, the reef model is able to maintain its stability in the currents and waves without sliding or rolling, indicating that the reef model has a reliable structure.

4. Discussion

This study took a trapezoidal reef as an example. A comparative analysis was conducted on changes in flow field around reefs with different disposal spacing based on the Fluent numerical simulation method, after which an analysis was carried out on the stability of the reef model in the currents.

To verify the reliability of the numerical simulation method and calculations, the flow velocity of the longitudinal dam body within the ecological engineering restoration area of the Yangtze River estuary was measured in September 2020. I used the LinkQuest ADCP (San Diego, CA, USA) Doppler flow velocity meter to map the flow velocity and flow pattern around the artificial reef in the artificial reef placement area, which can measure the flow velocity and direction of each vertical point of the cross section in the watershed, with an accuracy of 0.01 m/s, and analyze and obtain the maximum upwelling flow velocity according to the previous criteria on the definition of the upwelling. The experimental results show that the longitudinal artificial fish reef has a relatively small flow velocity, significant changes in direction, and complex flow patterns. Analytical calculation results are shown in Table 3. Comparing and analyzing the experimental test results with the numerical calculation results in Table 1, it can be seen that the simulation results are basically consistent with the surveying results, with an error of less than 10%, indicating the effectiveness and high accuracy of the simulation of artificial fish reef flow patterns.

Table 3. Comparison between simulation and test data of upwelling velocity of an artificial reef.

Velocity of Flow/($m{\cdot}s^{-1}$)	Maximum Upwelling Velocity Simulation Value/($m{\cdot}s^{-1}$)	Maximum Upwelling Velocity Test Value/ ($m{\cdot}s^{-1}$)	Error Analysis of Flow Velocities/%
0.1	0.047	0.051	8.5
0.5	0.230	0.215	6.5
1.0	0.470	0.434	7.6
1.5	0.680	0.619	8.9
2.0	0.830	0.872	5.1

The simulation results show that, in the transverse combination mode, the scale and strength of the upwelling and back eddy were greatest at a reef spacing of 1.0 L, meaning the maximum scale and strength of the upwelling and back eddy were obtained when the disposal spacing between two unit reefs was 1.0 L. In longitudinal combination mode, the scale and strength of the upwelling reached a maximum at 1.5 L; the scale of the back eddy on the second unit reef was significantly smaller than that for the first reef. This is possibly because the first unit reef weakens the back eddy on the second one; this means that a longitudinal layout has a large influence on the back eddy. This is basically consistent with Liu et al. [29], who analyzed the maximum change of the flow field when the model spacing is between 1.0 and 1.5 L. However, because they only analyzed the change characteristics of the flow field, without detailed analysis of the upwelling velocity and the volume of the upwelling, they were unable to judge the scale of the upwelling of the artificial reefs, and they also did not analyze the different velocities, which could not accurately respond to the characteristics of the upwelling at different velocities and under different layouts. Zheng et al. [25] also simulated the flow field effects of three stacked forms of circular tube-type artificial reefs numerically. When the number of stacked reefs is fixed, the maximum upwelling velocity and upwelling area increase with the increase in approaching velocity, which is consistent with the conclusion of this paper, but there is no analysis of the back eddy. As the simulation is mainly used to evaluate the maximum upwelling flow velocity and upwelling area, the simulation error value is around 20%, which is a large error, and at the same time, their study did not analyze the layout spacing and flow velocity of the artificial reefs, and the effect of the reefs was not investigated practically, so that the ecological significance of the reefs could not be determined. Liu et al. [17] also simulated and analyzed porous square-shaped artificial reefs, but only analyzed the height and area of the upwelling region of a single reef, with a large error value, and did not quantitatively and in detail analyze the strength of upwelling changes under different reef layouts and velocities. Yang et al. analyzed the effect of different fabrication methods on the flow field of permeable artificial reefs, but the evaluation method was simpler and did not analyze the strength of upwelling under different layouts, while the experimental results were not verified. In this study, we analyzed small trapezoidal reefs and applied the ratio method to compare the flow field effects under different velocities and different layouts, and the results had small errors, which can provide valuable references for the placement and design of artificial reefs. And stability analysis was conducted on the trapezoidal artificial reef model. It was found that it is able to maintain its stability in currents and waves, meaning stability risks are relatively low for the transverse and longitudinal reef combinations discussed in this study. This finding is of great significance for the actual planning and deployment of artificial reefs.

The survey and monitoring of the project area was carried out in September of the year before and after the ecological restoration project and at the same location according to the principle of spatial and temporal unity, and in terms of fish species, gillnets with the same mesh size of 40 mm were used to catch fish in the area where the artificial reefs were put in, as shown in Figure 9. The results showed that the number of fish species in the project area were from 40 to 49 after the construction of artificial fish reefs. Fish density is mainly used to assess the number of fish, with the American Bio Sonics DT-X (Seattle, WA, USA) finding that echo sounders (split beam, detection distance: 0.5–500 m, echo detection limit: −140 dB, sounding frequency: 0.01–30 times/second, pulse frequency: 0.1–1 milliseconds, dynamic range: more than 160 dB) are used to detect fish in the longitudinally combined artificial reef waters before and after the completion of the project; the fish density was about 0.314 fish per m^2 before the project. After the completion of the project, a fish echo detector was used to detect the fish density within the engineering area. The results showed that the fish density after the project construction was about 1.197 fish per m^2 after the construction; the fish density after project completion was about four times that before the restoration. The results are shown in Figure 9. This is mainly because the construction of artificial fish reefs changed the flow characteristics and produced more eddies, with significant changes in flow patterns and gradients, and allowed fish species to spawn in faster-flowing areas and rest and feed in still-water areas after spawning. Artificial fish reef construction provided better conditions for fish habitats. In summary, it was found that the fish species and quantities in the test area increased during these investigations, indicating that the longitudinally combined trapezoidal artificial reefs have a very good impact on the recovery of fishery resources. This study can thus provide a theoretical basis for the design, deployment, and ecological restoration in waters featuring trapezoidal artificial reefs. Figure 10 shows the trapezoidal artificial reef deployed on site.

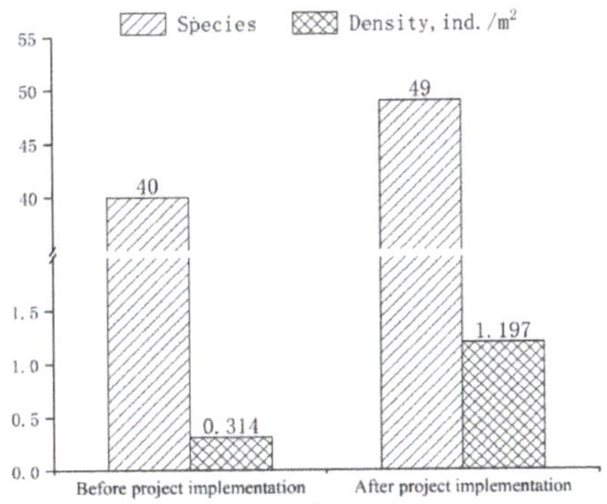

Figure 9. Comparison of fish before and after the implementation of the project.

Figure 10. Artificial reef before deployment.

5. Conclusions

This paper uses numerical simulation to simulate the strength of upwelling and back eddy around the flow field of small trapezoidal reefs under different deployment spacing and velocities, and compares and analyzes the optimal spacing of reefs under different velocities using the ratio method, and compares the results with the existing research results, which shows that the numerical simulation can better respond to the influence of the deployment spacing on the effect of the flow field around the trapezoidal reefs, and makes the analysis of the physical stability, which can provide a certain theoretical basis for the layout of artificial reefs in the actual sea area. This paper also investigates and analyzes the ecological benefits of the reefs, and the results also show that the trapezoidal reefs under this layout have better stability and ecological benefits, and also provide valuable references for the placement and design of artificial reefs.

This study analyzed the flow field effects of trapezoidal reefs under different flow velocities and spacing using numerical simulation and field survey methods. However, the flow field of reefs with different layouts was not verified in the experiments. Additionally, no optimization or improvement was made to the structure of the trapezoidal reefs, and there was no comparative analysis with other artificial reefs. In future research, it is necessary to conduct field verification experiments for reefs with different layouts and optimize the structure of trapezoidal reefs to provide a basis for their design and selection.

Author Contributions: Conceptualization, X.C. (Xiaolong Chen) and X.C. (Xuan Che); methodology, X.C. (Xiaolong Chen) and X.C. (Xuan Che); software, X.C. (Xiaolong Chen) and X.L.; validation, C.T. and Y.Z.; writing, X.C. (Xiaolong Chen). All authors have read and agreed to the published version of the manuscript.

Funding: This study was supported by Science & Technology Fundamental Resources Investigation Program (Grant No. 2022FY100404).

Data Availability Statement: The data presented in this study are available on request from the corresponding author.

Conflicts of Interest: The authors declare no conflict of interest.

References

1. Wang, H.; Chen, P.; Zhang, S.; Jia, X. Effect on fishery resources multiplication of artificial reefs. *Guangdong Agric. Sci.* **2009**, *8*, 18–21.
2. Gao, Y.; Chen, X.; Meng, S.; Hu, G.; Li, D.; Qiu, L.; Song, C.; Fan, L.; Xu, H.; XU, P. Research Progress of Artificial Reef Construction and Its Mechanism. *Chin. Agric. Sci. Bull.* **2023**, *39*, 138–144.
3. Yang, H. Construction of marine ranching in China: Reviews and prospects. *J. Fish. China* **2016**, *40*, 1133–1140.
4. Wang, Y.; Guo, D. Design and construction of marine ranch in my country. *China Fish.* **2011**, *4*, 25–27.
5. Pan, P. Marine ranch: Carrying new hope for China's fishery transformation. *China Fish.* **2016**, *1*, 47–49.
6. Yu, L. Study on Topography and Hydrodynamic Characteristics of Typical Spawning Grounds of Four Great Fishes in the Middle Reaches of the Yangtze River. Ph.D. Thesis, Southwest University, Chongqing, China, 2018.
7. Zhang, J.; Che, X.; Jiang, G.; Tian, C.; Chen, X. Effects of artificial dams on hydrodynamic characteristics of fish habitat in the upper reaches of the Yangtze River. *Chin. Soc. Agric. Eng.* **2021**, *5*, 140–146.
8. Fan, J.; Chen, P.; Feng, X.; Chen, G. Research Progress on hydrodynamics of artificial reefs. *Guangdong Agric. Sci.* **2013**, *2*, 185–188.
9. Yang, Y.; Yan, Z.; Qiao, Y. Characterization and review of hydraulic conditions of river fish habitat. *J. Hohai Univ. (Nat. Sci.)* **2007**, *2*, 125–130.
10. Sun, Y.; Niu, T.; Wang, Y.; Wang, R.; Guo, X. Simulation and restoration design of fish habitat based on terrain remodeling. *Environ. Ecol. Three Gorges* **2015**, *37*, 29–32.
11. Sun, J. Simulation of Fish Habitat in Heishui River, a Backwater Tributary of Baihetan Reservoir. Master's Thesis, Zhejiang University, Hangzhou, China, 2013.
12. Wu, J.; Wu, Z. Turbulence field of spur dike and its engineering significance. *J. Zhengzhou Inst. Eng.* **1994**, *2*, 22–27.
13. Jiang, Z. Numerical Simulation of Hydrodynamics for Artificial Reefs. Ph.D. Thesis, Ocean University of China, Qingdao, China, 2009.
14. Zhang, S.; Sun, M.; Chen, Y. Quantitative analysis of upwelling current features ofa artificial concrete reef with different height. *J. Dalian Fish. Univ.* **2008**, *23*, 353–358.
15. Pan, L.; Lin, J.; Zang, S. A numerical experiment of the effects of artificialreef on vertical 2-dimensional steady flow field. *J. Shanghai Fish. Univ.* **2005**, *14*, 406–412.
16. Li, J.; Zhang, S. The comparison between numerical simulation and water channel experiment on an Mi-zi artificial reef. *J. Fish. China* **2010**, *34*, 1587–1594. [CrossRef]
17. Liu, Y.; Guan, C.; Zhao, Y.; Cui, Y.; Dong, G. Experimental study on two-dimensional flow field of the star artificial reef in the water stream with PIV. *Chin. J. Hydrodyn.* **2010**, *25*, 777–783.
18. Guan, C.; Liu, Y.; Zhao, Y.; Cui, Y.; Li, J. Experimental study on two dimensional flow field of the compound Martificial reef with Particle Image Velocimetry (PIV). *Fish. Mod.* **2010**, *37*, 15–19.
19. Liu, Y.; Guan, C.; Zhao, Y. Numerical simulation and PIV study of unsteady flow around hollow cube artificialreef with free water surface. *Eng. Appl. Comput. Fluid Mech.* **2012**, *6*, 527–540.
20. Jiang, Z.; Liang, Z.; Tang, Y. Numerical simulation and experimental study of the hydrodynamics of a modeled reef located within a current. *Chin. J. Oceanol. Limnol.* **2010**, *28*, 267–273. [CrossRef]
21. Yakhot, V.; Orzag, S. A Renormalization group analysis of turbulence: Basic theory. *J. Sci. Comput.* **1986**, *1*, 3–11. [CrossRef]
22. Yu, D.; Yang, Y.; Li, Y. Research on hydrodynamic characteristics and stability of artificial reefs with different opening ratios. *Period. Ocean Univ. China* **2019**, *49*, 128–136.
23. Wang, J.; Liu, L.; Cai, X. Numerical simulation study on influence of disposal space on effects of flow field around porous square artificial reefs. *Prog. Fish. Sci.* **2020**, *41*, 40–48.
24. Wang, F. *Computational Fluid Dynamics Analysis*; Tsinghua University Press: Beijing, China, 2004; pp. 113–143.
25. Zheng, Y.; Liang, Z.; Guan, C. Numerical simulation and experimental study on flow field of artificial reefs in three tube-stacking layouts. *Oceanol. Limnol. Sin.* **2014**, *45*, 11–19.
26. Liu, H.; Ma, X.; Zhang, S.; Yu, H.; Huang, H. Research on model experiments of effects of artificial reefs on flow field. *J. Fish. China* **2009**, *33*, 229–236.

27. Guan, C.; Li, M.; Heng, Y.; Li, J.; Cui, Y.; Li, Z.; Wang, T. Numerical Simulation of Disposal Space and Analysis on Physical Stability of Three-Tube Artificial Reefs. *Periddical Ocean Univ. China* **2016**, *46*, 9–17.
28. Cui, Y.; Guan, C.; Wan, R. Numerical simulateon on influence of disposal space on effects of flow field around artificial reefs. *Trans. Oceanol. Limnol.* **2011**, *51*, 59–65.
29. Liu, H.; Ma, X.; Zhang, S.; Lin, j.; Yu, H.; Huang, H. Validation and comparison between wind tunnel experiments and numerical simulation of flow field around artificial reefs. *J. Fish. Sci. China* **2009**, *16*, 365–371.
30. Huggins, D.; Piedrahita, R.; Rumsey, T. Analysis of sediment transport modeling using computational fluid dynamics (CFD) for aquaculture raceways. *Aquac. Eng.* **2004**, *31*, 277–293. [CrossRef]
31. Huggins, D.; Piedrahita, R.; Rumsey, T. Use of computational fluid dynamics (CFD) for aquaculture raceway design to increase settling effectiveness. *Aquac. Eng.* **2005**, *33*, 167–180. [CrossRef]
32. Jiang, Z.; Liang, Z.; Liu, Y. Particle image velocimetry and numerical simulations of the hydrodynamic characteristics of an artificial reef. *Chin. J. Oceanol. Limnol.* **2013**, *31*, 949–956. [CrossRef]

Journal of
Marine Science and Engineering

MDPI

Article

Role of Shape and Kinematics in the Hydrodynamics of a Fish-like Oscillating Hydrofoil

Siddharth Gupta [1,2,3], Atul Sharma [2], Amit Agrawal [2], Mark C. Thompson [1] and Kerry Hourigan [1,*]

[1] Fluids Laboratory for Aeronautical and Industrial Research (FLAIR), Department of Mechanical and Aerospace Engineering, Monash University, Melbourne 3800, Australia; siddharth.gupta1@monash.edu (S.G.); mark.thompson@monash.edu (M.C.T.)
[2] Department of Mechanical Engineering, Indian Institute of Technology Bombay, Mumbai 400076, India; atulsharma@iitb.ac.in (A.S.); amit.agrawal@iitb.ac.in (A.A.)
[3] IITB-Monash Research Academy, IIT Bombay, Mumbai 400076, India
* Correspondence: kerry.hourigan@monash.edu

Abstract: In the present two-dimensional numerical study, we investigate the roles of geometrical parameters of a hydrofoil (shape/curvature of the leading and trailing edges and thickness) and kinematic parameters (phase difference between heave and pitch (ϕ)) on the propulsive performance of different-shaped hydrofoils oscillating at maximum angles of attack up to $\alpha_{\max} = 30°$. The study was carried out at a fixed non-dimensional maximum heave to chord ratio $h_\circ/C = 0.75$, Strouhal number $St = 0.25$, and Reynolds number $Re = 5000$. Our findings reveal that hydrofoil performance and stability improve with leading and trailing edge curvatures but decline as thickness increases. By analyzing the near-wake structure, we establish that even minimal flow separation increases power consumption while moderate flow separation enhances thrust. Over the range of different-shaped hydrofoils at different $\alpha_{\max}$ and ϕ, maximum propulsion efficiency occurs for those parameters for which there is a small degree of flow separation but with no roll-up of a separating vortex. In comparison, maximum thrust generation occurs when there is a moderately strong flow separation but without induction of a significant amount of fluid around the trailing edge. These insights offer valuable knowledge for understanding fish propulsion efficiency and have applications in designing autonomous underwater vehicles (AUVs) and micro-air vehicles (MAVs).

Keywords: fluid–structure interaction; fish-like propulsion; computational fluid mechanics

1. Introduction

The problems pertaining to fluid–structure interaction (FSI) can be classified into two primary categories: one-way or weakly coupled, and two-way or strongly coupled problems. In one-way or weakly coupled problems, only the fluid flow field is influenced by the structure/body motion. On the other hand, in two-way or strongly coupled problems, both the structure/body and the fluid flow field exert mutual influence upon each other [1]. The present work is on a one-way coupled FSI problem of an oscillating hydrofoil.

Especially over the past decade, the hydrodynamics of an oscillating hydrofoil has drawn the attention of many engineers, mathematicians, and biologists due to its association with understanding the biology of fish swimming and designing new innovative solutions for modern engineering applications [2]. Some of these applications include micro-air and water propulsion systems [3–5], and energy-harvesting systems [6–9]. The motivation of the present study is to understand the role of hydrofoil-shape on its propulsive performance. This knowledge is fundamental and holds significant potential for enhancing the precision and appropriateness of engineering designs for autonomous underwater vehicles (AUVs) and micro-air vehicles (MAVs).

For the present problem of a thrust-generating oscillating hydrofoil, the non-dimensional governing parameters can be categorized as two geometric parameters: the curvature (of leading and trailing edges) and thickness of the hydrofoil; four kinematic parameters: heave to chord ratio $h_\circ/C$, Strouhal number $St = 2h_\circ f/u_\infty$ (f is the frequency of oscillation), maximum angle of attack $\alpha_{\max}$ (Equation (2)), and the phase difference between heave and pitch ϕ (Equation (1)); and Reynolds number $Re = \rho u_\infty C/\mu$ as the flow parameter based on free stream velocity u_∞, chord length C, and dynamic viscosity μ. Studies on oscillating hydrofoils in the literature use a range of different NACA hydrofoils. To provide context, we first sequentially review previous experimental and numerical studies in this area.

In terms of studies focusing on experiments, Triantafyllou et al. [10,11] undertook a stability analysis of the wake of an oscillating hydrofoil and found that its stability is governed by a non-dimensional parameter, the Strouhal number, St, defined above. The wake was found to be stable for the frequency range of $0.25 < St < 0.4$. Further, they determined the operational St for different fish and found that most propel themselves within this narrow St range. Later, through a parameter study, Anderson et al. [12] determined the optimal parameters for efficient oscillation of a NACA0012 hydrofoil, which replicates the caudal fin motion for carangiform fish. Their results showed that the optimal efficiency with high thrust also lies within this narrow St range. In addition to oscillation frequency, they also examined optimal values for three more non-dimensional governing parameters—heave amplitude $h_\circ/C$, maximum angle of attack $\alpha_{\max}$, and the phase difference between heave and pitch ϕ. For efficient propulsion, they proposed that the values of these parameters should be $0.25 < St < 0.4$, $h_\circ/C \approx 1$, $15^\circ < \alpha_{\max} < 25^\circ$, and $\phi = 75^\circ$. Following this, Read et al. [13] also undertook a parameter study for the same set of governing parameters as Anderson et al. [12]. However, for almost all cases, they observed that the oscillating hydrofoil achieved high efficiency together with high thrust for $90^\circ < \phi < 100^\circ$. These two contradictory findings led to numerous further experimental and numerical studies with the prime objective of exploring the effect of ϕ on propulsive performance. Motivated by these results, Hover et al. [14] compared the effect of different angle of attack profiles on the hydrodynamic performance of a flapping hydrofoil. They observed that a cosine angle of attack profile provides an optimal combination of thrust and propulsive efficiency, as compared to a harmonic angle of attack profile. Recently, Van Buren et al. [15] also performed an experimental investigation on a pitching and heaving teardrop-shaped hydrofoil. They observed that a phase difference of 90° between heave and pitch is required to achieve maximum propulsive efficiency. Additionally, they proposed scaling laws based on lift and added mass forces that align well with the experimental measurements.

Many numerical studies have also been undertaken on the problem of flow across oscillating hydrofoils. Isogai et al. [16] were among the first who performed two-dimensional simulations to understand the effect of ϕ on the propulsive performance of a NACA0012 oscillating hydrofoil. They concluded that the highest propulsive efficiency occurs at $\phi = 90^\circ$. A similar conclusion was reached in other studies [17,18]. Numerical studies have also been conducted to understand the effect of near- and far-wake vortices on the propulsive performance. For near-wake vortices, Tuncer and Platzer [19] concluded that the formation of even a small leading-edge vortex (LEV) should be prevented for efficient propulsion. Similar conclusions have resulted from other studies, including Tuncer and Kaya [20] and Young and Lai [21]. In direct contrast, Guglielmini and Blondeaux [22] concluded that a small to moderately strong magnitude LEV is beneficial for propulsive efficiency. Thus, the role of LEV in efficient propulsion is contradictory in the literature. For the far-wake structures, Bose and Sarkar [23] investigated the transition in the flow pattern from periodic to chaotic through a quasi-periodic route by increasing the heave amplitude at $Re = 1000$. Numerical studies have also been conducted to understand the effect of various geometrical parameters on propulsive performance. Zhang et al. [24] studied the effect of chord and thickness ratio of an elliptical foil, while Ashraf et al. [25] studied the effect of thickness and camber using a variety of NACA hydrofoils.

The above brief literature survey shows that most previous studies focused on understanding the role of kinematic parameters on the performance of an oscillating hydrofoil, and few published works discuss the role of the hydrofoil shape on performance parameters. However, the effect of the hydrofoil curvature at the leading and trailing edges, along with its thickness, remains to be explored. Thus, one of the primary aims of the present study is to elucidate the effect of these geometrical parameters on the propulsive performance of an oscillating hydrofoil at various angles of attack, α_{max}. Through this analysis, we are also able to comment on the role of LEV on the performance of a thrust-producing oscillating hydrofoil. Next, the phase difference between the heave and pitch ϕ is a crucial parameter that governs the propulsive performance. Hence, its role in the propulsive performance of a NACA00XX hydrofoil has been studied extensively, as discussed in the above paragraphs. Those studies reach similar conclusions that a phase difference of 90° provides the maximum propulsive efficiency with an optimal magnitude of thrust. A change in the phase difference may enhance the thrust generation; however, it significantly reduces propulsion efficiency. Note that the motion of an oscillating hydrofoil at $\phi = 90°$ seems to be the most fish-like, cleanly slicing through the water with the lowest angles of attack. Further, fish are also known for maintaining a phase difference of $\phi = 90°$ between pitch and heave during swimming [26]. Thus, the next objective is to understand whether $\phi = 90°$ is a universal phase difference for attaining maximum efficiency or whether it depends on hydrofoil shape. Being the primary mechanism used by various fish for swimming and other engineering applications discussed above, the present study contributes from both fundamental and engineering points of view.

The paper structure is as follows. The methodology is detailed in Section 2, which provides a description of different hydrofoils considered for the study, oscillation kinematics, governing equations, validation and grid independence studies, and definitions of the input and output governing flow parameters. This is followed by a discussion of the results in Section 3. The main conclusions are summarized in Section 4.

2. Methods: Modeling and Simulation

2.1. Mapping the Hydrofoil Shape

For the study on the role of curvature, specifically the front and rear shape, and thickness of the hydrofoil, Figure 1 depicts various shapes that are considered in the present study. The different shapes are chosen in such a way that they connect the two extreme shapes (a NACA0012 airfoil and a thin flat plate) by considering the importance of the front and rear geometry, and thickness. Here, shape 1 corresponds to a NACA0012 hydrofoil with a bluff trailing edge; shape 2 corresponds to a NACA0012 hydrofoil with a bluff leading edge; the rectangular plate (effectively) corresponds to a NACA0012 hydrofoil with bluff leading as well as trailing edges; and the thin plate (effectively) corresponds to a NACA0012 hydrofoil or a rectangular plate with negligible thickness. Thus, comparing the NACA0012 hydrofoil with shape 1 explores the role of a tapering tail, a NACA0012 hydrofoil with shape 2 helps us understand the role of a rounded leading edge, a NACA0012 hydrofoil transformed into a rectangular plate helps to understand the combined role of a rounded edge and tapered tail, and finally a NACA0012 airfoil transformed into a thin plate investigates the role of thickness on the propulsive performance.

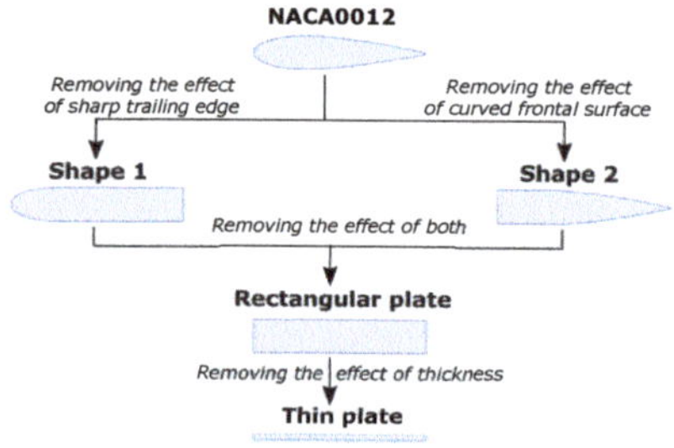

Figure 1. The different body shapes considered to understand the effects on propulsion and efficiency of curvature and thickness of hydrofoil-like geometries.

2.2. Oscillation Kinematics and Other Parameters of Interest

The oscillatory motion of the caudal (i.e., rear) fin of a fish can be modeled as a coupled pitching and heaving symmetric hydrofoil with a high aspect ratio [11,27]. This high aspect ratio allows the derivation of the three-dimensional flow characteristics around the foil from the two-dimensional flow considerations [12]. Figure 2 shows the schematic figure of a carangiform/thunniform fish about the medio-frontal plane. The hatched region shows the modeled caudal fin performing coupled pitching-and-heaving motions about the pivot point P in a freestream flow of velocity u_∞. The corresponding motion can be replicated by sinusoidal motion for both the heave and pitch [13], given as

$$\begin{aligned} h(t) &= h_0 \sin(2\pi f t), \\ \theta(t) &= \theta_0 \sin(2\pi f t + \phi), \end{aligned} \tag{1}$$

where h_0 is the maximum heave amplitude, f is the frequency, θ_0 is the maximum pitching amplitude, and ϕ is the phase-angle difference between heave and pitch. Notice from the figure that due to heaving motions, the hydrofoil moves at an additional velocity, the heave velocity $u_{heave} = dh/dt$, which results in a change in the resultant velocity (u_{res}) of the foil. Thus, the instantaneous angle of attack $\alpha(t)$ is no longer equal to pitching angle $\theta(t)$ [12] and can be given as:

$$\begin{aligned} \tan[\alpha(t) + \theta(t)] &= \frac{1}{u_\infty}\frac{dh(t)}{dt}, \\ \alpha_{\max} &= \max\left[\tan^{-1}\left(\frac{1}{u_\infty}\frac{dh(t)}{dt}\right) - \theta(t)\right]. \end{aligned} \tag{2}$$

For analyzing the output performance, the non-dimensional parameters considered in the study are the thrust coefficient C_T, lateral force coefficient C_L, input power coefficient C_P, and propulsive efficiency η_P, defined as, respectively,

$$C_T = \frac{F_T}{1/2\rho u_\infty^2 C}, \quad C_L = \frac{F_L}{1/2\rho u_\infty^2 C}, \quad C_P = \frac{\int c_L V_{body} dS}{1/2\rho u_\infty^3 C}, \tag{3}$$

$$\eta_p = \frac{P_{out}}{P_{in}} = \frac{\overline{F}_T\, u_\infty}{\int c_L V_{body} dS} = \frac{\overline{C}_T}{\overline{C}_P}. \tag{4}$$

Here, F_T and F_L are the net forces acting in the streamwise and lateral direction, respectively, and are calculated from the integration of pressure and viscous forces over the hydrofoil. Further, ρ is the density of the surrounding fluid, c_L is the local lateral force coefficient per unit surface area of the hydrofoil, and V_{body} is the lateral velocity of an oscillating hydrofoil. The propulsive efficiency η_P is defined as the ratio of the net output power to the input power [28,29], where an overbar represents time-averaged values obtained by using at least twenty oscillation cycles following statistical convergence of the simulations.

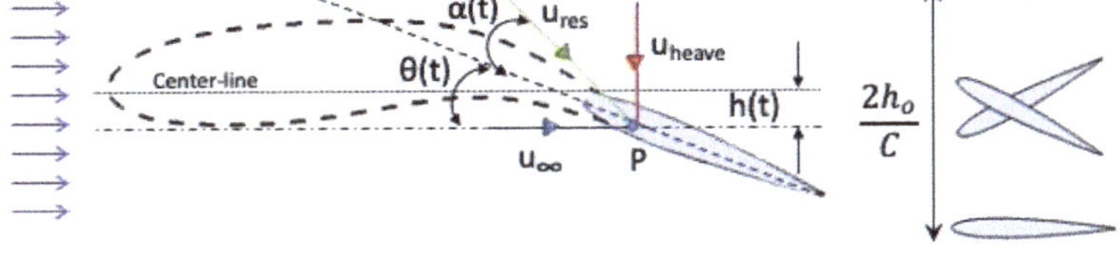

Figure 2. Schematics of a caudal-fin-like oscillating NACA0012 hydrofoil showing various governing parameters.

2.3. Governing Equations and Numerical Details

For an oscillating hydrofoil, Zurman et al. [30] found that the flow and forces are effectively two-dimensional for a NACA0016 hydrofoil at $St = 0.15$–0.45 and $Re = 5300$. Thus, for the present cases at $St = 0.25$ and $Re = 5000$, 2D numerical simulations are performed using an in-house code based on a level-set function-based immersed interface method (LS-IIM). The LS-IIM-based code was developed in the CFD lab at IIT Bombay by Thekkethil [31], in which the temporal variation in the solid–fluid interface is tracked by using a normal signed distance-based level set function, Φ. At each time step, the value of Φ is calculated using the minimum distance and winding algorithms, governed by the predefined kinematics of oscillation (discussed in Section 2.2). The positive Φ values represent fluid cells, whereas negative values represent solid cells. The fluid dynamics of the present problem are governed by the incompressible Navier–Stokes equation. In non-dimensional form, these are given by

$$\text{Continuity}: \nabla \cdot \vec{U} = 0, \tag{5}$$

$$\text{Momentum}: \frac{\partial \vec{U}}{\partial \tau} + \nabla \cdot (\vec{U}\vec{U}) = -\nabla P + \frac{1}{Re}\nabla^2 \vec{U}, \tag{6}$$

where $\vec{U}\,(\equiv \vec{u}/u_\infty, u_\infty$ is the freestream velocity) is the non-dimensional velocity vector and $P\,(\equiv p/(\rho u_\infty^2), \rho$ is the density of the surrounding fluid) is the non-dimensional pressure.

The LS-IIM uses the finite-volume method to discretize the above equations in space. Further, it is based on a fully implicit pressure projection method on a collocated grid. The non-linear advection term of the Navier–Stokes equation is integrated using the QUICK (Quadratic Upstream Interpolation for Convective Kinematics) scheme, while the CDS (Central Difference Scheme) is used for the diffusion term. More details of LS-IIM can be found in Thekkethil and Sharma [1].

2.4. Computational Details

Figure 3 shows the non-dimensionalized computational domain and the boundary conditions for the present problem on flow past an oscillating hydrofoil. The chord length C and the freestream velocity u_∞ are taken as the length and the velocity scale for the problem. The x-direction is aligned with the streamwise direction, while the y-direction is aligned with the lateral direction of the flow. The domain measures 16C in the streamwise direction, with the leading edge of the hydrofoil placed 5C from the inlet. The lateral dimension of the domain is 10C, with the hydrofoil located at the center. This provides a maximum lateral blockage ($0.12C/10C$) of 1.2%, ensuring a minimal effect of the lateral boundaries on the hydrodynamics of oscillation.

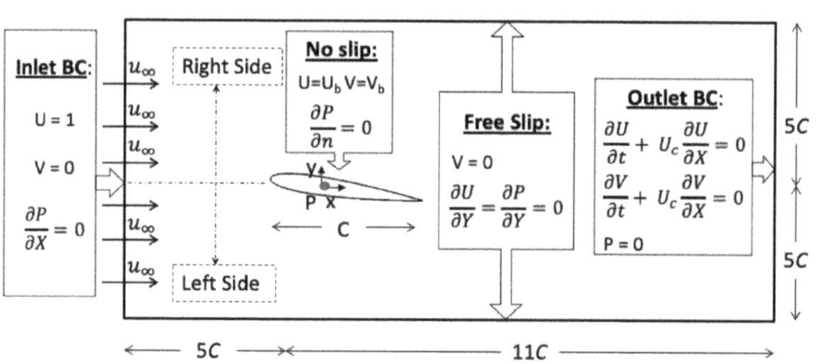

Figure 3. Non-dimensional computational setup along with boundary conditions for a coupled pitching-and-heaving NACA0012 hydrofoil in freestream flow. Here, U_b and V_b are the local velocity components of the hydrofoil-shaped body.

In Figure 3, U and V correspond to the non-dimensional velocities in the streamwise and lateral direction, respectively. The boundary condition at the inlet corresponds to a constant velocity (equal to the freestream velocity, u_∞). A convective boundary condition, with a convective velocity U_c equal to one, has been used at the outlet. A free-slip or symmetry boundary condition is used for the lateral boundaries. The hydrofoil is assumed to be non-deforming, and a no-slip (for the velocity) and a Neumann boundary condition (for pressure) have been employed at its surface.

For the grid independence study, we considered three non-homogeneous structured grids of sizes: 382×288, 768×552, and 1192×954. For each grid, a uniform size fine cell of size $\Delta x = \Delta y = 0.01, 0.05$, and 0.0025 is used in the region closer to the hydrofoil, and a uniform coarse cell size of $\Delta x = \Delta y = 0.5, 0.25$ and 0.125 is used in the far field. The finest and coarsest grid zones are connected using a hyperbolic stretching function, with a maximum expansion ratio of less than 1.01. For $\alpha_{\max} = 30°$, $h_\circ/C = 0.75$, $St = 0.25$, and $Re = 5000$, Figure 4a shows no significant visual difference in the temporal variation in the lateral force coefficient C_L between the two finest grids. Therefore, all further simulations for the present problem are carried out using the intermediate 768×552 grid size, noting that it has 200 points across the hydrofoil chord. The simulations are carried out in series on the local system with 32 GB RAM for a total of 48 simulation hours per case.

The numerical method used for the present study has already been applied successfully to various similar fluid–structure interaction (FSI) problems [1,32–37]. However, a further validation study has been carried out by comparing our results with that documented by Boiron et al. [38] and is presented in Figure 4b. The figure shows an excellent agreement between the present and published results for the temporal variation in the thrust and lift coefficients at $\alpha_{\max} = 20°$, $h_\circ/C = 0.75$, $St = 0.3$ and $Re = 4000$. The percentage of difference is within 2% and 5% for the lift and thrust coefficients, respectively.

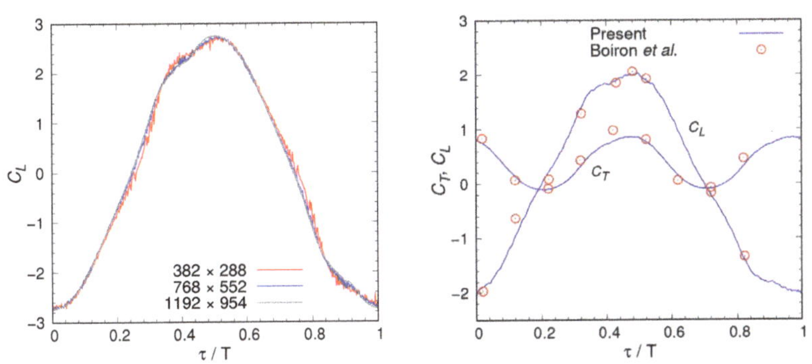

(a) (b)

Figure 4. Validation of the computer code for an oscillating NACA0012 hydrofoil. (**a**) Grid independence study. Temporal variation of the lift coefficient C_L at $\alpha_{\max} = 30°, h_\circ/C = 0.75, St = 0.25$ and $Re = 5000$. (**b**) Code verification study comparing predictions against those of Boiron et al. [38]. Temporal variation in thrust coefficient C_T and lift coefficient C_L at $\alpha_{\max} = 20°, h_\circ/C = 0.75, St = 0.3$ and $Re = 4000$. The intermediate size grid, 768×552, is used.

3. Results

The results section is divided into three subsections. Section 3.1 discusses the effect of varying the maximum angle of attack, $\alpha_{\max}$, on the propulsive performance of different shaped hydrofoils. Section 3.2 discusses the correlation between the time-averaged engineering parameters and instantaneous flow parameters using a cause-and-effect-based study and instantaneous vorticity in the vicinity of the hydrofoils at the extreme rightward lateral position. Section 3.3 examines hydrofoil cases at maximum efficiency to study the role of the phase difference ϕ between heave and pitch. In both Sections 3.1 and 3.3, the time-averaged performance parameters are calculated over twenty oscillation cycles following the statistical convergence of the simulations.

3.1. Effect of Curvature Variations on Hydrodynamic Performance

Considering the five different shapes of the hydrofoil (Figure 1), Figure 5 presents the effect of curvature and thickness on the time-averaged propulsive performance of a hydrofoil for various angles of attack, $\alpha_{\max} = 5°–30°$, at $St = 0.25$, $h_0/C = 0.75$, $\phi = 90°$ and $Re = 5000$. This leads to the pitching amplitude $\theta_0 \sim 34°–8°$ (Equation (1)).

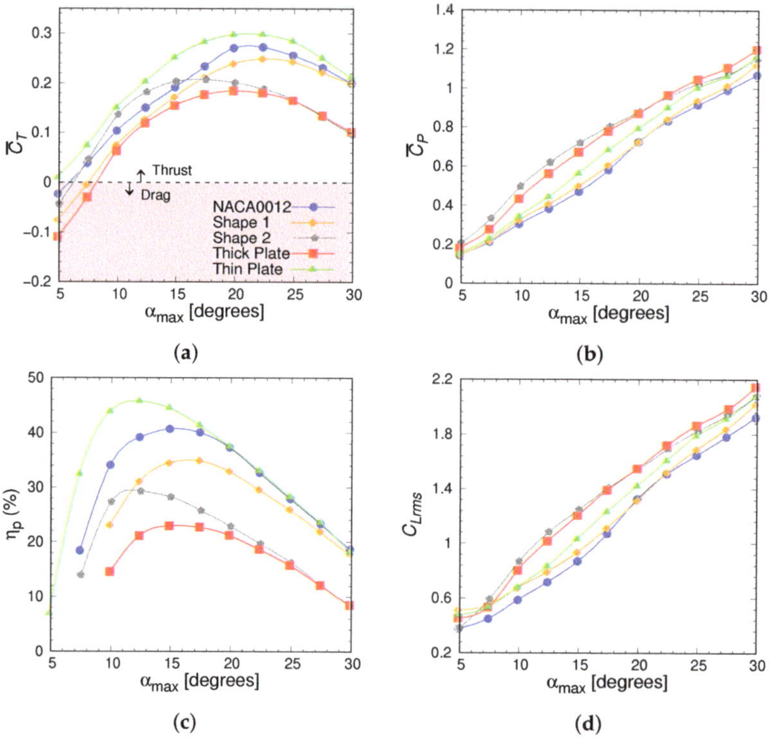

Figure 5. Variation of the time-averaged (**a**) thrust coefficient $\overline{C}_T$, (**b**) power coefficient $\overline{C}_P$, (**c**) propulsive efficiency η_P, and (**d**) rms of lift coefficient $C_{L,\text{rms}}$, as a function of maximum angle of attack $\alpha_{\max}$ for the five different shapes considered.

For the mean thrust coefficient $\overline{C}_T$, Figure 5a shows that the magnitudes of $\overline{C}_T$ for the NACA0012 and those with a bluff leading/trailing edge (shape 1 and shape 2) are bounded *within* the magnitude of $\overline{C}_T$ for the thin plate and thick plate. With an increase in $\alpha_{\max}$, the figure shows an increasing–decreasing trend for $\overline{C}_T$ with its maximum close to $\alpha_{\max} = 22.5°$ for the different shapes; the exception is shape 2, which achieves its maximum $\overline{C}_T$ at $\alpha_{\max} = 17.5°$. Further, at lower values of $\alpha_{\max}$, $\overline{C}_T$ for bluff frontal edge-based shape 2 approaches that of a thin plate, while $\overline{C}_T$ for bluff trailing edge-based shape 1 approaches that of a thick plate. However, at higher values of $\alpha_{\max}$, the reverse applies; that is, $\overline{C}_T$ for shape 1 approaches $\overline{C}_T$ of a thin plate, while $\overline{C}_T$ for shape 2 approaches that of a thick plate. The figure also shows the drag-to-thrust transition for different shapes with increasing $\alpha_{\max}$, except for the thin plate with a net thrust even at a very low $\alpha_{\max} = 5°$. For other hydrofoils, the figure shows that the transition occurs first for hydrofoils with a tapered trailing edge (NACA0012 and shape 2), compared with hydrofoils with a bluff trailing edge (shape 1 and the thick plate).

For the mean input power coefficient $\overline{C}_P$, Figure 5b shows an apparent increasing $\overline{C}_P$ with increasing α_{max}, with $\overline{C}_P$ for the different shapes bounded within the maximum $\overline{C}_P$ for shape 2 and minimum $\overline{C}_P$ for the NACA0012 airfoil. Further, the figure shows that $\overline{C}_P$ is smaller for shapes with a rounded frontal edge (NACA0012 airfoil, shape 1) compared with bluff frontal edges (the flat plate, shape 2 and the thick plate), and the trend is independent of α_{max} for the hydrofoil. Moreover, for the bluff frontal-edge hydrofoils, $\overline{C}_P$ increases with frontal surface area, i.e., $\overline{C}_P$ is larger for shape 2 and the thick plate compared to that for the thin plate at any α_{max}.

Similar to the variation with $\overline{C}_T$, Figure 5c shows that η_P for the different shapes is again bounded between the values for the thin and thick plate for any α_{max}. Note that the efficiency curve contains only those cases that resulted in a net mean thrust. Furthermore, the figure shows that the nature of the η_P curve is also of increasing–decreasing type. However, for three of the foil shapes, the maximum η_P occurs at $\alpha_{max} = 15°–17.5°$; this is not the case for the thin plate and shape 2, which have maximum η_P at $\alpha_{max} = 12.5°$. At larger angles of attack, $\alpha_{max} \geq 20.0°$, the figure shows that η_P for the NACA0012 matches that of a thin plate, whereas the η_P for shape 2 matches that of a thick plate. Further, by comparing shape 1 with the NACA0012 airfoil and shape 2 with the thick plate, it can be seen that η_P is greater for shapes that possess a tapered end. However, the benefits of possessing a tapered end decrease with increasing α_{max}.

For the rms variation in the lift coefficient, $C_{L,rms}$, Figure 5d shows that the variation is similar to the variation in $\overline{C}_P$ with α_{max}, as discussed above. For instance, the magnitude of $C_{L,rms}$ increases with α_{max}; for any α_{max}, the magnitude of $C_{L,rms}$ for different shapes is bounded within the maximum $C_{L,rms}$ for shape 2 and the minimum $C_{L,rms}$ for the NACA0012 airfoil, showing that shape 2 is the least stable and the NACA0012 airfoil is the most stable hydrofoil for propulsion. Further, it is important to notice that the $C_{L,rms}$ of shapes with a rounded frontal shape is smaller compared to shapes with a bluff frontal surface, which leads to a more stable propulsion for shapes with a rounded frontal area.

The above-discussed increasing–decreasing trend for $\overline{C}_T$ and the increasing trends for $\overline{C}_P$ and $C_{L,rms}$ with increasing α_{max} are at least partially understandable as these parameters are directly proportional to the surrounding fluid displaced by the hydrofoil. Thus, these parameters are directly proportional to the projected lateral surface area ($C\sin(\alpha)$) of the hydrofoil. To discuss this, Figure 6 shows instantaneous schematics of a hydrofoil at three values of α_{max} ($\alpha_1 < \alpha_2 < \alpha_3$). For simplicity, we have assumed that the hydrofoil is oscillating in a still fluid, resulting in the vertical relative velocity u_{res}. With increasing α_{max}, the figure shows that the projected lateral surface area of the hydrofoil increases, resulting in an increased volume of displaced fluid. Thus, a larger $\overline{C}_P$ is required for an oscillating hydrofoil to perform its movement at a larger α_{max}.

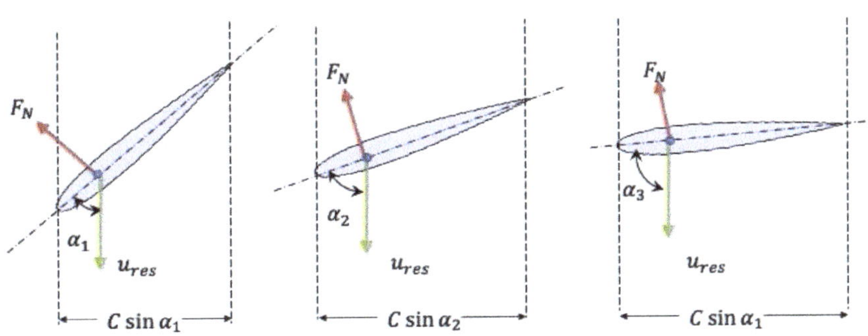

Figure 6. Schematics representation of NACA0012 hydrofoil oscillating in a still fluid at different angles of attack ($\alpha_1 < \alpha_2 < \alpha_3$) and time instant t.

Figure 6 also shows the normal direction in which the hydrodynamic force acts from the fluid to the hydrofoil. The axial component of this force provides the instantaneous thrust force, while the lateral component provides the instantaneous lift force. It can be seen from the figure that with increasing $\alpha_{\max}$, both the projected lateral surface area and lift component increase, resulting in an almost linear enhancement of $C_{L,\text{rms}}$ with $\alpha_{\max}$. In contrast, for C_T, there is a trade-off between increasing surface area and decreasing axial component, resulting in an increasing–decreasing variation. However, it is important to notice from Figure 5 that the magnitude of engineering parameters is different for different shapes. Further, the $\alpha_{\max}$ at which maximum $\overline{C}_T$ and η_p occur is different for different shapes. Thus, a more detailed study is needed; this is attempted in the next section (Section 3.2) by correlating the engineering parameters with the flow structures near the vicinity of the hydrofoil.

3.2. Flow Characteristics

The correlation between the above-discussed time-averaged engineering parameters with the instantaneous flow parameters is accomplished in two steps. In the first step, the instantaneous pressure contours in the vicinity of the hydrofoil are correlated with the corresponding instantaneous vortex structure and engineering parameters. For ease of understanding, the study is conducted only for an oscillating NACA0012 hydrofoil for a range of $\alpha_{\max}$. The corresponding study is known as a unified cause-and-effect-based CFD analysis, and has been proposed by Sharma [39]. The study helps to correlate the time-averaged engineering parameters with the instantaneous vortex structure when the NACA0012 hydrofoil is at the rightmost lateral position. In the second step, the correlation developed in the first step is applied to differently shaped hydrofoils.

Over one oscillation period, Figures 7 and 8 show a unified cause-and-effect-based analysis by plotting the time variation in 2D plots for flow properties and 1D line-plots for propulsive performance parameters for $\alpha_{\max}$ in the range $5° \leq \alpha_{\max} \leq 30°$. The 2D contour plots Figures 7a–i and 8a–i are for the vortex structure and pressure contours in the vicinity of the hydrofoil, respectively, while the 1D plots, Figure 7j,k, are for the thrust C_T and power C_P coefficients.

At $\tau/T = 0$, Figure 7 shows the onset of the leftward movement of the hydrofoil from its extreme right lateral position. At this instant of time, a net drag acts on all the hydrofoils independent of $\alpha_{\max}$; however, the magnitude of drag is larger for hydrofoils with a higher $\alpha_{\max}$, as shown in Figure 7j. The reason behind this increase in drag with $\alpha_{\max}$ can be seen in the instantaneous vortex contours of Figure 7(a1–a6). For $\alpha_{\max} = 5°$, the figure shows a fully attached flow across the hydrofoil, while the flow separates for hydrofoils with $\alpha_{\max} > 5°$—in this case, the flow starts to separate from the posterior region of the foil, and the point of separation moves towards the leading edge and becomes stronger with increasing $\alpha_{\max}$. The stronger flow separation causes a stronger pressure gradient along the chord (Figure 8(a1–a6)), resulting in higher drag.

Figure 7(b1–b6) show that the hydrofoil rotates in the counter-clockwise (CCW) direction with further leftward movement. With this movement to $\tau/T = 0.1$, Figure 7j shows the crossover from instantaneous drag to thrust for all $\alpha_{\max}$, except for $\alpha_{\max} = 5°$ for which crossover occurs later at $\tau/T = 0.25$. Further, the figure shows that the magnitude of instantaneous C_T at $\tau/T = 0.1$ increases with increasing $\alpha_{\max}$, reaches its maximum at $\alpha_{\max} = 20°$, and then decreases. In contrast, Figure 7k shows that the instantaneous power coefficient C_P increases with $\alpha_{\max}$. The corresponding vortex structure in Figure 7(b1–b6) shows that the increase in instantaneous C_T and C_P with $\alpha_{\max}$ is due to an increase in the strength of flow separation. However, above $\alpha_{\max} = 20°$, the strength of the flow separation is too large such that it draws the flow around the trailing edge of the hydrofoil, resulting in a reduction in net instantaneous thrust. With the further leftward movement to $\tau/T = 0.25$, Figure 7j,k show that the hydrofoils reach their maximum instantaneous C_T and C_P, except for $\alpha_{\max} = 5°$. Further, the figure shows that the largest instantaneous C_T occurs for $\alpha_{\max} = 25°$, in contrast to $\tau/T = 0.1$. The change in trend is due to the flow remaining attached up to $\alpha_{\max} = 20°$ (Figure 7(c1–c4)), a small separated flow at $\alpha_{\max} = 25°$ (Figure 7(c5)), and a larger separation for $\alpha_{\max} = 30°$ (Figure 7(c6)).

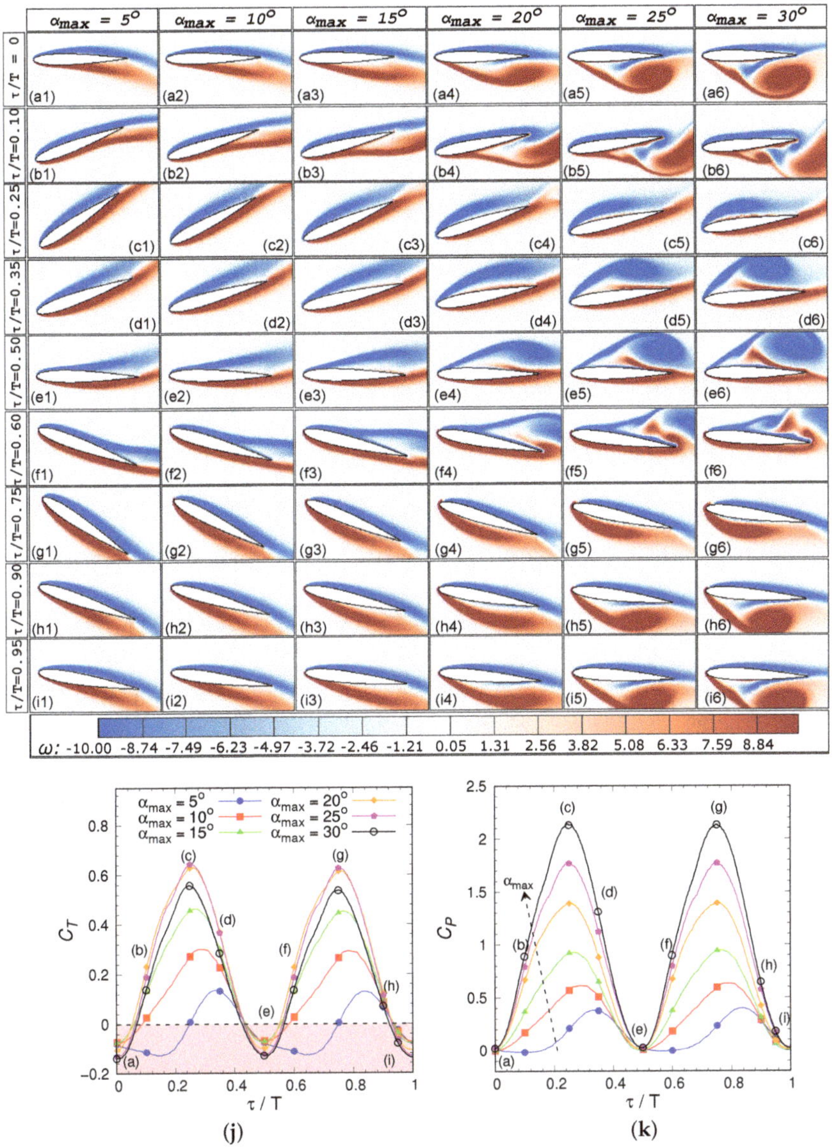

Figure 7. (**a–i**) Temporal variation in instantaneous vorticity contours, (**j**) thrust coefficient C_T, and (**k**) power coefficient C_P over one time period of an oscillating NACA0012 hydrofoil at various $\alpha_{\max} = 5°$–$30°$ (marked), for $St = 0.25$, $h_o/C = 0.75$, $\phi = 90°$ and $Re = 5000$. In the figure, column (**a1–i1**) represents variation in instantaneous vorticity contours for $\alpha_{\max} = 5°$, similarly column (**a2–i2**) is for $\alpha_{\max} = 10°$, column (**a3–i3**) is for $\alpha_{\max} = 15°$, column (**a4–i4**) is for $\alpha_{\max} = 20°$, column (**a5–i5**) is for $\alpha_{\max} = 25°$, and column (**a6–i6**) are for $\alpha_{\max} = 30°$. The multimedia movie in Supplementary Materials Video S1.

With further leftward movement, Figure 7(d1–d5) show that the CCW rotation of the hydrofoil switches to clockwise (CW) rotation, and the hydrofoils decelerate, reaching the extreme leftmost position. This deceleration of the hydrofoil reduces the instantaneous positive leftward pressure for all the cases, except for $\alpha_{\max} = 5°$—compare Figure 8(c1–c5) with Figure 8(d1–d5). The reduction in pressure reduces the pressure gradient, resulting in a reduction in the instantaneous thrust and power for all cases, except for $\alpha_{max} = 5°$, as shown in Figure 7j,k, respectively. Further, the switch in rotation from CCW to CW enhances the flow separation from the hydrofoil's rightward side, resulting in the continuation of the trend of increasing instantaneous C_T with α_{max} up to $\alpha_{max} = 25°$ and then decreasing, as observed for $\tau/T = 0.25$. The retardation continues as the hydrofoil approaches the extreme leftmost position (completing half a cycle), resulting in a further decrease in instantaneous thrust and power, as shown in Figure 7j,k.

For the subsequent half cycle, where the hydrofoil moves from the extreme leftmost to the rightmost position, Figures 7 and 8 show an opposite nature of periodic variation in vortex structures and pressure on the two sides of the foil, resulting in similar instantaneous thrust and power curve.

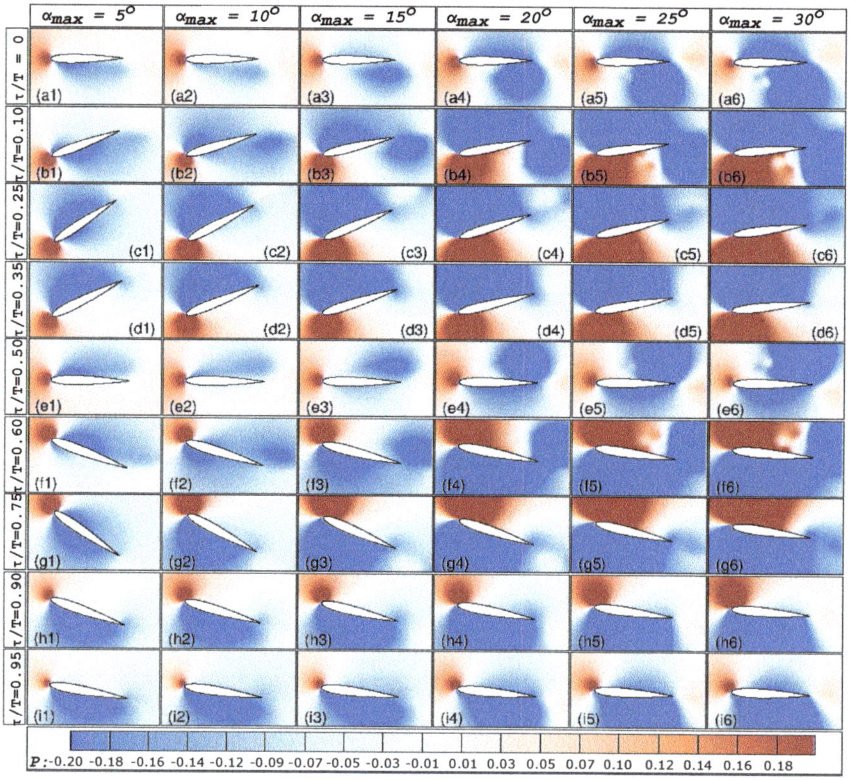

Figure 8. **(a–i)** Temporal variation in the instantaneous pressure contours over one time period of an oscillating NACA0012 hydrofoil at various $\alpha_{max} = 5°$–$30°$ (marked), for $St = 0.25$, $h_o/C = 0.75$, $\phi = 90°$ and $Re = 5000$. In the figure, column **(a1–i1)** represents variation in instantaneous pressure contours for $\alpha_{max} = 5°$, similarly column **(a2–i2)** is for $\alpha_{max} = 10°$, column **(a3–i3)** is for $\alpha_{max} = 15°$, column **(a4–i4)** is for $\alpha_{max} = 20°$, column **(a5–i5)** is for $\alpha_{max} = 25°$, and column **(a6–i6)** is for $\alpha_{max} = 30°$.

Therefore, by correlating the instantaneous vortex structure with the thrust and power coefficients for a cycle of oscillation of the NACA0012 hydrofoil at different α_{max}, it can be concluded that any amount of flow separation enhances the power consumed by the hydrofoil for performing its oscillation motion. In contrast, a moderately strong degree of flow separation (until it becomes so strong that it starts pulling a significant amount of fluid around the trailing edge) enhances the thrust. For further investigation, Figure 9 compares the instantaneous vortex structure in the vicinity of different shaped hydrofoils at their extreme rightward lateral position for various $\alpha_{\max}$. It can be seen from the highlighted rectangular regions in the figure that the $\eta_{\max}$ for any shape occurs at that $\alpha_{\max}$, for which there is a small degree of flow separation with almost no roll-up into a separating vortex. On the other hand, the highlighted cases for $\overline{C}_{T,\max}$ are those at which moderately strong flow separation occurs with no induction of a significant amount of fluid transferring around the trailing edge.

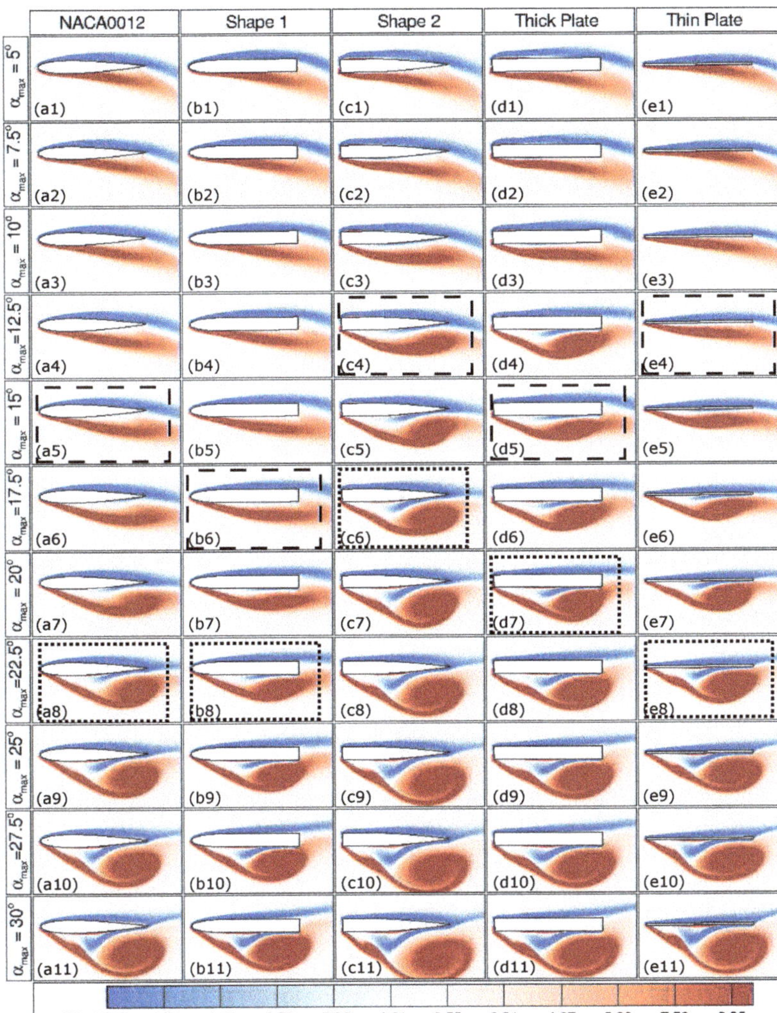

Figure 9. Time-instantaneous vortex structures (at maximum heave position) in the vicinity of the hydrofoils at various angles of attack $\alpha_{\max} = 5°$–$30°$, for $St = 0.25$, $h_o/C = 0.75$, $\phi = 90°$, and $Re = 5000$. Here, dashed and dotted rectangular regions correspond to cases for maximum η_p and $\overline{C}_T$, respectively. In the figure, column (**a1**–**a11**) represents instantaneous vorticity contours for NACA0012 hydrofoil, similarly column (**b1**–**b11**) is for Shape 1, column (**c1**–**c11**) is for Shape 2, column (**d1**–**d11**) is for thick plate, and column (**e1**–**e11**) is for thin plate.

Figure 9 also shows the effect of $\alpha_{\max}$ and geometrical parameters on the flow separation. For any shape, an increasing $\alpha_{\max}$ increases the magnitude of flow separation and brings the point of separation close to the leading edge. Geometrical features such as a bluff leading edge, tapered tail, and high thickness enhance the flow separation, while those such as a rounded leading edge and a flattened tail reduce the magnitude of flow separation.

3.3. Effect of Phase Difference between Heave and Pitch ϕ on Hydrodynamic Performance

In this section, the effect of the phase difference between heave and pitch, ϕ, on the propulsive performance of the hydrofoils is presented for cases at $\eta_{p,\max}$, as observed in Section 3.1. This corresponds to $\alpha_{\max} = 15°$ for the NACA0012 airfoil and thick plate, $\alpha_{\max} = 17.5°$ for shape 1, and $\alpha_{\max} = 12.5°$ for shape 2 and the thin plate for phase differences $\phi = 75°$–$105°$, all at constant $St = 0.25$, $h_o/C = 0.75$ and $Re = 5000$.

Figure 10 shows the variation in the normalized $\overline{C}_T$ and η_p with phase difference ϕ for different shaped hydrofoils at $\eta_{p,\max}$, i.e., $\alpha_{\max} = 15°$ for the NACA0012 airfoil and thick plate, $\alpha_{\max} = 17.5°$ for shape 1, and $\alpha_{\max} = 12.5°$ for shape 2 and the thin plate. Here, the normalized values are obtained by normalizing with corresponding values at $\phi = 90°$. For all cases, the figure shows that both the maximum thrust and efficiency occur in a narrow range of phase difference $90° \leq \phi \leq 100°$; this indicates that the optimal phase difference for propulsion is independent of the shape and shows the hydrodynamic (and evolutionary) reason behind its adoption by different body–caudal fin (BCF) fish in nature.

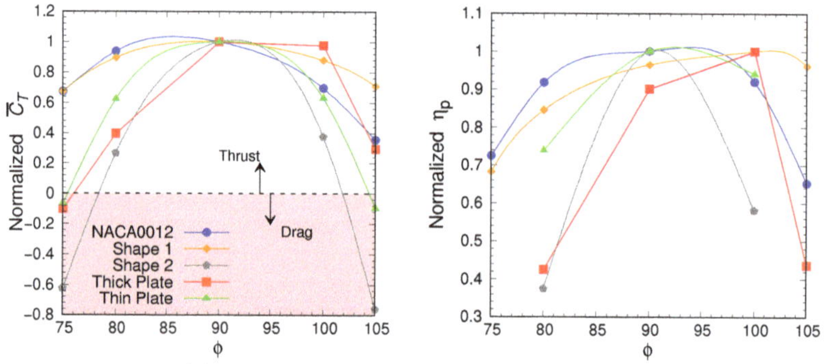

(a) (b)

Figure 10. Variation of normalized (**a**) thrust coefficient $\overline{C}_T$, and (**b**) efficiency η_p, with the phase difference between heave and pitch ϕ for cases at $\eta_{p,\max}$.

The occurrence of maximum efficiency and thrust coefficient at phase angles $90° \leq \phi \leq 100°$ is explained here using the analogy developed in Section 3.1 between the instantaneous wake structure in the vicinity of hydrofoil (at maximum heave position) and the engineering parameters. Similar to Figure 9, Figure 11 also shows that the maximum efficiency occurs at those ϕ for which there is a small degree of flow separation with no roll-up of a separating vortex. Similarly, the maximum $\overline{C}_{T,\max}$ occurs at those ϕ for which moderately strong flow separation occurs with no significant induction of fluid around the trailing edge.

Figure 11 also shows the effect of the phase difference ϕ and geometrical parameters on the instantaneous vorticity in the vicinity of the hydrofoils. It is seen that the flow separation for all the shapes reduces by decreasing the phase difference below $\phi = 90°$. However, the flow separation enhances, and the point of separation travels posterior (towards the tail) by increasing the phase difference above $\phi = 90°$. Further, by increasing the phase difference above $\phi = 90°$, secondary flow separation occurs from the rightward leading edge for shapes with a bluff leading edge (shape 2, the thick plate and the thin plate).

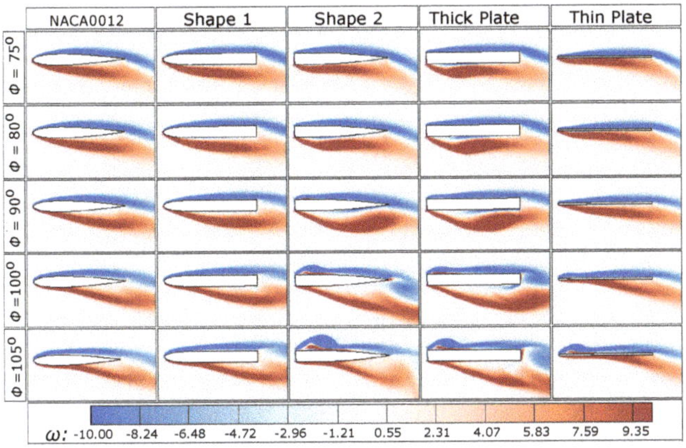

Figure 11. Time-instantaneous vortex structure (at maximum heave position) in the vicinity of the different hydrofoils at various ϕ for cases at $\eta_{p,\max}$.

4. Conclusions

The present study documents and contributes towards understanding the role of an oscillating hydrofoil's curvature/shape at the leading and trailing edges, along with its thickness, in the propulsive performance at various maximum angles of attack in the range $\alpha_{\max} = 5°–30°$, and for phase differences between heave and pitch of $\phi = 75°–105°$, at constant $St = 0.25$, $h_o/C = 0.75$, and $Re = 5000$. The conclusions drawn are as follows.

In terms of the role of leading- and trailing-edge curvatures, a tapered tail provides (compare NACA0012 with shape 1) a more significant thrust, consumes slightly lower power, and generates a lower $C_{L,\text{rms}}$. On the other hand, a rounded leading edge (compare the NACA0012 airfoil with shape 2) provides a smaller and larger thrust for $\alpha_{\max} < 15°$ and $\alpha_{\max} > 15°$, respectively. However, the power consumption and $C_{L,\text{rms}}$ reduce significantly. Thus, these leading and trailing curvatures enhance the oscillating hydrofoil's overall propulsive performance and stability. Therefore, it is evident that the combination of the rounded leading edge and tapered tail (compare the NACA0012 airfoil and thick plate) provides better thrust and lower power consumption and $C_{L,\text{rms}}$, providing overall enhanced propulsion. For the role of hydrofoil thickness (compare the thick and thin plates), a greater thickness reduces the net thrust and enhances the power consumption and $C_{L,\text{rms}}$, resulting in degradation of the overall propulsive performance and stability of an oscillating hydrofoil. Finally, for the role of the phase difference between heave and pitch, it was found that both the maximum thrust and efficiency occur over a narrow range $90° \leq \phi \leq 100°$ independent of the hydrofoil geometry.

The hydrodynamic reason behind the above variation is explained by correlating both the time-varying and time-instantaneous vortex structures in the vicinity of the hydrofoil with the corresponding engineering parameters. It is found that any amount of flow separation enhances the power consumed by the hydrofoil for performing its oscillation motion. In contrast, a moderate flow separation enhances the thrust generated. For all the different-shaped hydrofoils examined, the maximum efficiency occurs at those $\alpha_{\max}$ at which there is a small degree of flow separation with no roll-up of separating vortex. In comparison, maximum thrust generation occurs when there is a moderately strong flow separation with no induction of a significant quantity of fluid around the trailing edge. For any $\alpha_{\max}$, flow separation can be controlled by the geometrical parameters of the hydrofoil. A rounded leading edge and flattened tail reduce the magnitude of flow separation, while a bluff leading edge, tapered tail, and increased thickness enhance it.

We anticipate that by contributing to an understanding of the correlation between different hydrofoil geometrical parameters with engineering parameters and flow separation, our results can aid in designing and developing more efficient modern autonomous underwater and micro air vehicles. Our results also have implications for understanding fish swimming. A fish possesses a symmetrical caudal fin with a somewhat rounded leading edge and tapered tail, and maintains a phase difference of 90° while swimming. The results from the present study show that their NACA-like caudal fin generates high thrust, as high as that for a very thin plate, with minimum power requirement, resulting in higher propulsive efficiency. Moreover, their propulsion is very stable. A greater thickness without losing propulsive performance can give them a caudal fin with more structural strength, which is essential for directing water around it. The present study also shows that a phase difference of 90° for any shape provides high thrust with high efficiency and stability of oscillation. Thus, an oscillating caudal fin of NACA-like shape with a 90° phase difference between heave and pitch provides fish with an ideal combination for achieving high thrust, efficiency, and stability for propulsion.

In the current preliminary investigation, we have compared the hydrodynamics of hydrofoils with smooth, rounded leading edges and tapered tails to those with sharp leading and trailing edges. In the future, we plan to conduct a more comprehensive comparison using a non-dimensional geometric parameter that can assess the roundness and tapering of any hydrofoil shape to develop a flow law for optimization of thrust and propulsive efficiency.

Supplementary Materials: The following are available online at https://www.mdpi.com/article/10.3390/jmse11101923/s1, Video S1: Temporal variation in instantaneous vorticity contours, thrust coefficient C_T, and power coefficient C_P over one time period of an oscillating NACA0012 hydrofoil at various $\alpha_{\max} = 5°–30°$ (marked), for $St = 0.25$, $h_o/C = 0.75$, $\phi = 90°$ and $Re = 5000$.

Author Contributions: Conceptualization, S.G., A.S., A.A., M.C.T. and K.H.; methodology, S.G., M.C.T. and K.H.; software, A.S.; validation, S.G. and A.S.; formal analysis, S.G.; investigation, S.G.; resources, K.H. and M.C.T.; data curation, S.G.; writing—original draft preparation, S.G.; writing—review and editing, A.S., A.A., M.C.T. and K.H.; supervision, A.S., A.A., M.C.T. and K.H.; project administration, A.S., M.C.T. and K.H.; funding acquisition, K.H. and M.C.T. All authors have read and agreed to the published version of the manuscript.

Funding: This work was funded by the Australian Government through the Australian Research Council's Discovery Projects funding scheme (projects DP190103388 and DP210100990).

Data Availability Statement: The data that support the findings of this study are available from the corresponding author upon reasonable request.

Conflicts of Interest: The authors declare no conflict of interest.

Nomenclature

Symbols	**Definitions**
C	Chord length
C_L	Lateral force coefficient
c_L	Local lateral force coefficient per unit surface area of the hydrofoil
$C_{L,\mathrm{rms}}$	Rms variation in the lift/lateral coefficient
C_P	Input power coefficient
C_T	Thrust coefficient
f	Oscillation frequency
F_L	Net force acting in the lateral direction
F_N	Net force acting in the normal direction
F_T	Net force acting in the streamwise direction
h	Heave amplitude
h_0	Maximum heave amplitude
P	Non-dimensional pressure
Re	Reynolds number
St	Strouhal Number
T	Non-dimensional time to complete one cycle of periodic oscillations.
t	Dimensional time

U	Non-dimensional velocity in streamwise direction
$\vec{U}$	Non-dimensional velocity vector
U_b	Local velocity component of the hydrofoil-shaped body in streamwise direction
u_{heave}	Heave velocity
u_{res}	Vertical relative velocity
u_∞	Free stream velocity
V	Non-dimensional velocity in lateral direction
V_b	Local velocity component of the hydrofoil-shaped body in lateral direction
V_{body}	Lateral velocity
X	Non-dimensional streamwise distance
x	Streamwise direction
Y	Non-dimensional lateral distance
y	Lateral direction
α	Angle of attack
α_{max}	Maximum angle of attack
$\Delta x, y$	Grid cell size in x,y directions, respectively
η_P	Propulsive efficiency
θ	Pitching amplitude
θ_0	Maximum pitching amplitude
μ	Dynamic viscosity
ρ	Fluid density
τ	Non-dimensional time
Φ	Level set function
ϕ	Phase difference between heave and pitch
ω	Vorticity
Acronyms	
CCW	Counter-clockwise
CW	Clockwise
LEV	Leading edge vortex
LS-IIM	Level-set function-based immersed interface method

References

1. Thekkethil, N.; Sharma, A. Level set function–based immersed interface method and benchmark solutions for fluid flexible-structure interaction. *Int. J. Numer. Methods Fluids* **2019**, *91*, 134–157. [CrossRef]
2. Smits, A.J. Undulatory and oscillatory swimming. *J. Fluid Mech.* **2019**, *874*, 1. [CrossRef]
3. Zdunich, P.; Bilyk, D.; MacMaster, M.; Loewen, D.; DeLaurier, J.; Kornbluh, R.; Low, T.; Stanford, S.; Holeman, D. Development and testing of the mentor flapping-wing micro air vehicle. *J. Aircr.* **2007**, *44*, 1701–1711. [CrossRef]
4. Techet, A.H. Propulsive performance of biologically inspired flapping foils at high Reynolds numbers. *J. Exp. Biol.* **2008**, *211*, 274–279. [CrossRef] [PubMed]
5. Du, R.; Li, Z.; Youcef-Toumi, K.; y Alvarado, P.V. *Robot Fish: Bio-Inspired Fishlike Underwater Robots*; Springer: Berlin/Heidelberg, Germany, 2015.
6. Kinsey, T.; Dumas, G. Parametric study of an oscillating airfoil in a power-extraction regime. *AIAA J.* **2008**, *46*, 1318–1330. [CrossRef]
7. Kinsey, T.; Dumas, G.; Lalande, G.; Ruel, J.; Mehut, A.; Viarouge, P.; Lemay, J.; Jean, Y. Prototype testing of a hydrokinetic turbine based on oscillating hydrofoils. *Renew. Energy* **2011**, *36*, 1710–1718. [CrossRef]
8. Zhu, Q. Optimal frequency for flow energy harvesting of a flapping foil. *J. Fluid Mech.* **2011**, *675*, 495–517. [CrossRef]
9. Wang, Z.; Du, L.; Zhao, J.; Thompson, M.; Sun, X. Flow-induced vibrations of a pitching and plunging airfoil. *J. Fluid Mech.* **2020**, *885*, A36. [CrossRef]
10. Triantafyllou, M.; Triantafyllou, G.; Gopalkrishnan, R. Wake mechanics for thrust generation in oscillating foils. *Phys. Fluids A Fluid Dyn.* **1991**, *3*, 2835–2837. [CrossRef]
11. Triantafyllou, G.S.; Triantafyllou, M.; Grosenbaugh, M. Optimal thrust development in oscillating foils with application to fish propulsion. *J. Fluids Struct.* **1993**, *7*, 205–224. [CrossRef]
12. Anderson, J.; Streitlien, K.; Barrett, D.; Triantafyllou, M. Oscillating foils of high propulsive efficiency. *J. Fluid Mech.* **1998**, *360*, 41–72. [CrossRef]
13. Read, D.A.; Hover, F.; Triantafyllou, M. Forces on oscillating foils for propulsion and maneuvering. *J. Fluids Struct.* **2003**, *17*, 163–183. [CrossRef]
14. Hover, F.; Haugsdal, Ø.; Triantafyllou, M. Effect of angle of attack profiles in flapping foil propulsion. *J. Fluids Struct.* **2004**, *19*, 37–47. [CrossRef]

15. Van Buren, T.; Floryan, D.; Smits, A.J. Scaling and performance of simultaneously heaving and pitching foils. *AIAA J.* **2019**, *57*, 3666–3677. [CrossRef]
16. Isogai, K.; Shinmoto, Y.; Watanabe, Y. Effects of dynamic stall on propulsive efficiency and thrust of flapping airfoil. *AIAA J.* **1999**, *37*, 1145–1151. [CrossRef]
17. Ramamurti, R.; Sandberg, W. Simulation of flow about flapping airfoils using finite element incompressible flow solver. *AIAA J.* **2001**, *39*, 253–260. [CrossRef]
18. Soueid, H.; Guglielmini, L.; Airiau, C.; Bottaro, A. Optimization of the motion of a flapping airfoil using sensitivity functions. *Comput. Fluids* **2009**, *38*, 861–874. [CrossRef]
19. Tuncer, I.H.; Platzer, M.F. Computational study of flapping airfoil aerodynamics. *J. Aircr.* **2000**, *37*, 514–520. [CrossRef]
20. Tuncer, I.H.; Kaya, M. Optimization of flapping airfoils for maximum thrust and propulsive efficiency. *AIAA J.* **2005**, *43*, 2329–2336. [CrossRef]
21. Young, J.; Lai, J.C. Mechanisms influencing the efficiency of oscillating airfoil propulsion. *AIAA J.* **2007**, *45*, 1695–1702. [CrossRef]
22. Guglielmini, L.; Blondeaux, P. Propulsive efficiency of oscillating foils. *Eur. J. Mech. B/Fluids* **2004**, *23*, 255–278. [CrossRef]
23. Bose, C.; Sarkar, S. Investigating chaotic wake dynamics past a flapping airfoil and the role of vortex interactions behind the chaotic transition. *Phys. Fluids* **2018**, *30*, 047101. [CrossRef]
24. Zhang, X.; Ni, S.; Wang, S.; He, G. Effects of geometric shape on the hydrodynamics of a self-propelled flapping foil. *Phys. Fluids* **2009**, *21*, 103302. [CrossRef]
25. Ashraf, M.; Young, J.; Lai, J. Reynolds number, thickness and camber effects on flapping airfoil propulsion. *J. Fluids Struct.* **2011**, *27*, 145–160. [CrossRef]
26. Müller, U.K.; Stamhuis, E.J.; Videler, J.J. Riding the waves: The role of the body wave in undulatory fish swimming. *Integr. Comp. Biol.* **2002**, *42*, 981–987. [CrossRef]
27. Anderson, J.M. *Vorticity Control for Efficient Propulsion*; Technical Report; Massachusetts Institute of Technology: Cambridge, MA, USA, 1996.
28. Verma, S.; Hemmati, A. Evolution of wake structures behind oscillating hydrofoils with combined heaving and pitching motion. *J. Fluid Mech.* **2021**, *927*, A23. [CrossRef]
29. Zurman-Nasution, A.N.; Ganapathisubramani, B.; Weymouth, G.D. Fin sweep angle does not determine flapping propulsive performance. *J. R. Soc. Interface* **2021**, *18*, 20210174. [CrossRef] [PubMed]
30. Zurman-Nasution, A.; Ganapathisubramani, B.; Weymouth, G. Influence of three-dimensionality on propulsive flapping. *J. Fluid Mech.* **2020**, *886*, A25. [CrossRef]
31. Thekkethil, N. CFSD Development and Its Application for Hydrodynamic Analysis on Various Types of Fishes-Like Kinematics, Propulsion, and Flexibility of 2D and 3D NACA0012 Hydrofoils. Ph.D. Thesis, Indian Institute of Technology Bombay, Mumbai, India, 2019.
32. Thekkethil, N.; Sharma, A.; Agrawal, A. Unified hydrodynamics study for various types of fishes-like undulating rigid hydrofoil in a free stream flow. *Phys. Fluids* **2018**, *30*, 077107. [CrossRef]
33. Thekkethil, N.; Sharma, A.; Agrawal, A. Self-propulsion of fishes-like undulating hydrofoil: A unified kinematics based unsteady hydrodynamics study. *J. Fluids Struct.* **2020**, *93*, 102875. [CrossRef]
34. Gupta, S.; Thekkethil, N.; Agrawal, A.; Hourigan, K.; Thompson, M.C.; Sharma, A. Body-caudal fin fish-inspired self-propulsion study on burst-and-coast and continuous swimming of a hydrofoil model. *Phys. Fluids* **2021**, *33*, 091905. [CrossRef]
35. Gupta, S.; Sharma, A.; Agrawal, A.; Thompson, M.C.; Hourigan, K. Hydrodynamics of a fish-like body undulation mechanism: Scaling laws and regimes for vortex wake modes. *Phys. Fluids* **2021**, *33*, 101904. [CrossRef]
36. Gupta, S.; Agrawal, A.; Hourigan, K.; Thompson, M.C.; Sharma, A. Anguilliform and carangiform fish-inspired hydrodynamic study for an undulating hydrofoil: Effect of shape and adaptive kinematics. *Phys. Rev. Fluids* **2022**, *7*, 094102. [CrossRef]
37. Gupta, S.; Zhao, J.; Sharma, A.; Agrawal, A.; Hourigan, K.; Thompson, M.C. Two-and three-dimensional wake transitions of a NACA0012 airfoil. *J. Fluid Mech.* **2023**, *954*, A26. [CrossRef]
38. Boiron, O.; Guivier-Curien, C.; Bertrand, E. Study of the hydrodynamic of a flapping foil at moderate angle of attack. *Comput. Fluids* **2012**, *59*, 117–124. [CrossRef]
39. Sharma, A. *Introduction to Computational Fluid Dynamics: Development, Application and Analysis*; John Wiley & Sons: Hoboken, NJ, USA, 2016.

Journal of
Marine Science and Engineering

MDPI

Article

Computational Study on Influence Factors and Vortical Structures in Static Drift Tests

Chen Yang [1,*], Ke Zeng [1], Jilong Chu [1], Shuxia Bu [1,2] and Zhang Zhu [1]

1 China Ship Scientific Research Center, Wuxi 214082, China; cssrc702zk@163.com (K.Z.); long8616767@163.com (J.C.); bushuxia@cssrc.com.cn (S.B.); zhuzhang@cssrc.com.cn (Z.Z.)
2 Taihu Laboratory of Deep-Sea Technological Science, Wuxi 214082, China
* Correspondence: yangchen@cssrc.com.cn

Abstract: This paper conducted a computational study on the KCS benchmark model at static drift conditions. At the first instance, the roles played by the grid size, turbulence model, and time step are qualitatively and quantitatively analyzed with the orthogonal experimental method (OEM). After the verification of simulated results compared with experimental data in a Static Oblique Towing Test (OTT), hydrodynamic performance is obtained with the employment of the SST κ-ω turbulence model. The grid size is set as 0.07 m while the time step as 0.01 s. The characteristics of the wake field are illustrated in different forms, such as contours of the free surface, distribution of pressure and hydrodynamic forces, variation of turbulent kinetic energy (TKE), and so on. For a deep insight into the physical mechanisms of the asymmetrical flow field, the Detached Eddy Simulation (DES) method is also utilized to capture vortical structures occurring around the hull, in comparison with results obtained through the Reynolds Averaged Navier Stokes (RANS) model. With the aim of a hydrodynamic derivative estimation or detailed flow characteristics analysis, corresponding selections of the computational method are disparate.

Keywords: static oblique towing test (OTT); CFD; turbulence model; vortical structure

1. Introduction

Identified as a significant factor for navigational safety, an accurate prediction of ship maneuverability is of great necessity [1]. Experimental measurements are usually carried out for the estimation of hydrodynamic derivatives, and serve as input of mathematical models to predict ship motion, which is in demand of expenditure, labor, and experience [2,3]. The determination of hydrodynamic derivatives is fundamental for the establishment of a model. Taking both rotational and viscous effects of fluid into consideration [4], compared with simplified potential flow theory, Computational Fluid Dynamics (CFD) methods have the advantage in the description of the viscous effect of fluid, resulting in higher accuracy in predicting. No matter a full scale or model scale [5], CFD methods show great promise to work as a substitute for experimental approaches [6].

Compared to results from experimental measurements and potential flow computations, there is a better agreement in the ship response between results from the URANS codes CFDSHIP-IOWA and experimental measurements, especially in calm water. Carrica et al. [7] utilized the advanced version of CFDSHIP-IOWA V4.5 to calculate ship motions in head waves for a model-scaled KCS with a rudder in close proximity to the propeller. The wake field close to the vessel is obtained through the DES turbulence approach. Due to the lack of an appropriate maneuvering basin with a current generator, as well as relevant control units and measurement devices, there are few studies focused on free-running model tests in currents. Instead of highly laborious experiments, Kim et al. [8] deployed a CFD method with high fidelity to analyze the effect of currents on ship maneuverability.

After decades of development in computer technology, direct CFD simulation [9] has undisputed competence in reproducing a flow field, especially details. Owing to tremendous technological progress, researchers have conducted virtual maneuvering model tests for ships by solving equations of motion for fluid and rigid bodies. To save computational resources, RANS equations [10] are widely used instead of direct solving methods, in modeling viscous turbulent flow. Large amounts of successful cases involved S175 [11], DTMB 5415 [12], KVLCC2 [13], KRISO Container Ship (KCS) [14], and so on.

Zhang et al. [15] applied the viscous method to predict the roll motion of KCS in calm water and then converted it to equivalent linearized roll damping. Subsequently, the occurrence of parametric rolling in regular head waves could be achieved. Yu et al. [16] selected one discretized propeller model (DPM) for simulating the turning circle test at a rudder angle of 35° under the condition of Fr = 0.26. Three body force models [17–19] were also applied and an analysis was made on ship turning circles. Based on results derived from four propulsion models, differences are noticeable with a drift angle exceeding 10°. Simonsen et al. [20] carried out a study by the CFD method to calculate the added resistance and motions of a ship with an appendage in regular head seas. Served as a cost-effective method for obtaining self-propulsion factors, Can et al. [21] applied a CFD method in combination with Telfer's GEOSIM method to investigate the effective wave factor. Three model scales, as well as the full-scale condition, were investigated, and an accurate effective wake fraction of the full-scaled KCS was extrapolated.

Despite the prediction of the combined seakeeping and maneuvering performance, Kim et al. [22] introduced a CFD approach that could be used to study the course-keeping ability of the KCS. Simultaneously, the effect of wave directions on turning maneuvers could also be investigated. Generally speaking, the application of CFD methods to maneuvering is less mature compared to the solution of resistance [23]. The existence of an unsteady flow, ship motion, and variable environment increases the complexity of numerical simulation [24]. Yuan [25] found that the asymmetric flow passing on two sides of a hull results in a lifting force, which contributes to the yaw moment.

A suitable pattern of computational settings for virtual static captive model tests has not been established in the existing literature due to limited research efforts. In this work, great emphasis is placed on a systematic analysis concerning the OTT. An orthogonal experimental method (OEM) is applied to discuss the effects of different influence factors, including the turbulence model, grid size, and time step. Additionally, an uncertainty analysis was performed during the calculation of hydrodynamic forces and moment, with the adoption of the Grid Convergence Index (GCI) method. On this basis, a satisfactory computational setup for a virtual oblique towing test is achieved, and simulations are conducted at drift angles ranging from 0° to 12°. In addition to non-uniform force distribution and free surface elevation, vortical structures are captured to provide a comprehensive understanding of the asymmetric flow field. This work contributes to advancing the numerical solutions for virtual static captive model tests.

2. Numerical Method

2.1. Governing Equations

In this study, the commercial software STAR-CCM+ 16.06 is utilized for the simulation of three-dimensional viscous flow around the hull. Taking calculating resource and computing time into account, this work applied RANS equations instead of solving NS equations directly, which include Reynolds stress terms characterizing the turbulence effect. Under the assumption of impressible flow, relevant equations for the description of two-phase turbulent flow can be presented as follows:

$$\frac{\partial(\varphi_m\rho_m)}{\partial t}+\frac{\partial}{\partial x_i}(\varphi_m\rho_m\overline{u_i})=0 \tag{1}$$

$$\frac{\partial\rho_m\overline{u_i}}{\partial t}+\frac{\partial}{\partial x_j}(\rho_m\overline{u_i u_j})=-\frac{\partial\overline{P}}{\partial x_i}+\frac{\partial}{\partial x_j}\left[\mu_m\left(\frac{\partial\overline{u_i}}{\partial x_j}+\frac{\partial\overline{u_j}}{\partial x_i}\right)\right]+\frac{\partial}{\partial x_j}\left(-\rho_m\overline{u_i' u_j'}\right)+\rho_m\overline{f_i} \tag{2}$$

where m represents the phase id concerning water and air; i and j are coordinate components of velocity; u denotes velocity; ρ stands for density; μ is viscosity; P means pressure; f refers to gravity in this work; and $\overline{u_i' u_j'}$ indicates the Reynolds stress term based on the Boussinesq hypothesis, and a specific explanation can be found in reference [4].

2.2. Ship Geometry

The well-known benchmarking KCS is studied for simulation at a scale factor of 1/75.5. It is worth mentioning that a fixed rudder is located at the stern of the ship model, at a distance of nearly three meters from the bow, as displayed in Figure 1. Principal characteristics of the ship geometry are obtained from SIMMAN (2014) [26] and listed in Table 1. For the sake of a validation analysis, experimental results were released by the National Maritime Research Institute (NMRI).

Figure 1. The side view of the KCS model.

Table 1. Dimensions of KCS including the full scale and model scale.

Description	Parameter	Unit	Full Scale	Model Scale
Length between perpendiculars	L_{pp}	m	230	3.046
Beam at waterline	B_{wl}	m	32.2	0.427
Draught	T	m	10.8	0.143
Volume of displacement	∇	m^3	52,030	0.121
Block coefficient	C_B	-	0.651	0.651
Wetted surface area	S	m^2	9530	1.688
Metacentric height	GM	m	0.600	0.097

2.3. Case Description of Simulation

From Figure 2, a Cartesian coordinate system is fixed on the ship, with positive x, y, and z axes pointing to the upstream, portside, and upward directions in sequence. The origin is set as the intersection of the central longitudinal section, the midship section, and the water plane, and the vessel is upright without any trim at the beginning. The length of the computational domain starts from 2.7 L_{pp} to 4.0 L_{pp} in the longitudinal direction, while the width from 3.2 L_{pp} to 2.5 L_{pp} in the horizontal direction due to the symmetrical incoming flow. The construction of the virtual tank in the height direction is defined as consistent with the depth of the maneuvering basin in the NMRI. The setup of boundary conditions is summarized in Table 2, including the velocity inlet, the pressure outlet, and no-slip walls. To eliminate the effect of boundaries, a wave damping zone is specified near the rear part of the virtual tank.

Table 2. Summary of boundary conditions involved in this work.

Name of Boundary	Type of Boundary Condition
Inlet	Velocity inlet
Outlet	Pressure outlet
Side_R	Velocity inlet
Side_L	Velocity inlet
Top	Velocity inlet
Bottom	Velocity inlet
Hull surface	No-slip wall

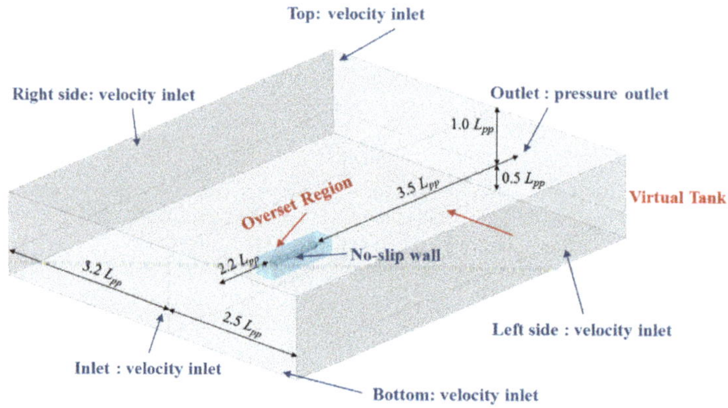

Figure 2. The setup of the computational domain.

Figure 3 displays the mesh structure of simulation. The trimmed mesher and the surface remesher are applied for mesh discretization. The latter is employed to generate volume meshes, while the former is used for surface meshes. Specifically, the prism mesh was generated at the boundary layer close to hull surfaces, with a layer number of six. To satisfy the requirement of $30 < y+ < 100$, the total thickness of the boundary layer is limited to 0.042 m. For a reasonable mesh distribution instead of a uniform grid, three refinement levels are identified from inside out. The interface of the two-phase flow is another refinement region to capture the elevation in the free surface precisely. Additionally, the VOF method, one of the multiphase flow models, is adopted to identify phases of water and air. Taking the complexity of specific geometric structures into consideration, local mesh refinement is also implemented in parts including the bow, the stern, and the rudder. A three-degree-of freedom vessel is investigated by solving the equations for rigid body motion, including heave, pitch, and roll motions during the simulation.

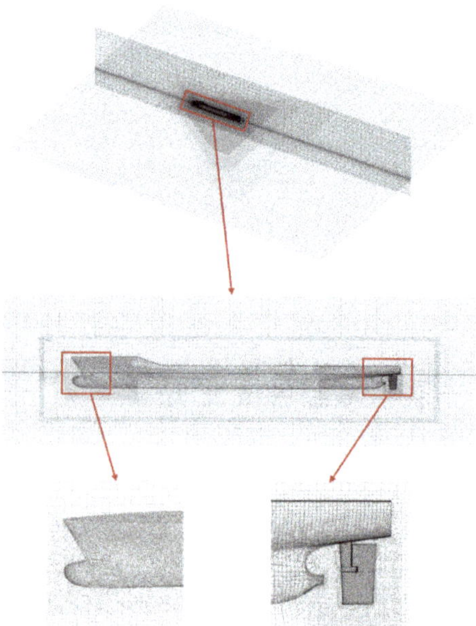

Figure 3. Mesh structure of simulation.

Virtual oblique towing tests contain conditions of drift angles ranging from 0° to 12°. Table 3 introduces specific working conditions of simulation. The velocity v_m of the ship model is constant, corresponding to a Froude number (Fr) of 0.20.

Table 3. Computational case conditions.

Drift Angle (deg)	Velocity of Model (m/s)	*Fr*
0, 4, 8, 12	1.1	0.20

When conducting the simulation of the OTT, ship motions are confined to the transverse X-Y plane. Hence, particular attention is paid to longitudinal force X, lateral force Y, and yaw moment N. For simplification, non-dimensional longitudinal force X' and lateral force Y' can be derived by dividing $\frac{1}{2}\rho v_m{}^2 L_{pp} d$, while non-dimensional yaw moment $N' = \frac{N}{\frac{1}{2}\rho v_m{}^2 L_{pp}{}^2 d}$.

2.4. Orthogonal Uncertainty Analysis

During the use of numerical tools, great necessity is attached to an uncertainty analysis relating to multi-factors. When estimating numerical errors and uncertainties, at least three sets of data in relation to one variable are in demand for verification. Considering the efficiency of the analysis, the OEM is utilized for the simulation of the OTT. The application of the turbulence model is deemed as influence factor A, selection of grid base size is regarded as influence factor B, and setting of the time step is taken as influence factor C. Figure 4 illustrates the workflow of the orthogonal uncertainty analysis method employed in this work.

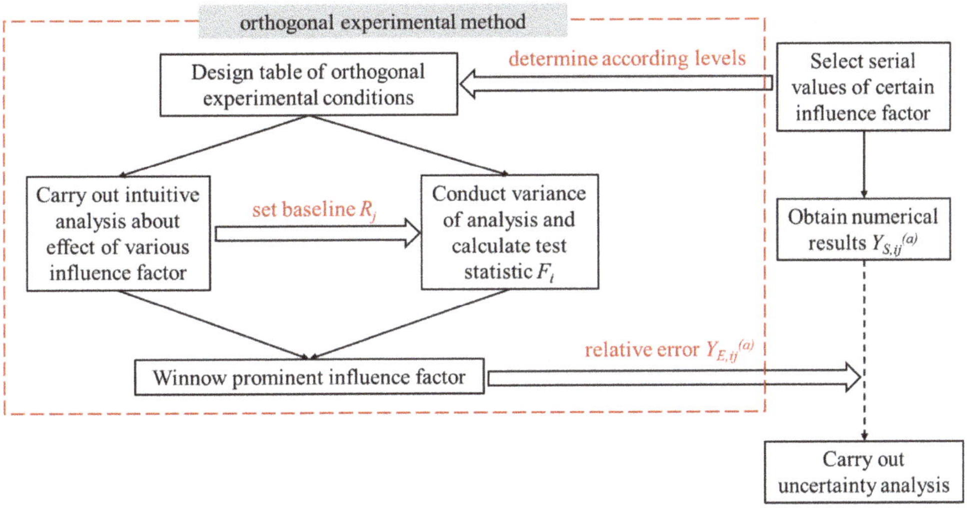

Figure 4. Workflow of orthogonal uncertainty analysis method.

During orthogonal experimental tests, the count of test conditions is denoted as n, while the number of influence factors as p and number of levels as q. For convenience, the relevant indicator of an experiment with a serial number of a is denoted by $Y_{ij}{}^{(a)}$, where i represents the order of the influence factor and j stands for the order of the level, respectively. Average values at targeted levels K_{ij} play a significant role in the OEM including an intuitive analysis and variance analysis, implemented as a qualitative analysis and quantitative analysis, respectively. Calculation equations are provided as follows:

$$K_{N,ij} = \frac{q}{n}\sum_{a=1}^{n/q} Y_{N,ij}^{(a)} \tag{3}$$

$$K_{E,ij} = \frac{q}{n}\sum_{a=1}^{n/q} Y_{E,ij}^{(a)} \tag{4}$$

where subscript 'N' represents data obtained by means of numerical simulation and 'E' refers to the relative error between results calculated from the numerical method and experimental measurement. To put it another way, ${K_{N,ij}}^{(a)}$ stands for numerical results of influence factor i at the level j (denoted as serial number of a). In this work, ${Y_{N,ij}}^{(a)}$ is deployed in the uncertainty analysis and ${Y_{E,ij}}^{(a)}$ is calculated for comparison between influence factors.

In a gesture to the winnow prominent influence factor, which has a great impact on the relative error, an analysis of variance is conducted alongside the intuitive analysis. The definition of extreme deviation R_j is shown as Equation (6), and the influence factor corresponding to the minimum value of extreme deviation will be taken as the baseline:

$$R_j = \max\left(K_{E,ij}\right) - \min\left(K_{E,ij}\right) \tag{5}$$

Take the relative error of influence factor i at the level j (denoted as serial number of a) ${Y_{E,ij}}^{(a)}$ as an indicator, and the average of results is as follows:

$$\overline{T} = \frac{1}{n}\sum Y_{E,ij}^{(a)} \tag{6}$$

Then, S_i stands for the sum of the standard deviation, which is concerned with all influence factors:

$$S_i = \frac{n}{q}\sum_{j=1}^{q}\left(K_{E,ij} - \overline{T}\right)^2 \tag{7}$$

After the summation of standard deviation obtained through Equation (7), mean square MS_i can be calculated as follows:

$$MS_i = S_i/f_i = \frac{n}{q}\sum_{i=1}^{q}\left(K_{E,ij} - \overline{T}\right)^2 \Big/ (p-1) \tag{8}$$

where f_i indicates the degree of freedom.

Aiming at quantifying the effect of the influence factor, test statistic F_i is applied to compare impact with the baseline, which is written as follows:

$$F_i = S_i/S_e \tag{9}$$

where S_e corresponds to the sum of the standard deviation for the influence factor with the minimum extreme deviation.

A smaller value of the test statistic indicates a less significant role. Hence, influence factors with a test statistic below the reference value will not be taken into account in the subsequent uncertainty analysis. For space reasons, relevant procedures will not be introduced in this paper, which can be found in published study [27].

3. Numerical Results

3.1. Effect of Selected Influence Factors

During the simulation of an OTT, an OEM is utilized for a comprehensive analysis on the effects of various computational settings. Definitions of influence factors A, B, and C are stated in Section 2.4. For each influence factor, three different levels are chosen for the analysis, as presented in Table 4. With regard to the turbulence model, level 1 and 2 stand for the widely used two-equation models SST κ-ω model and Realizable κ-ε model, respectively. A Reynolds Stress Model (RSM) is represented by level 3, which is another effective tool to address complex flow phenomena. Taking the grid discretization method and computing time into account, the DES model or LES model is excluded from the design of an orthogonal array. Concerning grid convergence research, a refinement ratio of $\sqrt{2}$ is chosen according to reference [27] in maneuvering tasks.

Table 4. Relative error of test conditions by OEM.

No.	Influence Factor A	Influence Factor B	Influence Factor C	State of Conditions	Relative Error (%)
1	SST κ-ω	0.05 m	5×10^{-3} s	$A_1B_1C_1$	−12.54
2	SST κ-ω	0.07 m	0.01 s	$A_1B_2C_2$	−10.81
3	SST κ-ω	0.10 m	0.02 s	$A_1B_3C_3$	−6.74
4	Realizable κ-ε	0.05 m	0.01 s	$A_2B_1C_2$	−12.43
5	Realizable κ-ε	0.07 m	0.02 s	$A_2B_2C_3$	−9.64
6	Realizable κ-ε	0.10 m	5×10^{-3} s	$A_2B_3C_1$	−6.02
7	RST	0.05 m	0.02 s	$A_3B_1C_3$	−10.45
8	RST	0.07 m	5×10^{-3} s	$A_3B_2C_1$	−7.82
9	RST	0.10 m	0.01 s	$A_3B_3C_2$	−1.72

The orthogonal array is also illustrated in Table 4 as the column of 'State of Conditions', indicating the parameter configuration for the numerical simulation. Two principles for the design of test conditions are the uniform distribution and homogeneous design. To guarantee the orthogonality of the test design, the occurrence frequencies of level 1, 2, and 3 for the influence factor are identical. In addition to this, the combination of level orders concerning either of the two influence factors is also the same. Values of relative errors related to each state of experimental conditions are also provided. Among all conditions, it can be observed that data obtained through the CFD method are smaller than experimental results. The peak value of the relative error comes to −12.54% under the experimental condition of $A_1B_1C_1$, while the minimum value is about −1.72%.

Average values at different levels are listed in Table 5, together with extreme deviations of three influence factors, which reflect the extent of significance. From the results, the impact exerted by the influence factor C is negligible. With the variation of the time step from 0.01 s to 0.02 s, the extreme deviation is only 0.62%. As a result, the time step is chosen as the baseline factor for the variance analysis. Values of the test statistic are consistent with the extent of significance obtained from extreme deviation. The impact of grid base size imposed on the numerical results cannot be ignored. Compared with the time step, the test statistic of grid size reaches 32.32, which is larger than the reference value of $F_{0.95} = 19$. Among the three influence factors, grid size plays the most significant role in numerical simulation, followed by the turbulence model.

In order to carry out an intuitive comparison of results obtained with different turbulence models, the longitudinal force and lateral force are plotted under an Fr of 0.20 as displayed in Figure 5. During post-processing, the ship hull is divided into 100 parts at an interval of $\Delta x / L_{pp} = 0.01$. Values of each part can be obtained by means of integration according to the active area. The force induced by pressure and viscous effect are designated as subscripts p and s, respectively. For instance, physical quantity X_p represents the part of longitudinal force deriving from pressure while Y_s denotes the part of lateral force resulting from friction. The drift angle is chosen as 12° for a more conspicuous separate flow in the virtual test. In the figure, the ordinate represents the variation in the longitudinal direction with a non-dimensional parameter x/L_{pp} for convenience, while the abscissa represents the values of forces. From the gradient of value, the pressure component makes a major contribution, particularly in lateral force. The grey line indicates data related to the SST κ-ω model, while the red line and blue line correspond to level 2 and level 3 of turbulence models, respectively. The diversity of the turbulence model can be observed facilely in forces stemming from the viscous effect of water, especially in the range of x/L_{pp} = 0.45~0.70. To put it another way, the discrepancy between curves of the pressure component in different colors is not as distinct as that of the shear component, which is consistent with results in Table 5.

Table 5. Intuitive analysis results and variance analysis of non-dimensional sway force Y'.

	Influence Factor A	Influence Factor B	Influence Factor C	
$K_{E,i1}$	−10.03%	−11.81%	−8.79%	
$K_{E,i2}$	−9.36%	−9.42%	−8.32%	
$K_{E,i3}$	−6.67%	−4.83%	−8.94%	
R_i	3.36%	6.98%	0.62%	
Extent of significance		B > A > C (influence factor B ranks first)		
Sum of standard deviation S_i	1.905×10^{-3}	7.550×10^{-3}	0.064×10^{-3}	
Degree of freedom f_i	2	2	2	$F_{0.95}(2,8) = 19$ $F_{0.99}(2,8) = 99$
Mean square MS_i	0.952×10^{-3}	3.775×10^{-3}	0.032×10^{-3}	
Test statistic F_i	8.154	32.32	0.272	

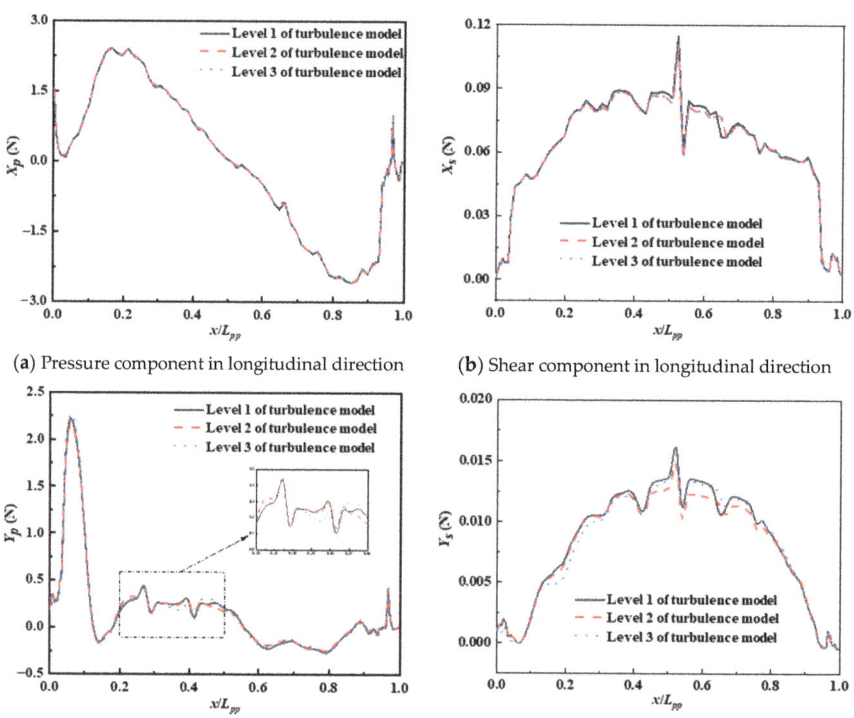

(**a**) Pressure component in longitudinal direction (**b**) Shear component in longitudinal direction

(**c**) Pressure component in transverse direction (**d**) Shear component in transverse direction

Figure 5. The distribution of forces along the longitudinal and the transverse directions by using different turbulence models.

To implement the requisite quantitative analysis, the GCI is applied to estimate the uncertainty caused by the grid size, which is a quantifiable influence factor. It is worth mentioning that the average value at different levels of $K_{S,ij}$ is taken as a substitute for numerical results in the grid convergence study, which can be found in Table 6. Utilizing the SST κ-ω model at a time step of 0.01 s, it only takes the influence exerted by grid size into consideration. Surge force is opposite to the forward direction, manifested as negative values in the table. For the sake of convenience, number 1 represents the fine grid, while numbers 2 and 3 stand for medium and coarse grids, respectively. Corresponding grid numbers are 4.15×10^6, 1.64×10^6, and 0.66×10^6 in sequence. To guarantee a monotonic convergence condition, hydrodynamic forces and moment are non-dimensionalized to fall in the range of 0 to 1, including dimensionless forms of drag force X', transverse force Y', and yaw moment N'. When comparing the results of adjacent convergence models with symbols 1 and 2, the approximate relative error $e_a{}^{21}$ is below the acceptable upper limit of 5%, as well as the GCI_{21} index.

Table 6. Results of grid convergence analysis.

Parameter	Dimensionless Surge Force X'	Dimensionless Sway Force Y'	Dimensionless Yaw Moment N'
$K_{S,1}$ ($\times10^{-3}$)	−20.246	60.802	27.925
$K_{S,2}$ ($\times10^{-3}$)	−20.260	62.447	28.157
$K_{S,3}$ ($\times10^{-3}$)	−20.658	65.614	28.621
ε_{21}	-1.420×10^{-5}	1.6453×10^{-3}	2.319×10^{-4}
ε_{32}	-3.978×10^{-4}	3.1664×10^{-3}	4.635×10^{-4}
R	0.0357	0.5196	0.5003
p_{GCI}	3.7033	0.7524	0.7928
$e_a{}^{21}$	0.070%	2.706%	0.830%
GCI_{21}	0.003%	3.332%	0.948%
$e_a{}^{32}$	1.963%	5.071%	1.646%
GCI_{32}	0.091%	6.537%	1.970%

Figure 6 displays the distribution of sway force on both sides along the longitudinal direction under different grid sizes, which is also integrated by uniformly distributed sections with the same Δx. For an intuitive comparison, it has to be stressed that negative values on the portside are plotted. Under the condition of a 12° drift with v_m = 1.1 m/s, a smoother transition can be noticed with the fine grid. In other words, an increase in the grid number can bring about an improvement in accuracy to some extent. Nevertheless, when the grid number reaches a certain value, the promotion of accuracy may be negligible regarding the consumed computing resources. For convenience, we take the bow of the hull as the position of $x = 0$, and stern of the model as $x = 1$. From lateral force acting on the portside, peak values emerge at the distance of x/L_{pp} = 0.139 and 0.522, which is independent of grid size. Such a phenomenon also takes place when it comes to the starboard side, with a different peak point at x/L_{pp} = 0.064. In addition to the location of the peak value, discrepancy also exists between the magnitude of lateral force on two sides, which can be attributed to the influence of the drift angle, irrelevant to the refinement of the grid in this research. Combined with GCI values, the intermediate set of grid size, which equals 0.07 m, could satisfy the calculation accuracy of hydrodynamic force.

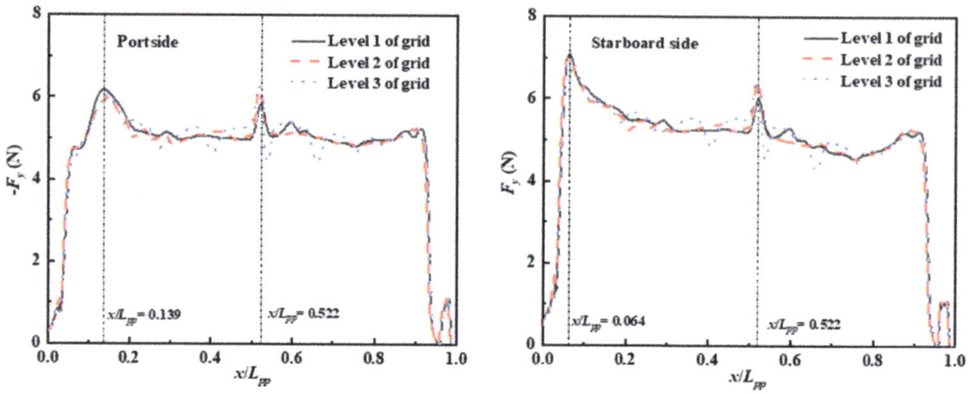

(**a**) Inverse value of lateral force at the portside (**b**) Lateral force at the starboard side

Figure 6. Lateral force at two sides under diverse grid sizes.

3.2. Numerical Results under Different Test Conditions

Based on the analysis undertaken in Section 3.1, the effects of three influence factors at various levels are investigated, and suitable computational settings for a virtual static drift test are received. The combination of both the grid size and time step at level 2, together with the RST turbulence model, is a proper selection. Apart from the simulation accuracy, the stability of the solution and the difficulty of convergence should also be considered. Hence, the SST κ-ω model is employed for simulation under several drift angles. Table 7 provides a comparison between solving results and experimental data given by the NMRI under drift angles of 6° and 12°. It can be observed that dimensionless surge force and sway force calculated through numerical simulation are smaller than those obtained by experiments, while the numerical result of the dimensionless yaw moment is larger. From the results, the maximum value of the relative error occurs when the drift angle reaches 12°, which is about 11.37% for the non-dimensional yaw moment. It can be attributed to many reasons. For a static drift test, hydrodynamic forces and moment are affected by flow separation, especially that taking place around the sonar dome. The difficulty in capturing the onset and progression increases at a larger drift angle. However, the RANS turbulence model has its limitation in shielding of the boundary layer, which may result in a large error. Additionally, the more unsteady behavior of the free surface under the condition of a 12° drift test poses a greater challenge for the grid discretization method. Inalterable grid refinement regions as the condition of small-drift-angle tests may underperform during the simulation of interaction between the fluid and ship hull. The average relative error of hydrodynamic forces and moment is nearly 10% within the acceptable range. Generally speaking, adopted computational settings can meet the requirement for the simulation of an OTT.

Table 7. Validation of the CFD method compared with experimental data provided by NMRI.

Drift Angle	Physical Quantity	Experimental Data ($\times 10^{-2}$)	Numerical Result ($\times 10^{-2}$)	Relative Error (%)
$\beta = 6°$	X'	−1.938	−1.900	−1.95
	Y'	2.338	2.220	−5.06
	N'	1.271	1.328	4.55
$\beta = 12°$	X'	−2.211	−2.072	−6.27
	Y'	6.894	6.480	−6.02
	N'	2.455	2.734	11.37

With the transformation of the drift angle, the incident direction of incoming flow alters while positioning of the vessel remains. To illustrate the interaction between the vessel and fluid, Figure 7 depicts the volume fraction of water at the bow. Under the drift angle of 0°, the distribution of water is the same on two sides of the hull. When the drift angle increases from 0° to 12°, the discrepancy between two sides can be easily observed, especially in the case of a 12° drift angle. The larger the drift angle is, the larger the difference between crests at the portside and the starboard side will be.

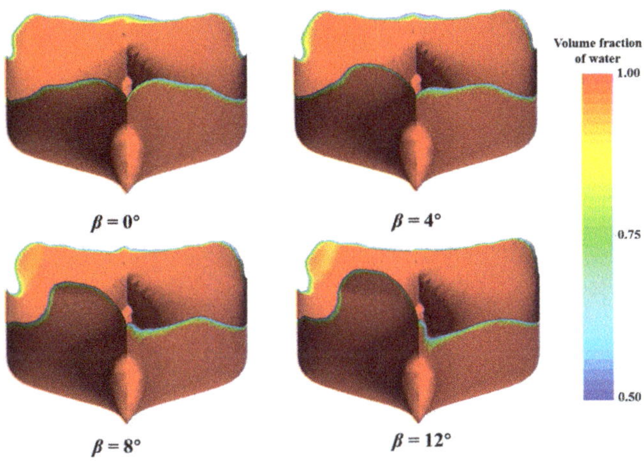

Figure 7. The general view of waterlines with a drift angle in the range of 0~12°.

For a clear representation of wave elevation, Figure 8 displays the top view of the free surface at an interval of 4° for the drift angle. The wave height at the portside and the starboard side correlates to the distribution of pressure on the hull, which is closely bound up with the acting force attributed by fluid. Apart from this, contours of the wave pattern are also displayed in Figure 8 as supplementation. The grey line denotes the portside while the red line represents the starboard side, both of which are extracted at sections located 0.072 L_{pp} away from the longitudinal section. Under different drift angles, the diversity of wave height concentrates on the region of x/L_{pp} = 0~1. In particular, critical positions can be observed near x/L_{pp} = 0.177 and x/L_{pp} = 0.828, which correspond to the regions of the fore body and the aft body, respectively. Under the 4° drift condition, the wave surface ahead of the bow is conspicuously elevated at the starboard side, with a maximum difference in wave height of 0.027 m. Accompanied by the increase in the drift angle, more obscure deviation in wave height at two sides can be observed. The dispersed waves at the portside vanish gradually with the opposite situation observed at the other side, especially in the bow region.

During the oblique towing test, the onset of the included angle between the incoming flow and the vessel will give rise to pressure differences around the hull, which can be attributed to the asymmetry of the fluid at the portside and the starboard side. It is undeniable that hydrodynamic forces and moment exerted on the vessel are deeply dependent on the flow field. From Figure 9, divergence can be observed between longitudinal distributions of pressure in drift angle tests, especially in the region of x/L_{pp} = 0~0.2. Similarly, the difference of pressure distributed at two sides is more significant with an increase in the drift angle. The existence of a bulbous bow exaggerates the difference originating from the oblique flow. After development along the longitudinal direction, there is a minor discrepancy between pressures on the two sides.

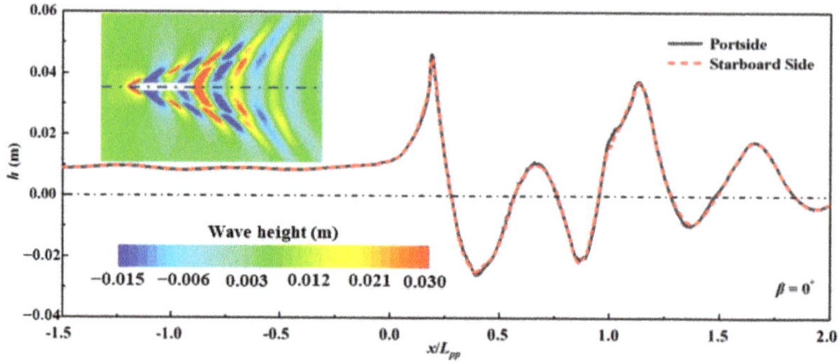

(**a**) drift angle of 0°

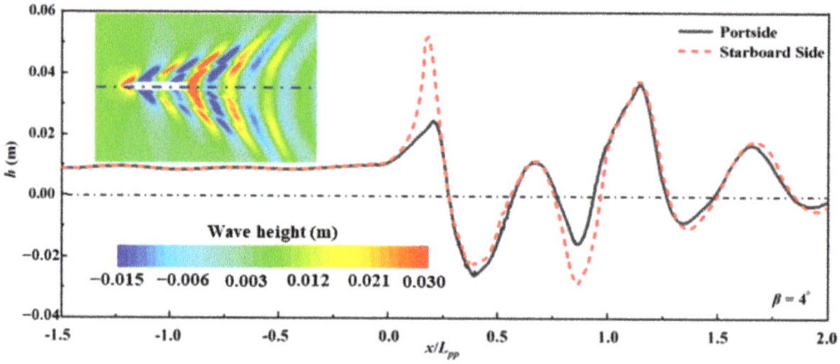

(**b**) drift angle of 4°

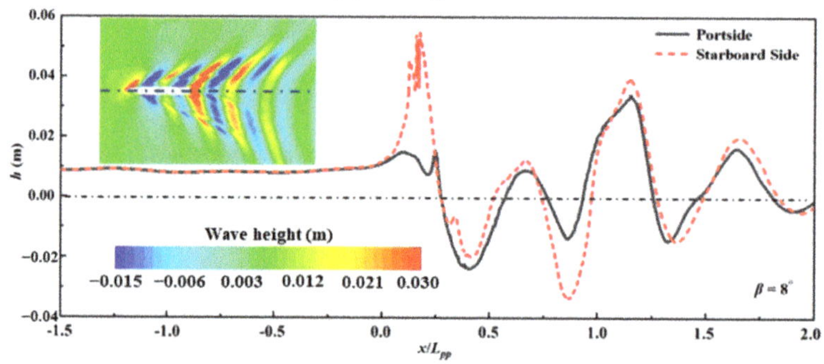

(**c**) drift angle of 8°

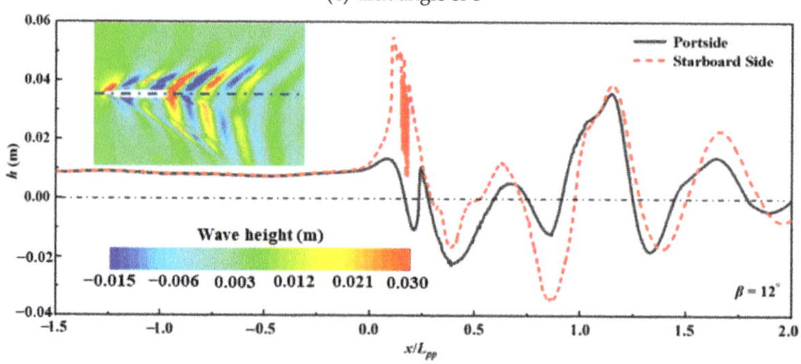

(**d**) drift angle of 12°

Figure 8. Comparison of wave height along longitudinal direction.

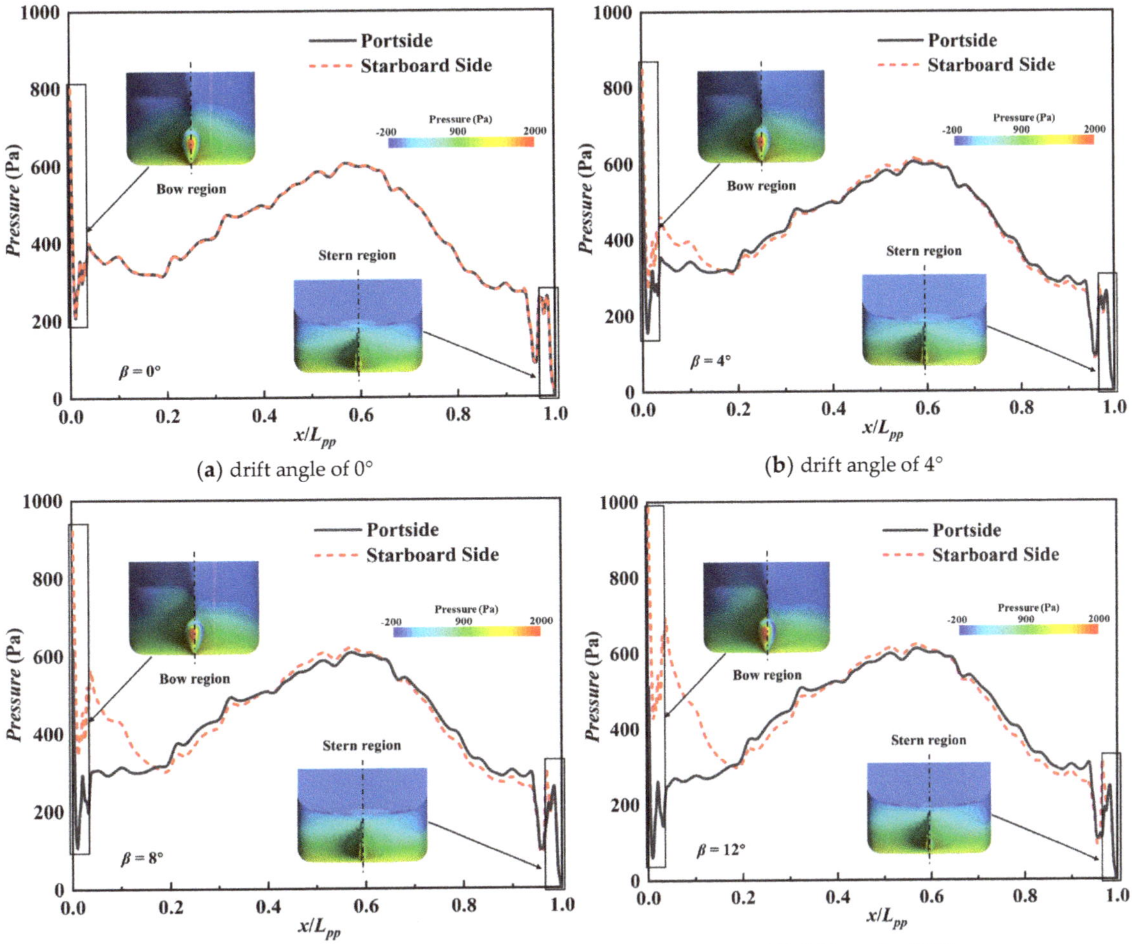

(**a**) drift angle of 0° (**b**) drift angle of 4°

(**c**) drift angle of 8° (**d**) drift angle of 12°

Figure 9. Pressure distribution on the hull along the longitudinal direction.

The pressure difference at two sides along the longitudinal section leads to disparity in the velocity component, which is a significant indicator of the flow field. The asymmetrical velocity field around the symmetrical vessel, displayed in Figure 10, results in the generation of diverse sway force, with the solid line corresponding to the portside and the dashed line for the starboard side. The distribution of fluid velocity is symmetrical under the condition of $\beta = 0°$, which is similar to Figures 7–9. It is noticeable in the difference of lateral force at two sides with an increase in the drift angle. The greatest disparity of $|F_y|$ emerges at the location of $x/L_{pp} = 0.055$ in the 12° drift angle test. At the starboard side, the maximum lateral force is 33% higher than that at the portside, which comes up to 11.94 N. For an intuitive analysis of local areas, velocity components on sections of $x/L_{pp} = 0.12$ and $x/L_{pp} = 0.88$ are also demonstrated in Figure 10, corresponding to bow and stern regions, respectively. The dimensionless parameter u' is selected for illustration, and contours are demonstrated from the perspective of the stern. The velocity component in the longitudinal direction experiences continuous reduction when the drift angle increases. In the bow region of the vessel (section $x/L_{pp} = 0.12$), fluid at the starboard side shows higher longitudinal velocity speed, especially in Figure 10d. On the contrary, the opposite appearance is perceived at the section of x/L_{pp} = 0.88. The asymmetry in velocity component arises from the lateral flow, which brings about a leaking vortex in the region near the bulbous bow. The deviation extent of fluid will be enhanced due to the interaction between the leaking vortex and the boundary layer at the starboard side of the vessel.

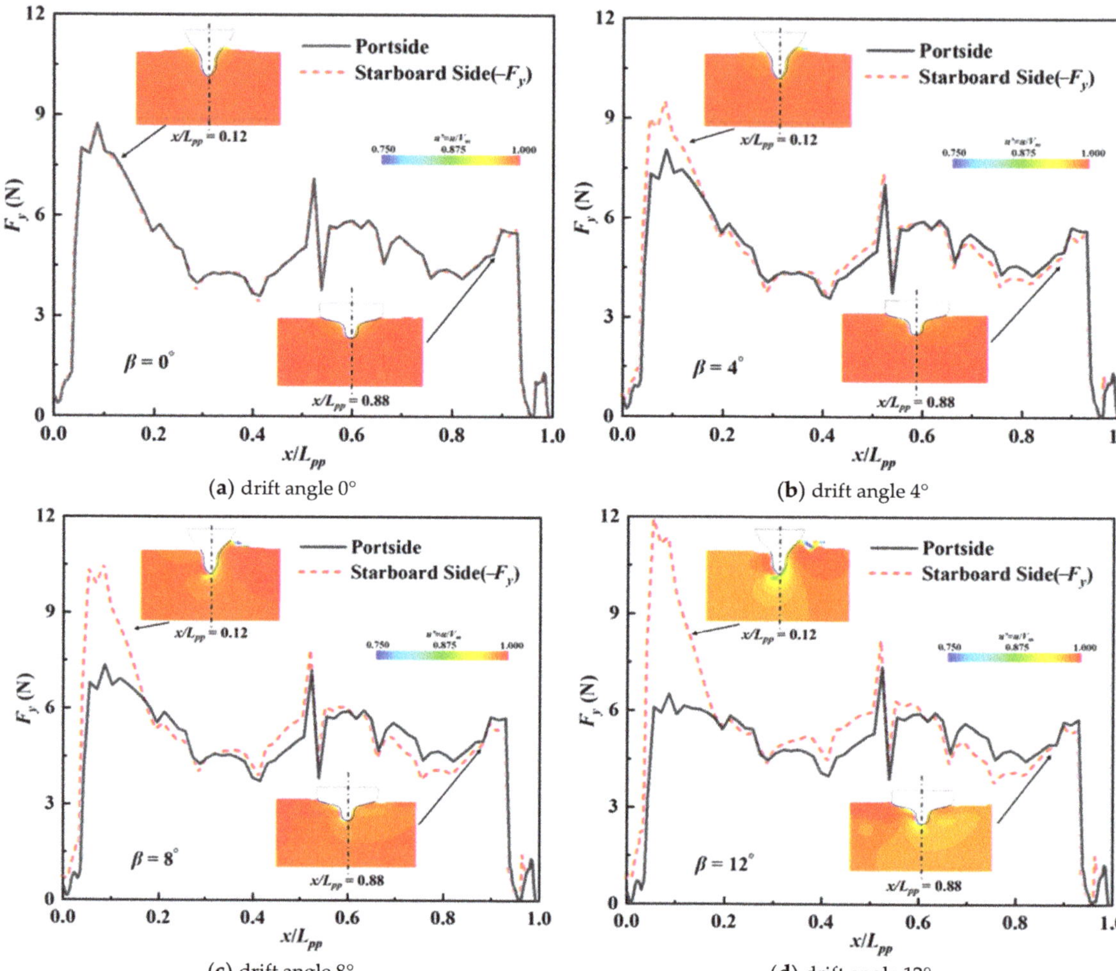

(**a**) drift angle 0° (**b**) drift angle 4°

(**c**) drift angle 8° (**d**) drift angle 12°

Figure 10. The comparison of lateral force on two sides of the hull.

The longitudinal evolution of turbulent kinetic energy is depicted in Figure 11. The asymmetry can be easily observed when the drift angle increases, which shows consistency with that of the longitudinal velocity component. Nevertheless, the magnitude is underestimated due to anisotropy in a Reynolds Stress Transport model [28]. As a result, these contours are merely appropriate for the reflection of a certain distribution pattern.

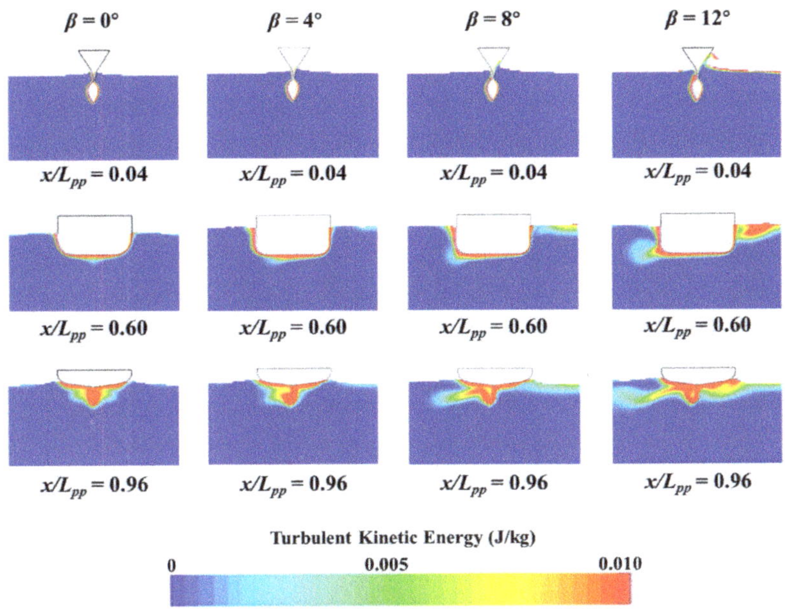

Figure 11. Turbulent kinetic energy at different sections.

3.3. Vortical Structures

With keen attention paid to detailed flow separation and realistic demand for an authentic representation of the real physics, researchers are in persistent pursuit of capturing vortex structures. In a gesture to look deeper into the physical mechanisms of the asymmetrical flow field, a comparison is made with the method of the RANS model and DES model [29]. The main time-averaged vortical structures occurring around the ship hull under a static drift condition of 12° are illustrated in Figure 12, together with the position of fluid in contrast to the free surface. For vortex identification, the iso-surface of the Q-criterion [30] is displayed with a threshold of 50, which is colored by the vorticity magnitude. The major vortex structure is the windward sonar dome tip vortex (WW-SDTV). It can be clearly distinguished from the sonar dome end tip. Under the premise of an identical grid refinement method at the same location, the discrepancy in the evolution of the WW-SDTV can be easily detected when utilizing two different models. In Figure 12b, the vortex stretches to a longer range along the hull. To put it another way, the progressive damping of the WW-SDTV simulated by the RANS method is faster. Concerning the stern vortex (SV), more detailed vortex structures are visible when utilizing the DES method. Apart from this, a smoother transition of the kelvin wave is observed, indicating a better solution over the wake field.

For the research on flow separation and vortex shedding, the development of the vortex in the direction of incoming fluid is demonstrated in Figure 13, with the extraction of six sections along the hull. From longitudinal vorticity contours, the deflection of the vortex is inconspicuous until the stern region in the 4° static drift test. Under high drift angles, the evolution of the vortex at the portside is more violent, especially for 12° DES simulation.

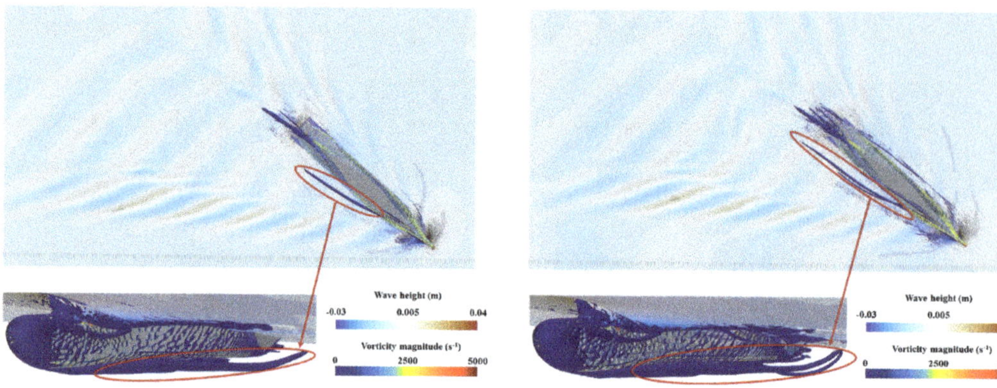

(**a**) results with the deployment of the RANS method (**b**) results with the deployment of the DES method

Figure 12. Vortex structures obtained through RANS model and DES model.

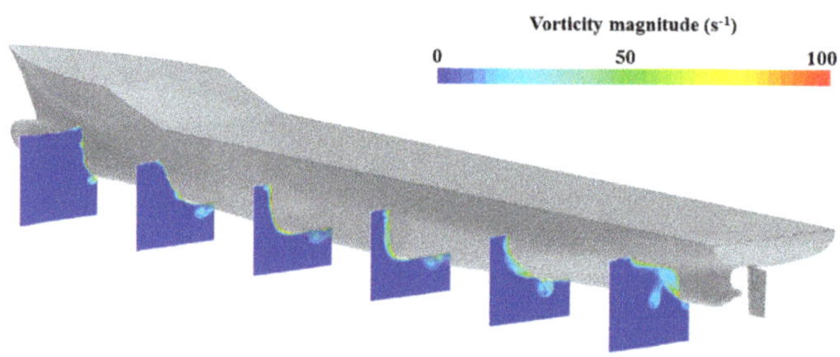

(**a**) drift angle of 4°

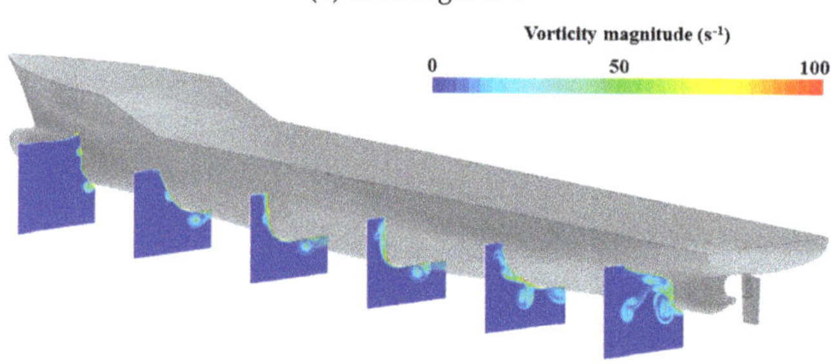

(**b**) drift angle of 6°

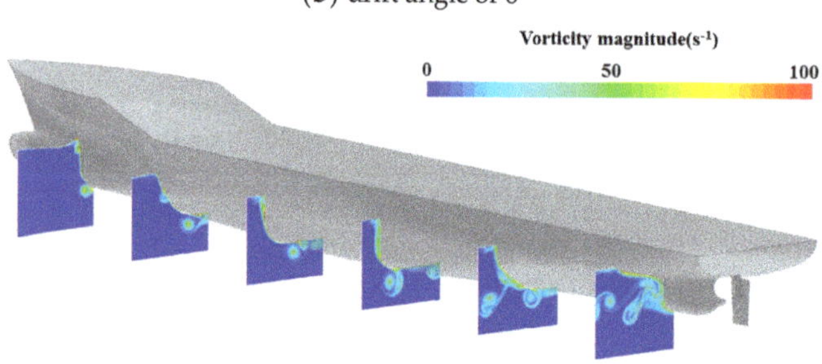

(**c**) drift angle of 8°

Figure 13. *Cont.*

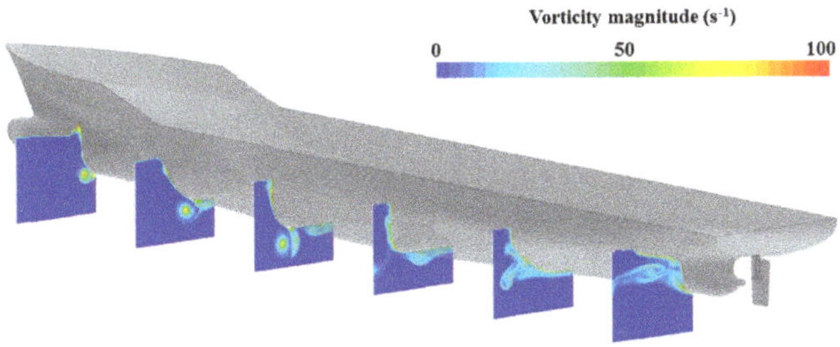

(**d**) drift angle of 12°

Figure 13. Vorticity magnitude in different static drift tests.

Figure 14 depicts the simulation results of lateral force using the RANS method and DES method. Two drift conditions are investigated, including drift angles of 6° and 12°. Subscripts of '*P*' and '*S*' in plots stand for the portside and starboard side, respectively. Disparities are mainly distributed at the portside, where incoming fluid originates. An abrupt variation of $F_{y,P}$ solved by the RANS method is visible around the center of the vessel in Figure 14a, with a maximum value and a minimum value. However, the distribution pattern obtained by the DES method, representing a smoother transition at the portside, does not match that of the RANS method. Concerning the starboard side, divergence can hardly be observed when the drift angle is 6°. When the drift angle increases to 12°, the variation tendency of $F_{y,S}$ calculated by DES is similar to that of the 6° static drift condition, showing distinct differentiation from the results of the RANS method. As illustrated in Table 8, the dimensionless sway force acting on the hull under the condition of $\beta = 6°$, solved by the DES method, is 2.412×10^{-2}. The corresponding relative error compared with experimental data is 3.17%. For the 12° static drift test, the value of Y' predicted by the DES method is 6.168×10^{-2}, which is close to the result achieved from the RANS method as shown in Table 7. The above analysis demonstrates the outperformance of the DES method in capturing details of vorticities, which is in accordance with the conclusion from predecessors' work [17]. For the sake of predicting hydrodynamic forces and moments, the utilization of the RANS model is preferable, taking efficiency and accuracy into account.

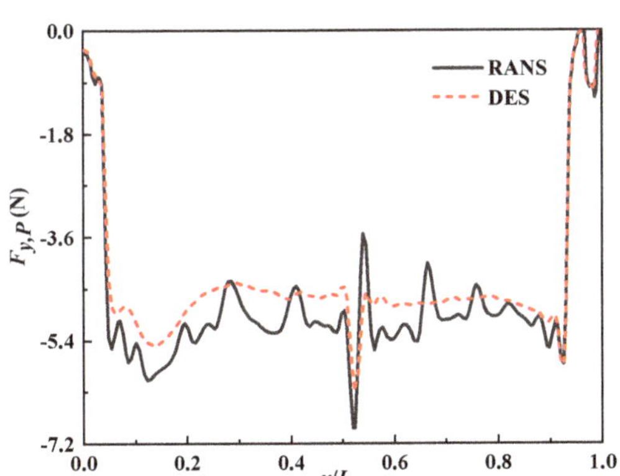

(**a**) portside at $\beta = 6°$

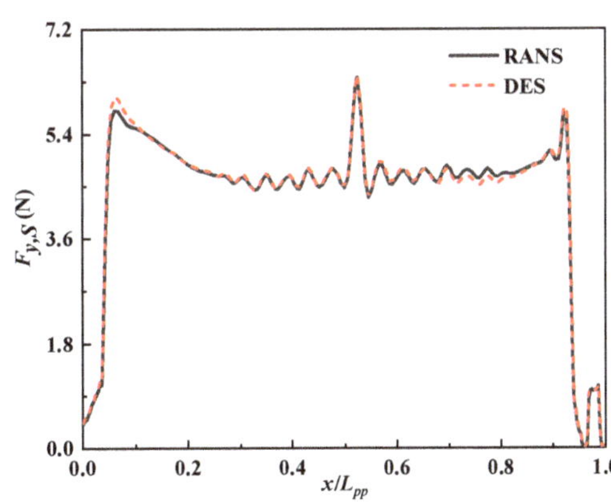

(**b**) starboard side at $\beta = 6°$

Figure 14. *Cont.*

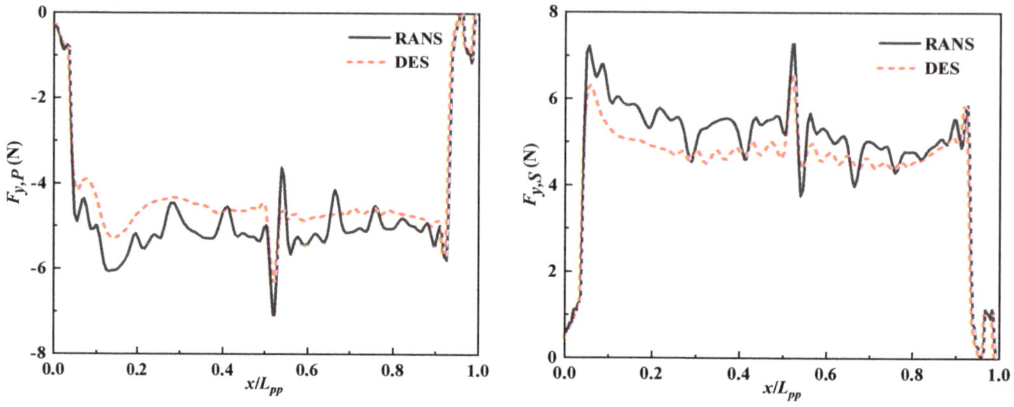

(**c**) portside at $\beta = 12°$ (**d**) starboard side at $\beta = 12°$

Figure 14. The comparison of lateral force through the RANS method and DES method.

Table 8. Validation of the DES model compared with experimental data provided by NMRI.

Drift Angle	Physical Quantity	Experimental Data ($\times 10^{-2}$)	Numerical Result ($\times 10^{-2}$)	Relative Error (%)
$\beta = 6°$	Y'	2.338	2.412	3.17
	N'	1.271	1.356	6.71
$\beta = 12°$	Y'	6.894	6.168	−10.53
	N'	2.455	2.744	12.99

4. Conclusions

A systematic analysis is conducted qualitatively and quantitatively to explore the roles of three influence factors in the computational study of the KCS. The aim of simulation determines the selection of the computational method. After the discussion of the oblique flow field in virtual static captive tests, some conclusions are addressed in the following:

Among influence factors of the grid size, turbulence model, and time step, grid size plays the most significant role in numerical simulation, followed by the turbulence model. According to the verification of numerical results, the relative error of the yaw moment is larger compared with longitudinal force and lateral force.

From the oblique flow field, the diversity between the lateral force at the portside and the starboard side along the longitudinal direction becomes more conspicuous with a larger drift angle, especially in the bow region.

The DES method has irreplaceable advantages in providing detailed information about vortex structures. Nevertheless, the utilization of the RANS model is preferable for the sake of predicting hydrodynamic forces and moments.

Author Contributions: Conceptualization, C.Y. and K.Z.; methodology, C.Y., K.Z. and Z.Z.; software, C.Y. and K.Z.; validation, J.C. and Z.Z.; formal analysis, C.Y. and K.Z.; investigation, C.Y., K.Z. and J.C.; resources, C.Y. and K.Z.; data curation, J.C. and Z.Z.; writing—original draft preparation, C.Y.; writing—review and editing, K.Z. and S.B.; supervision, K.Z. and J.C.; project administration, S.B.; funding acquisition, S.B. All authors have read and agreed to the published version of the manuscript.

Funding: This research was funded by the National Natural Science Foundation of China (Grant No. 52271336).

Institutional Review Board Statement: Not applicable.

Informed Consent Statement: Not applicable.

Data Availability Statement: The data that support the findings of this study are available within the article.

Conflicts of Interest: The authors declare no conflict of interest.

Nomenclature

B_{wl}	beam at waterline	m
C_B	block coefficient	
f	gravity	N
F_i	test statistic	
f_i	degree of freedom	
Fr	Froude number	
F_y	lateral force	N
GM	metacentric height	m
h	wave height	m
K_{ij}	average values at targeted levels	
L_{pp}	length between perpendiculars	m
MS_i	mean square	
N'	non-dimensional yaw moment	
P	pressure	Pa
S	wetted surface area	m^2
T	draught	m
R_j	extreme deviation	
S_i	sum of the standard deviation	
v_m	velocity of ship	m/s
u_i	x-coordinate components of velocity	m/s
u_j	y-coordinate components of velocity	m/s
X'	non-dimensional longitudinal force	
X_p	part of longitudinal force deriving from pressure	N
X_s	part of longitudinal force deriving from friction	N
Y'	non-dimensional lateral force	
${Y_{E,ij}}^{(a)}$	relative error of the orthogonal experimental test with a serial number of *a*	
${Y_{N,ij}}^{(a)}$	simulation result of the orthogonal experimental test with a serial number of *a*	
Y_p	part of longitudinal force deriving from pressure	N
Y_s	part of longitudinal force deriving from friction	N
$y+$	dimensionless wall distance	
Abbreviations		
CFD	Computational Fluid Dynamics	
DES	Detached Eddy Simulation	
DPM	Discretized Propeller Model	
GCI	Grid Convergence Index	
KCS	KRISO Container Ship	
OEM	Orthogonal Experimental Method	
OTT	Oblique Towing Test	
RANS	Reynolds Averaged Navier Stokes	
RSM	Reynolds Stress Model	
SV	Stern Vortex	
TKE	Turbulent Kinetic Energy	
WW-SDTV	Windward Sonar Dome Tip Vortex	
Greek symbols		
β	drift angle	°
ε	dissipation rate	m^3/s^2
κ	turbulent kinetic energy	J/kg
μ	viscosity	Pa·s
ρ	density	kg/m^3
ω	dissipation rate	m^3/s^2
∇	volume of displacement	m^3

Subscripts

i	influence factor
j	level
m	phase id
P	values of the portside
S	values of the starboard side

References

1. He, Z.; Liu, C.; Chu, X.; Negenborn, R.R.; Wu, Q. Dynamic anti-collision A-star algorithm for multi-ship encounter situations. *Appl. Ocean Res.* **2022**, *118*, 102995. [CrossRef]
2. Balagopalan, A.; Tiwari, K.; Rameesha, T.; Krishnankutty, P. Manoeuvring prediction of a container ship using the numerical PMM test and experimental validation using the free running model test. *Ships Offshore Struct.* **2020**, *15*, 852–865. [CrossRef]
3. Zhang, S.; Liu, J.; Li, S.; Yasukawa, H.; Wu, Q. Impact of bow shapes on hydrodynamic derivatives due to drifting conditions. *Ocean Eng.* **2022**, *245*, 110347. [CrossRef]
4. Bhushan, S.; Yoon, H.; Stern, F. Large grid simulations of surface combatant flow at straight-ahead and static drift conditions. *Int. J. Comput. Fluid D* **2016**, *30*, 356–362. [CrossRef]
5. Yang, J.; Feng, D.K.; Liu, L.W.; Wang, X.Z.; Yao, C.B. Research on the performance of pumpjet propulsor of different scales. *J. Mar. Sci. Eng.* **2022**, *10*, 78. [CrossRef]
6. Yoon, H.; Simonsen, C.D.; Benedetti, L.; Longo, T.; Toda, Y.; Stern, F. Benchmark CFD validation data for surface combatant 5415 in PMM maneuvers—Part I: Force/moment/motion measurements. *Ocean Eng.* **2015**, *109*, 705–734. [CrossRef]
7. Carrica, P.M.; Fu, H.P.; Stern, F. Computations of self-propulsion free to sink and trim and of motions in head waves of the KRISO Container Ship (KCS) model. *Appl. Ocean Res.* **2011**, *33*, 309–320. [CrossRef]
8. Kim, D.; Tezdogan, T.; Incecik, A. A high-fidelity CFD-based model for the prediction of ship manoeuvrability in currents. *Ocean Eng.* **2022**, *256*, 111492. [CrossRef]
9. Wang, J.H.; Wan, D.C. CFD investigations of ship maneuvering in waves using naoe-FOAM-SJTU solver. *J. Mar. Sci. Appl.* **2018**, *17*, 443–458. [CrossRef]
10. Cura-Hochbaum, A. Virtual PMM tests for manoeuvring prediction. In Proceedings of the 26th ONR Symposium on Naval Hydrodynamics, Rome, Italy, 17–22 September 2006.
11. Huang, S.X.; Jiao, J.L.; Chen, C.H. Numerical prediction of ship motion and slamming load characteristics in cross wave. *J. Mar. Sci. Technol.* **2022**, *27*, 104–124. [CrossRef]
12. Sun, H.B.; Yang, S.Q.; Xu, Y.F.; Xiao, J.F. Prediction of the pitch and heave motions in regular waves of the DTMB 5415 ship using CFD and MMG. *J. Mar. Sci. Eng.* **2022**, *10*, 1358. [CrossRef]
13. Kim, I.; Kim, C.; Kim, S. Estimation of the manoeuvrability of the KVLCC2 in calm water using free running simulation based on CFD. *Int. J. Nav. Archit. Ocean Eng.* **2021**, *13*, 466–477. [CrossRef]
14. Jin, Y.T.; Yiew, L.J.; Zheng, Y.Y.; Magee, A.R.; Duffy, J.; Chai, S.H. Dynamic manoeuvres of KCS with CFD free-running computation and system-based modelling. *Ocean Eng.* **2021**, *241*, 110043. [CrossRef]
15. Zhang, W.; Liu, Y.D.; Chen, C.; He, Y.P.; Tang, Y.Y.; Sun, J.J. Research on the parametric rolling of the KCS container ship. *J. Mar. Sci. Technol.* **2023**, *28*, 675–688. [CrossRef]
16. Yu, J.W.; Feng, D.K.; Liu, L.W.; Yao, C.B.; Wang, X.Z. Assessments of propulsion models for free running surface ship turning circle simulations. *Ocean Eng.* **2022**, *250*, 110967. [CrossRef]
17. Hough, G.R.; Ordway, D.E. The generalized actuator disk. *Dev. Theor. Appl. Mech.* **1964**, *2*, 317–336.
18. Tokgoz, E. A CFD Study on the Propeller-Hull Interaction Flow in Waves Using Body-Force Distribution Model. Ph.D. Thesis, Osaka University, Osaka, Japan, 2015.
19. Feng, D.; Yu, J.; He, R.; Zhang, Z.; Wang, X. Improved body force propulsion model for ship propeller simulation. *Appl. Ocean Res.* **2020**, *104*, 102328. [CrossRef]
20. Simonsen, C.D.; Otzen, J.F.; Joncquez, S.; Stern, F. EFD and CFD for KCS heaving and pitching in regular head waves. *J. Mar. Sci. Technol.* **2013**, *18*, 435–459. [CrossRef]
21. Can, U.; Delen, C.; Bal, S. Effective wake estimation of KCS hull at full-scale by GEOSIM method based on CFD. *Ocean Eng.* **2020**, *218*, 108052. [CrossRef]
22. Kim, D.; Song, S.; Tezdogan, T. Free running CFD simulations to investigate ship manoeuvrability in waves. *Ocean Eng.* **2021**, *236*, 109567. [CrossRef]
23. Feng, D.K.; Ye, B.; Zhang, Z.G.; Wang, X.Z. Numerical simulation of the ship resistance of KCS in different water depths for model-scale and full-scale. *J. Mar. Sci. Eng.* **2020**, *8*, 745. [CrossRef]
24. Di Mascio, A.; Broglia, R.; Muscari, R. Numerical simulations of viscous flow around a fully appended hull with enforced motion. In Proceedings of the 9th Numerical Ship Hydrodynamics Conference, Ann Arbor, MI, USA, 5–8 August 2007.
25. Yuan, Z.M. Ship hydrodynamics in confined waterways. *J. Ship Res.* **2019**, *63*, 16–29. [CrossRef]
26. SIMMAN. Workshop on Verification and Validation of Ship Manoeuvring Simulation Methods. 2014. Available online: https://simman2014.dk (accessed on 8 December 2014).
27. Islam, I.; Soares, C.G. Uncertainty analysis in ship resistance prediction using OpenFOAM. *Ocean Eng.* **2019**, *191*, 105805. [CrossRef]

28. Visonneau, M.; Guilmineau, E.; Rubino, G. Local Flow around a surface combatant at various static drift conditions: The role played by turbulence closures. In Proceedings of the 33rd Symposium on Naval Hydrodynamics, Osaka, Japan, 18–23 October 2020.
29. Durbin, P.A. Some Recent Developments in Turbulence Closure Modeling. *Annu. Rev. Fluid Mech.* **2018**, *50*, 77–103. [CrossRef]
30. Zhang, Y.N.; Wang, X.Y.; Zhang, Y.N.; Liu, C.Q. Comparisons and analyses of vortex identification between Omega method and Q criterion. *J. Hydrodyn.* **2019**, *31*, 224–230. [CrossRef]

Journal of
Marine Science and Engineering

MDPI

Article

Fishing Vessel Bulbous Bow Hydrodynamics—A Numerical Reverse Design Approach

Héctor Rubén Díaz Ojeda [1,*], Sebastian Oyuela [2], Roberto Sosa [3], Alejandro Daniel Otero [4] and Francisco Pérez Arribas [5,*]

1 Instituto Universitario de Sistemas Inteligentes y Aplicaciones Numéricas en Ingeniería, Universidad de Las Palmas de Gran Canaria, 35017 Las Palmas, Spain
2 Canal de Experiencias de Arquitectura Naval (CEAN), Facultad de Ingeniería, Universidad de Buenos Aires, Buenos Aires C1063, Argentina; soyuela@fi.uba.ar
3 Canal de Experiencias de Arquitectura Naval (CEAN), Instituto de Tecnologías y Ciencias de la Ingeniería "Hilario Fernández Long" (INTECIN), Buenos Aires C1063, Argentina; rsosa@fi.uba.ar
4 Centro de Simulación Computacional para Aplicaciones Tecnológicas (CSC), Consejo Nacional de Investigaciones Científicas y Técnicas (CONICET), Buenos Aires C1425, Argentina; aotero@fi.uba.ar
5 Departamento Arquitectura Construcción y Sistemas Oceánicos y Navales, Universidad Politécnica de Madrid, Calle Ramiro de Maeztu, 7, 28040 Madrid, Spain
* Correspondence: hectorruben.diaz@ulpgc.es (H.R.D.O.); francisco.perez.arribas@upm.es (F.P.A.)

Abstract: Naval hydrodynamics typically focus on reducing ship resistance, which can be achieved by incorporating a bulbous bow. This feature is commonly used in the merchant fleet and smaller vessels, such as fishing boats, to minimize wave-making resistance. However, it is important to note that the use of a bulbous bow may not always be necessary or effective in all ship designs. In some cases, fishing ship designs may include a bulbous bow that is not optimized due to the use of procedures and methods intended for larger merchant ships or based on past experience. This study examines the effect of different bow designs, including the bulbous bow, on ship resistance in calm water, with a focus on a typical Argentinian trawler fishing vessel. The objective of this research is to assess the hydrodynamics of various designs for a particular ship by modifying its vessel lines. Firstly, the bulbous bow is removed, and then the reduction in ship resistance achieved by the bulbous bow under different load conditions and speeds is evaluated by comparing the vessel with and without the bulbous bow. The numerical analysis is performed using OpenFOAM, and the results are validated through towing tank experiments. This research indicates that the performance of the bulbous bow varies under different conditions. Therefore, it is recommended to conduct an initial study and a full evaluation of the design and operation alternatives.

Keywords: ship resistance; EFD and CFD; fishing vessels; bulbous bow; ship design

1. Introduction

In naval hydrodynamics, one of the main evaluations that should be carried out in the early stages of ship design is the ship's resistance. This is related to the operational speed of the ship that is established in the ship's contract. An accurate prediction is basic for an appropriate ship use. In addition, this resistance will influence other areas such as engine selection, stability analysis, scantling, etc., without taking into account the huge impact that might be produced on the ship's operation, profitability, and performance.

Historically, when dealing with the early stages of ship design, nonexpensive and accessible procedures are used. These procedures are regression and statistical methods based on database information from previous towing tank experiments and sea trials, such as [1] or, one of the most used, the Holtrop and Mennen method [2]. These methods are used as guides, but they are also used with certain caution due to the uncertainties presented when compared with numerical or experimental studies, such as those presented by [3], where discrepancies of up to 30% in ship resistance were found for fishing vessels. In [4], there are also large differences in resistance found between regression methods and CFD. As it was presented, these differences present a high impact on environmental inefficiencies which might lead to a wrong design or even operational problems. The presented differences come, for instance, from the fact that regression and statistical methods are unable to evaluate the skills of a naval architect drawing lines or drawing a bulbous bow in different ways.

As the design stages progress, the needs change towards more reliable and accurate methods or procedures. These methods are required since a good capture of the drawing lines effects, viscous effects, or wave breaking, among others, is essential for a good resistance evaluation. Nowadays, two procedures are the most used and recognized by the ITTC (International Towing Tank Conference). On the one hand, experimental fluid dynamic (EFD) tests carried out in towing tanks are considered the most reliable procedure. On the other hand, there are computational fluid dynamics (CFD) simulations that allow many different trials and visualizations for those that do not have access to experimental results as a reliable alternative, as is presented in the text of [5]. This last alternative is commonly used in design as a tool for testing different design concepts during the last years, commonly validated and used for ship resistance determination in calm water.

For the research community and experts in the field, the ITTC publishes Recommended Procedures in order to establish guidelines to follow for a consistent methodology both numerical and experimental in order to obtain the most accurate and reproducible results with a minimum error.

In hull design, a bulbous bow is commonly incorporated below the waterline because it helps reduce total drag and installed power, improving the economics of the design by reducing fuel consumption. Assuming the vessel maintains a consistent speed, an appropriate bulbous bow design can lead to favorable interference between the bow waves generated by the bulb and those generated by the hull alone. According to [6,7], this can lead to a decrease in resistance of between 12% and 15%. Other types of bows, such as dihedral bulbous bows, have been confirmed in [8,9] as good alternatives for ship resistance. These bulbous bows present a resistance reduction for small fishing vessels under varying loads and speeds. The study was performed with model scale ships and calm water, numerically and experimentally noting that the use of a bulbous bow should be considered once a certain speed is reached.

However, an erroneous design may cause a detrimental interaction between these two wave systems, resulting in a substantial rise in ship resistance. This topic, for instance, is important, as shown in [10]. Due to the complex interactions between the bulbous bow waves and hull, ship designs cannot be universally efficient or standardized for all speeds and operating conditions. This is because different speeds and varying load conditions must be taken into consideration for each vessel's operation. When traveling to and from fishing grounds, trawler fishing vessels experience different displacements and speeds. It is important to note that the weight of the fish capture can also impact the displacement and speed of the vessel when returning to port. These can also be affected by factors such as weather conditions and competition with other fishing vessels. In this sense, the bulbous bow is placed in a fixed ship, which makes the ship have different efficiencies according to the specific operation condition that the ship has throughout its lifetime. For this reason, sometimes, the election of a bulbous bow is mostly a compromise election that is based on multiple considerations and evaluations.

Designing a bulbous bow is a complex and controversial task that lacks a standardized procedure. Model tests have been conducted to select an optimal bulbous bow design with minimal resistance, but they can be time-consuming and expensive [11]. Numerical methods can aid the design process when model tests are not feasible. The authors of [12] conducted one of the most well-known experimental studies on bulbous bow models. The study involved testing numerous models and using statistical analysis to develop design charts that are commonly used in bulbous bow design. The study involved testing numerous models and using statistical analysis to develop design tables that are commonly used in bow design, but the tables are not suitable for smaller fishing vessels, and if used by design engineers, the resultant bow may not be optimal due to the differences between the designed bow and those used in the trials designed for larger vessels.

Previous research is based on Kracht's work [13] or [14]. Some studies use CFD analysis to optimize design variables, such as the length, breadth, frontal area, and depth of a bulbous bow. CFD has been used more often than experimental studies [15–17].

Bulbous bows have been applied in Argentinian fishing vessels. In Argentina, the fishing industry has changed since 1960. Hake fishery started in 1961, changing into the incorporation of fresh fish ships in 1970, which brought an enhancement of the capture quality and freshness. In 1976, processing ships allowing freezing were introduced, while in the 1980s, an expansion of the shrimp fishery appeared [18]. Nowadays, Argentina has started a fleet modernization using more environmentally sustainable technologies to improve crew safety. Aligning with this modernization, studies of hydrodynamic optimizations might be interesting from a consumption and operation point of view, with whether to use or not use a bulbous bow being one of the most relevant topics.

This study aims to assess the effectiveness of a bulbous bow for a fishing trawler that has been in use for over 20 years. This research began with the design and validation approximation presented in [19,20]. The ship lines used are the typical lines of an Argentinian fishing vessel which was initially designed with a bulbous bow. To achieve this, the authors decided first to evaluate the hydrodynamics of the original vessel with a bulbous bow numerically and experimentally in order to quantify its resistance and hydrodynamic behavior. The plan was to remove the bulbous bow from the original ship design, which is a unique and innovative approach to ship design. This inverse procedure has no previous references. Using CFD, this study compares the hydrodynamic behavior of the original hull with the bulbous bow with the two hulls designed without it. The results provide valuable insights into the range of applications of the original bulbous bow and two typical fishing bow designs. This will lay the foundation for future bow designs using the knowledge gained from these studies. This research provides valuable data for different operational conditions of a fishing vessel and also provides useful information on bow behavior in calm water that can be used for consideration in different cases. The results are in a field where limited research is found, which makes it important to evaluate the forces' decomposition for each case that identify the determinant factors for each case of sailing, because this will provide some useful ideas for naval designers when this topic is considered.

This paper is organized as follows: A description of the model and full-scale ship, the towing tank, and the experimental setup is presented in Section 2. The numerical set-up and procedure are described in Section 3. The results are presented in Section 4, and finally, the main conclusions are given in Section 5.

2. Towing Tank Experiments

Towing tank experiments are the traditional way of obtaining the resistance–velocity curve for a specific ship. The accuracy provided by the experiments allows one to validate the numerical procedure with the obtained results. For this reason, in this work, the total model resistance is measured by Froude numbers (Fn) between 0.1 and 0.45. With Fn defined as

$$\mathrm{Fn} = \frac{V}{\sqrt{g L_{WL}}} \tag{1}$$

with V being the model ship velocity, L_{WL} the model length on the waterline, and g the acceleration of gravity. As mentioned in the introduction, towing tank experiments are highly costly. For that reason, it was decided to only experimentally study the original case, the ship with a bulbous bow, which will serve as the CFD validation.

2.1. Model Ship Description

In the present research, the ship lines used are similar to a typical Argentine fishing vessel. This ship is built with a model scale λ = 20 using Glass-Fiber Reinforced Plastic (GFRP), taking into account the ITTC procedures for manufacturing [21]. Special attention was paid to tolerances and surface finish manufacture, as well as the location and size of the turbulence stimulation wires.

The principal characteristics of the model and full-scale ship are presented in Table 1. This table represents the case with a bow of a typical Argentinian fishing vessel. Three load conditions (LC) are presented in Table 1 for the model scale, since a full evaluation of those three load conditions will be performed. These are identified as a full load condition (LC heavy), an intermediate load condition (LC medium), and a ballast condition (LC light). The principal characteristics of the full-scale ship are presented only for the ballast condition (LC light) for reference purposes.

Table 1. Main parameters of full-scale ship and ship model.

Main Particulars	Symbol	Unit	Full Scale LC Light	Model LC Light	Model LC Medium	Model LC Heavy
Model scale	λ	[-]	-	20	20	20
Length on waterline	L_{WL}	[m]	32.68	1.634	1.662	1.641
Length, overall submerged	L_{OS}	[m]	34.795	1.670	1.740	1.740
Breadth	B	[m]	9.28	0.464	0.464	0.464
Draught	T	[m]	3.30	0.165	0.180	0.195
Displacement volume	∇	[m^3]	599.40	0.075	0.085	0.095
Wetted surface area	S	[m^2]	392.67	0.982	1.058	1.124
Block Coefficient	C_B	[m]	0.60	0.60	0.61	0.64
Midship section coefficient	C_M	[m]	0.86	0.86	0.87	0.89

The three models used for this research are shown in Figures 1–3. The original model is shown in Figure 1, and then two alternatives were tested in order to evaluate their hydrodynamics in calm water. As presented previously, there is not a general rule for ship bulbous bow design. Therefore, the designs shown in Figures 2 and 3 were based on the authors' experience with similar ship designs. It was decided to study inclined and straight bow shapes since the straight bow is common in traditional designs without a bulbous bow, and the vertical bow was considered a limit case of the previous bow shape, which also maximizes the waterline length. For designing these models, the ship lines and all the ship characteristics were kept the same, with the only change being the bulbous that was removed. Then, the bow was closed in two different ways.

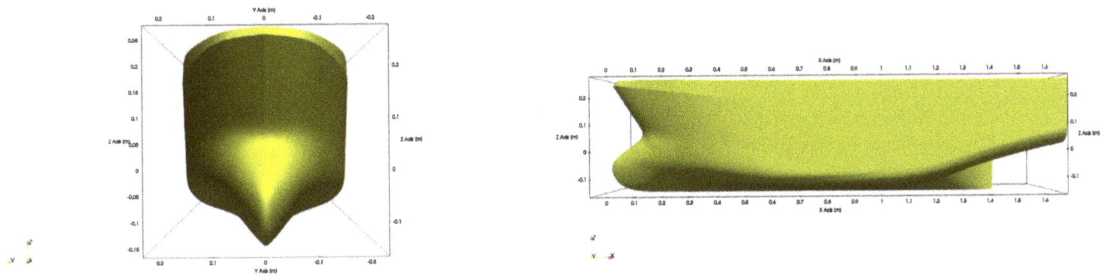

Figure 1. Bulbous bow model (BBM): front view (**left**) and side view (**right**).

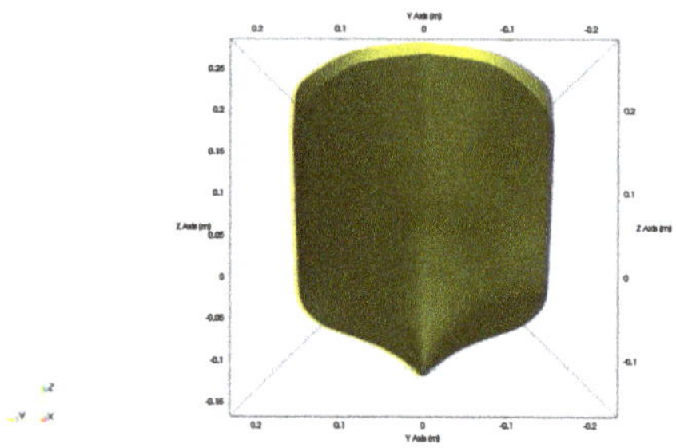
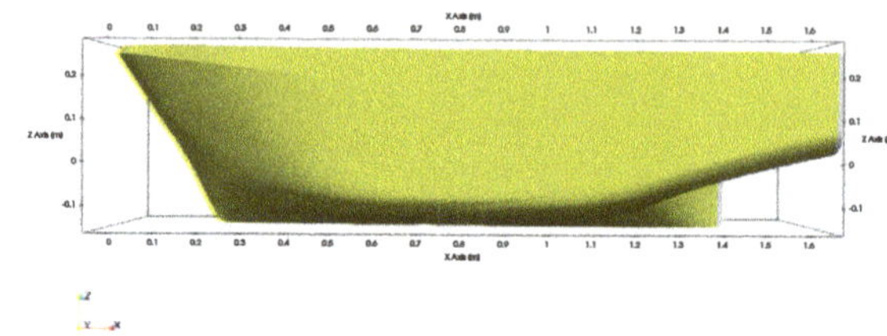

Figure 2. Inclined bow model (IBM): front view (**left**) and side view (**right**).

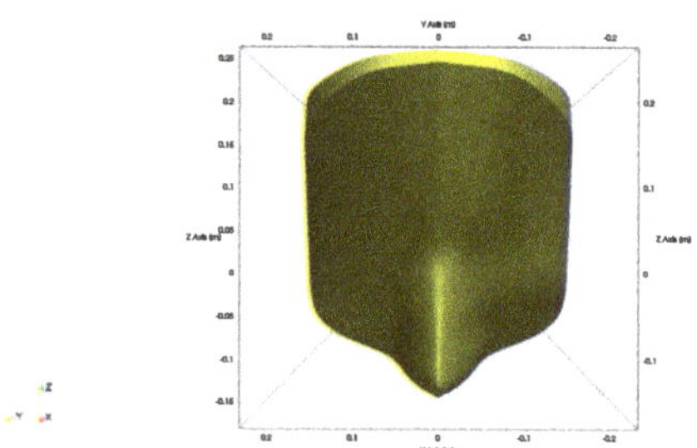
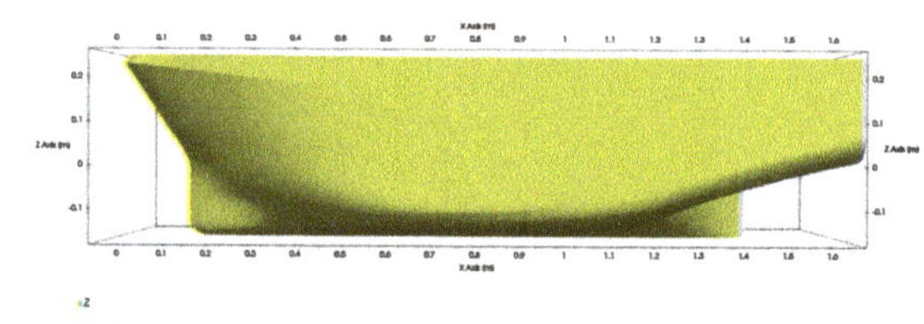

Figure 3. Straight bow model (SBM): front view (**left**) and side view (**right**).

2.2. Experimental Set-Up

The University of Buenos Aires towing tank was used to conduct experimental tests. The tank measures 72 m in length, 3.6 m in width, and has a depth of 2 m. Figure 4 shows a picture of the model. Two trip wires of 1.0 mm were employed to promote turbulence. One is located downstream from the fore perpendicular at a distance equal to five percent of L_{WL}. And another is downstream the bulb fore-end at a distance equal to one-third the bulb length.

The force measurement procedure consists of attaching the ship model, once ballasted, to a dynamometer. An R47 by Kempf & Remmers one-component force transducer capable of handling a full-scale load of ±100 N with a sensitivity of approximately ±1 mV/V was used. The model was only free to move in trim and sinkage.

The resistance measurements' uncertainty analysis considers several components, including hull geometry, towing speed, water temperature, dynamometer calibration, and repeat tests.

The standard uncertainty for each component is estimated individually and then combined using the Root Sum Square (RSS) method [22]. The best estimate of resistance is determined by adopting the mean of repeat resistance measurements. Table 2 shows the uncertainty component of the mean (with N = 4 repeat tests) for four different speeds As seen in Table 2, the dynamometer and the precision of measurement are the major sources of the uncertainties originating in the repeat tests. The expanded uncertainties presented in Table 2 correspond to a 95% confidence level.

Table 2. Combination of uncertainty in measurement for resistance at Fn = 0.14, Fn = 0.26, Fn = 0.37, and Fn = 0.45. Repeat test N = 4.

Uncertainty Components	Fn = 0.14	Fn = 0.26	Fn = 0.37	Fn = 0.45
Hull geometry	0.05	0.05	0.05	0.05
Speed	0.067	0.067	0.067	0.067
Water temp.	0.03	0.03	0.03	0.03
Dynamometer	4.73	1.04	0.35	0.12
Repeat test, Deviation [a]	5.00	3.50	1.50	1.10
Combined for single test	6.88	3.65	1.54	1.11
Repeat test, Deviation of mean	2.50	1.75	0.75	0.55
Combined for repeat mean	5.35	2.04	0.83	0.57
Expanded for repeat mean	10.70	4.08	1.66	1.14

[a] Repeat test, Deviation = (Repeat test, Deviation of mean) × $N^{1/2}$.

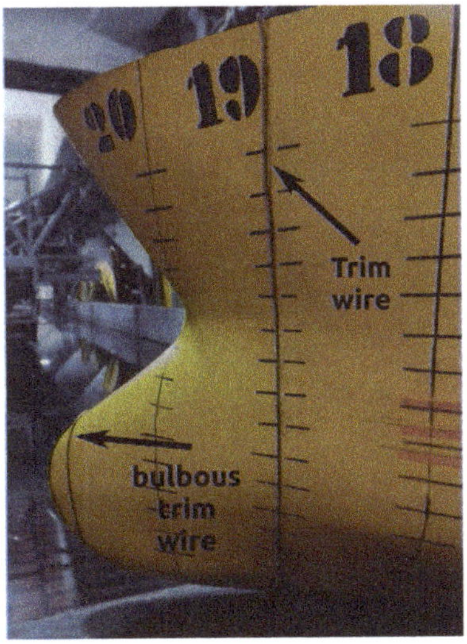

Figure 4. Top: experimental model. **Bottom**: tripwire detail.

3. CFD Description

Nowadays, computational fluid dynamics (CFD) is extended to a wide range of areas. In naval engineering, CFD is used in many topics, with resistance calculation being one of the most interesting ones [8]. OpenFOAM is a commonly used code for this task and in research. This one is an open tool based on the finite volume method that implements the equations for fluid dynamics, Navier–Stokes equations, different numerical schemes, and turbulence methods, among other equations. In this work, this tool is used for creating the mesh and solving the numerical equations. The software Rhinoceros 7 was used for creating the geometry and Paraview for visualizing the results.

We compared free heave and pitch CFD cases with constrained cases (fixed sinkage and trim) and, even for the highest Fn, we did not find differences of more than 1.5 % in the calculated resistance values. So, although trim and sinkage were monitored in the towing tank tests in this work, we do not compare ship attitudes, because the CFD calculations were carried out with fixed sinkage and trim in order to reduce the computational effort.

3.1. Numerical Equations

When the ship resistance is computed numerically, the equations to be used must imply a Newtonian, turbulent, incompressible, and viscous flow. The equations of the fluid used are then the momentum and mass equations of Navier–Stokes Equations (2) and (3).

$$\nabla \cdot \overline{U_f} = 0 \tag{2}$$

$$\frac{\partial(\rho_f \overline{U_f})}{\partial t} + \nabla(\rho_f \overline{U_f U_f}) = \rho_f g - \nabla p + \mu_f \nabla^2 \overline{U_f} \tag{3}$$

Since the turbulence cannot be solved in all the scales due to the computational resources, a modelization is needed. The Reynolds-Averaged Navier–Stokes (RANS) method is used with the SST k-ω model that provides good results, such as the ones presented in [8]. This model is based on two equations, where k is the turbulent kinetic energy, and ω is the specific dissipation rate.

On the other hand, these kinds of simulations imply the presence of two fluids, air and water. Therefore, an equation that can handle the two fluids is required, that is, the Equation (4) of the Volume of Fluid (VOF) method. This method distinguishes between 0 and 1 for a scalar α, indicating the proportion of fluid contained in the different domain cells.

$$\frac{\partial(\alpha)}{\partial t} + \overline{U_f} \cdot \nabla\alpha = 0 \tag{4}$$

3.2. Numerical Domain

A 3D domain was selected with the dimensions presented in Figure 5 after conducting some simulations verifying that no blockage effect was present. Therefore, the different domain walls are chosen far enough from the model to avoid any effect produced by those walls.

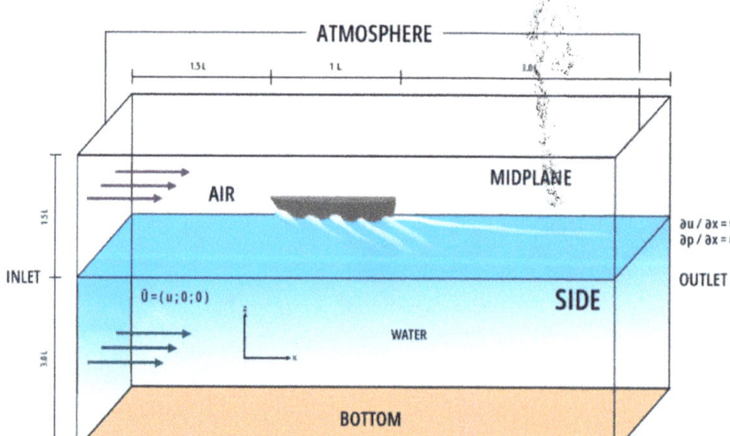

Figure 5. Problem domain.

3.3. Numerical Boundary Conditions

The numerical boundary conditions are presented in Table 3. It should be noted that only half of the ship is computed in order to optimize computational resources. This implies the use of the *symmetry* boundary condition.

Table 3. Numerical boundary conditions.

Patch	U	p	k	ω	μ Turbulent
Inlet	uniform	fixed flux	fixed value	fixed value	fixed value
Outlet	zero gradient	zero gradient	zero gradient	zero gradient	zero gradient
Atmosphere	zero gradient	zero gradient	zero gradient	zero gradient	zero gradient
Bottom	symmetry	symmetry	symmetry	symmetry	symmetry
Midpl/Side	symmetry	symmetry	symmetry	symmetry	symmetry
Ship	no slip	fixed flux	wall function	wall function	wall function

3.4. Numerical Mesh

The mesh used for the numerical analysis was created using blockMesh and snappyHexMesh, both of these tools are provided by OpenFOAM. The mesh is structured with 92% hexahedrons and divided into uniform and graded zones. Figures 6 and 7 present the final mesh, which has a horizontal expansion ratio of 1.3 and 1.2 in the 'x' and 'y' axes, respectively, and a vertical expansion ratio of 1.2, all towards the center zone. The vessel is contained within a block composed of hexahedrons with an aspect ratio of one, except for the interface zone, where the cells were reduced by 25% of their height to accurately capture wave height. The mesh is designed to ensure that the y+ value remains below 5 in all cases.

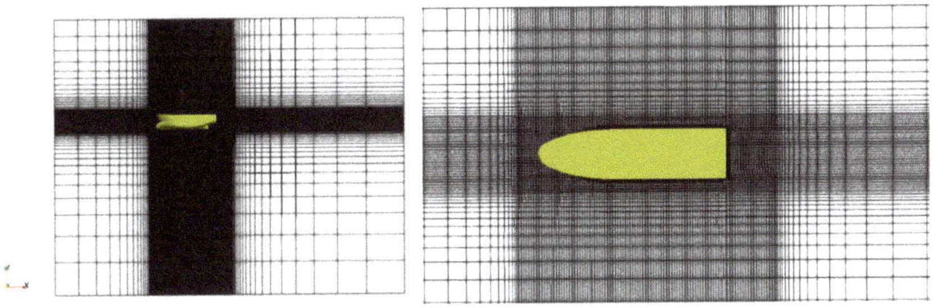

Figure 6. General mesh view. Mesh 4.

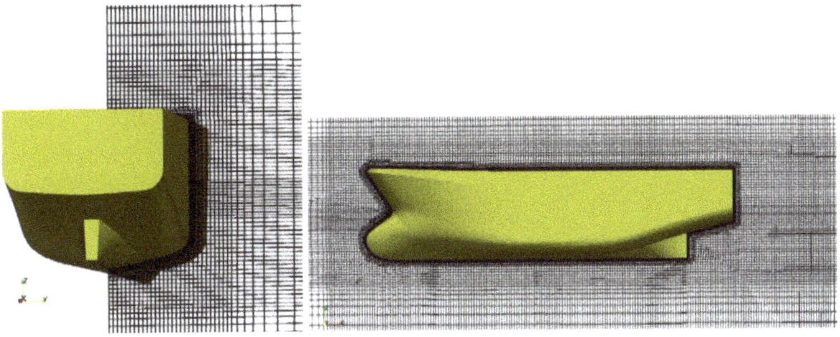

Figure 7. Mesh zoom for different orientations. Mesh 4.

A mesh validation was conducted in order to see that the results provided by the numerical procedure were obtained with a minimum error. Therefore, a convergence study was conducted using the heavy LC and Fn 0.45. The different meshes tested, with the cell size in the center box refined systematically by a ratio of 1.2, presented an oscillatory convergence, shown in Table 4. All the meshes presented show an error that might be acceptable since it is lower than 4%. Finally, the simulations were conducted using a first-order pseudotransient time scheme that employs a specific local time step. Therefore, there is no need for further numerical sensitivity analysis. The numerical set-up is validated.

Table 4. Mesh validation.

Case	Mesh Cells	Fn	F [N]	Error (%)
EFD	-	0.45	40.27	-
Mesh 1	633.938	0.45	40.26	0.02
Mesh 2	903.744	0.45	41.76	3.70
Mesh 3	1.213.419	0.45	40.34	0.17
Mesh 4	2.485166	0.45	41.16	2.21

4. Results

4.1. CFD Validation

Before analyzing the differences between the use of the bulbous bow and the two different cases without a bow, the comparison between CFD and EFD drag resistance for the case with bulbous was performed. Therefore, the drag forces obtained for the case with bulbous bow at different speeds and different load conditions were compared. In Figure 8, the drag forces are presented in a dimensionless way using the coefficient C_T, which is defined as

$$C_T = \frac{Drag}{\frac{1}{2}\rho S v^2} \tag{5}$$

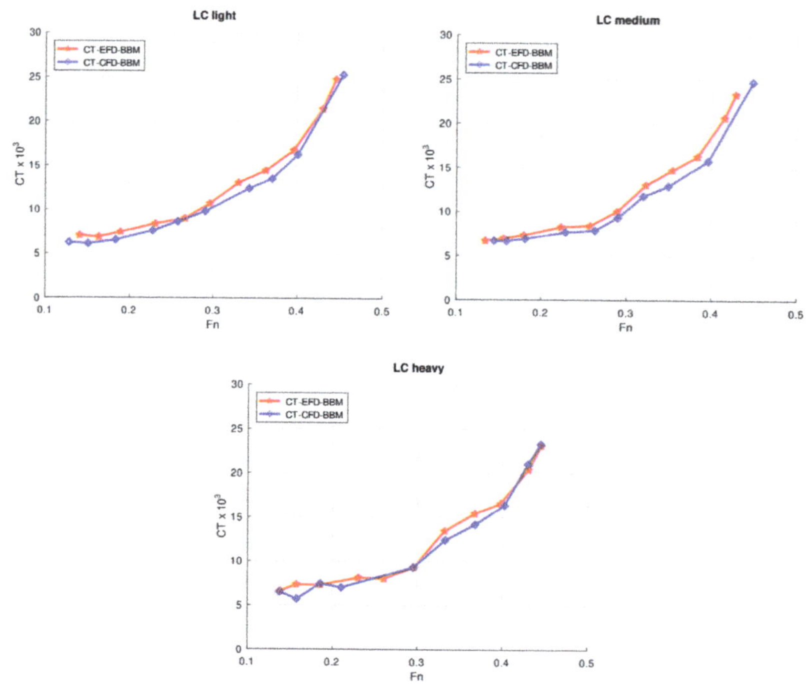

Figure 8. C_T vs Fn for LC light, LC medium, and LC heavy.

The numerical results present a good agreement with the experimental ones for all the speeds and load conditions. In addition, the wave pattern generated by the ship model is also compared in Figure 9, showing an apparently similar wake. According to these results, the numerical procedure for the wake evaluation and the drag forces estimation can be validated.

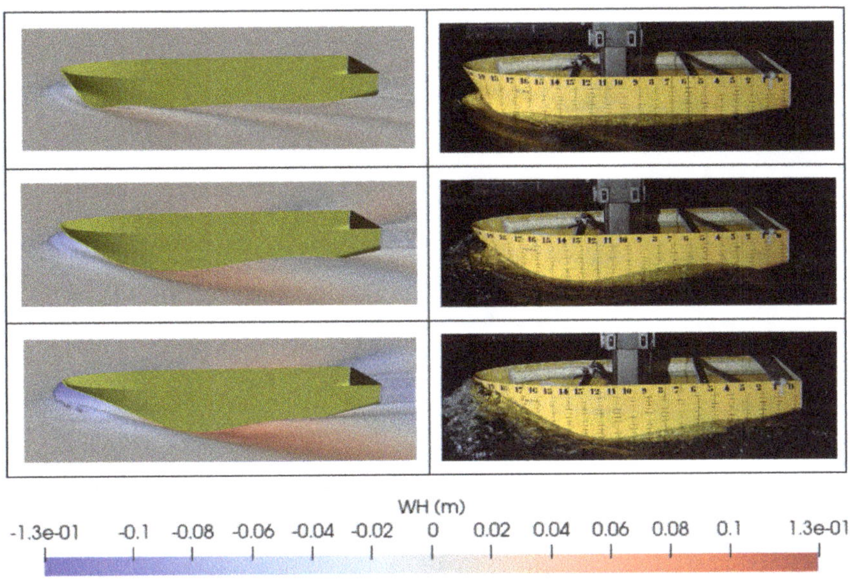

Figure 9. Wave pattern for different Fn and draught = LC heavy. Left: CFD. Right: EFD. First row: Fn = 0.26, second row: Fn = 0.37, third row: Fn = 0.45.

The evaluation of different drafts is because fishing ships sail under different situations depending on their activities and load. For instance, when the ship is fishing, it leaves the port with an empty load, and when it is coming back to port it has fish inside it. Also, the speed is determined by the sea conditions, weather, fishing restrictions, etc. In this work, the selected ship is designed for an operation of Fn 0.35. Lower velocities will be used in port and other fishing activities or even when that speed is unreachable due to weather conditions. The results presented in Figures 8 and 9 show the typical behavior of these kinds of hull ships. For Fn < 0.3, the major component of the resistance is the viscous resistance growing C_T linearly. This speed is the critical speed, since for higher velocities, the C_T increases in a exponential way. The main reason for this behavior is that at low speeds, the wave resistance has almost no influence on the total resistance, while becoming predominant at higher speeds.

Once the CFD has been validated, its use makes it possible to obtain data that are difficult or very complicated to measure experimentally, for instance, the pressure distribution around the hull. In Figure 10, the dimensionless pressure C_P around the hull for the BBM case is shown. This dimensionless parameter is defined as follows:

$$C_P = \frac{p - p_\infty}{\frac{1}{2}\rho_\infty V_\infty^2} \tag{6}$$

where $p - p_\infty$ is the dynamic pressure.

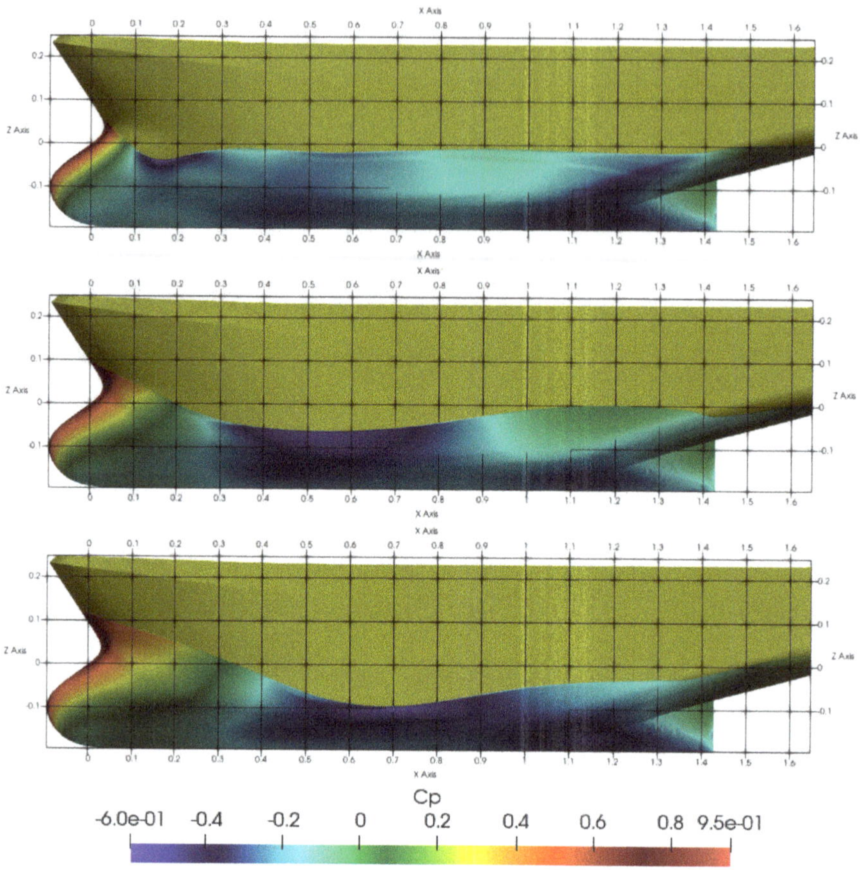

Figure 10. Cp distribution for Fn = 0.26, Fn = 0.37, and Fn = 0.45, T = LC Heavy.

When studying wave generation, it is essential to consider the distribution of pressure over the hull. The pressure distribution over the hull indicates the size and shape of the waves. In addition, this pressure distribution may provide insight into the distribution of forces around the hull, highlighting areas that are more susceptible to higher forces. In Figures 10–12, different pressure zones are presented for the three cases of study. It is noticeable that these contours are presented for different Fn numbers and load conditions. The pressure distribution and wave formation change with the Fn number, as shown in Figure 10 for the bulbous bow case where the wake formation is bigger for higher Fn numbers. Additionally, the maximum pressure distribution expands to a larger area as Fn increases. Figures 11 and 12 show that hulls without a bulbous bow have a very similar pressure distribution and wake formation. In contrast, hulls with a bulbous bow have a more linear wake formation and do not present a hollow in the shoulders of the ship, as is the case with hulls without a bulbous bow.

Based on these contours, it can be concluded that for optimal efficiency and reduced drag, a smooth pressure distribution over the hull is necessary. An even pressure distribution reduces disturbances in the water, resulting in less energy loss due to wave generation and therefore less resistance. Conversely, an uneven pressure distribution may produce significant disturbances in the water, resulting in more waves and greater resistance. The distributions are highly dependent on various factors, and the behavior changes once the bulbous bow is introduced.

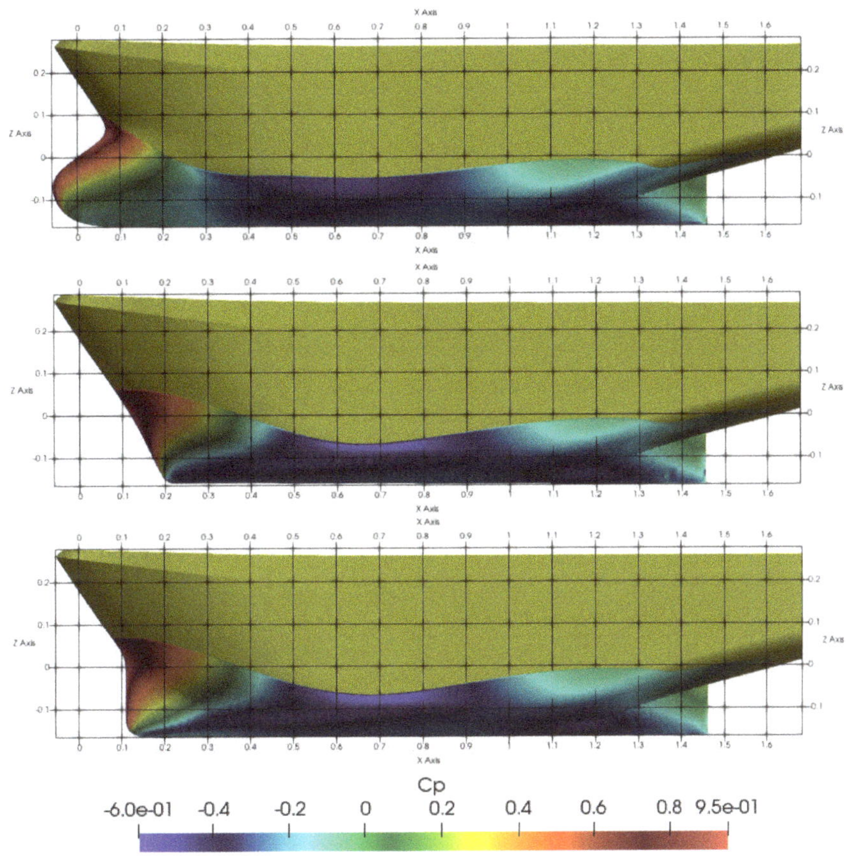

Figure 11. Cp distribution for the different bulbous bows for Fn = 0.37, T = LC light.

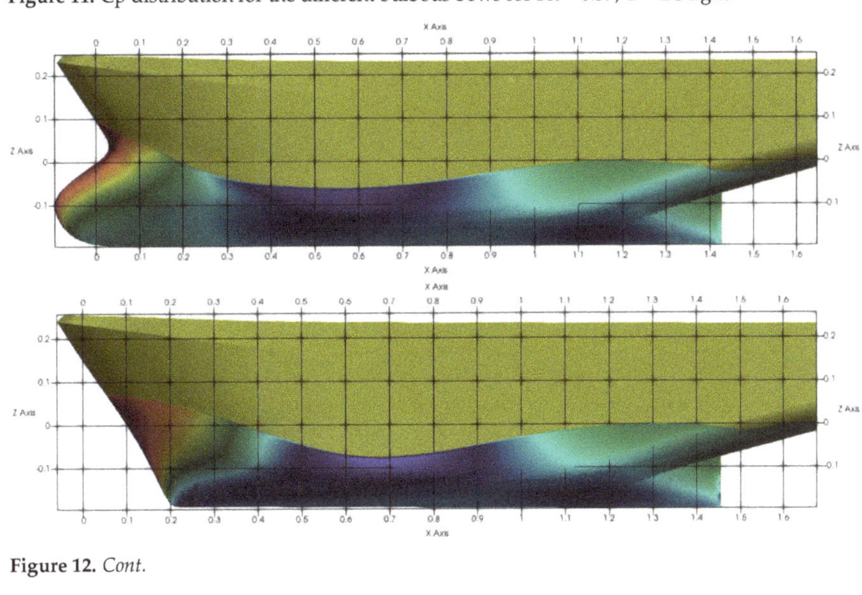

Figure 12. *Cont.*

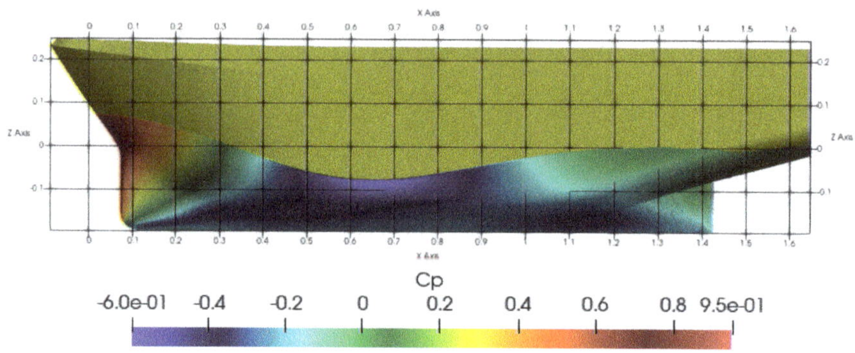

Figure 12. Cp distribution for the different bulbous bows for Fn = 0.37, T = LC heavy.

As fishing vessels operate at a wide range of speeds and loads, i.e., when fishing, leaving port, or returning from the ground with fish, the hull shape and resistance calculation should be carefully evaluated, as one hull shape may work for one load condition and speed range but not for the rest. In this situation, it must be the ship owner who decides the optimum range of operation.

4.2. Bows Designs Analysis

After evaluating the ship model resistance numerically and experimentally for the case with a bulbous bow, the model is redesigned removing the bulbous bow. Two different configurations for this kind of vessel will be implemented and evaluated. Those cases will bring more details about the hydrodynamic behavior, providing valuable information for this kind of ship with limited research. The evaluation between the two new designs without bulbous bows and the original one with bulbous bows will be performed only numerically, keeping the same mesh, range of velocities, and load conditions. In order to better appreciate the results and the differences between hulls, the resistance will be split in two, for moderate speeds, $Fn < 0.3$ in Figure 13, and for higher speeds, $Fn > 0.3$ in Figure 14.

In Figure 13 and for $Fn < 0.3$, we can appreciate that there is not a clear tendency or any hull that works better for the three loads conditions at this range of speed. For LC light, the hull BBM produces less drag until Fn 0.15. From this speed, there is a linear growth of resistance, and it is the worst hull in terms of resistance for Fn between 0.2 and 0.25. For this LC, the hull SBM always offers less resistance than the IBM hull. Indeed, it is the most efficient hull for the range of speed 0.2–0.25. The IBM model seems to be the less efficient hull for all the range of speed.

When LC medium is evaluated, in Figure 13, we can appreciate that the hull BBM is less efficient for velocities lower than Fn < 0.25. On the contrary, and in comparison with the LC light, the IBM hull is the most efficient. However, this hull seems to be efficient for a speed lower than Fn < 0.25, since from this velocity onwards, the resistance grows exponentially and it becomes larger than the BBM case.

For the LC heavy, there is not a clear tendency that allows one to determine which hull is better. Only when the speed is higher than 0.25, the BBM becomes a more efficient hull. For the different load conditions and hull shapes, we can appreciate that Fn 0.25 is a critical value where the ship resistance curve behavior changes.

In Figure 14, we analyze the resistance produced by the three hulls for higher speeds. In this range of speeds, the wave resistance is the most relevant part of the total resistance. For the three load conditions, we can appreciate an exponential growth in the total resistance for all of the hulls. For the three load conditions, the hull BBM performs better for the whole range of speed. The IBM and the SBM have almost identical drag for the load conditions of medium and heavy. For the LC light, the SBM hull has less drag than the IBM for Froude numbers lower than 0.4, demonstrating almost the same for higher speeds. It is remarkable that for Fn > 0.35, the increment of resistance makes the bulbous bow the best.

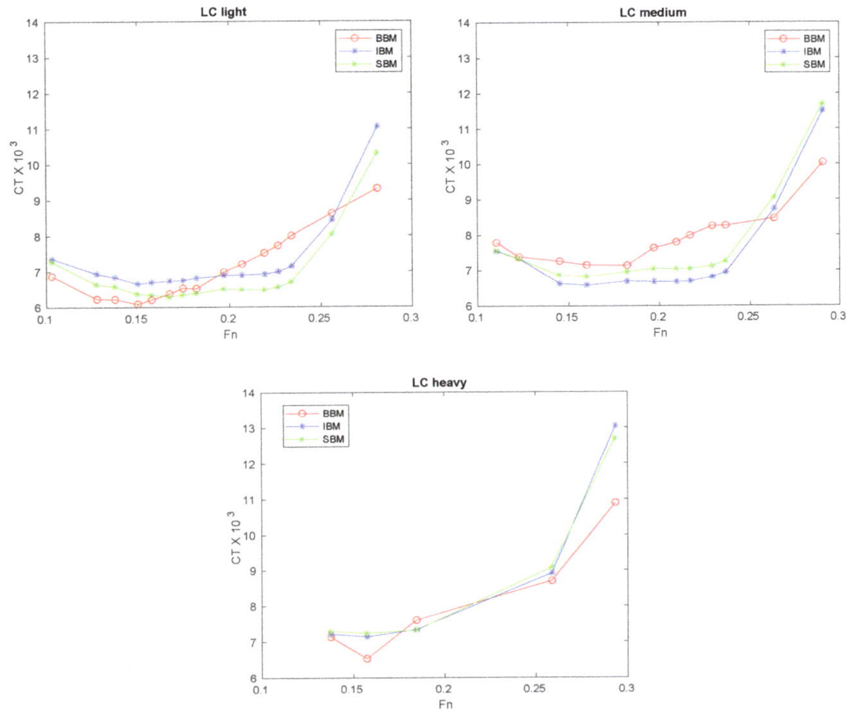

Figure 13. C_T vs. Fn comparison for the three hulls and for Fn < 0.3.

Based on the previous results, it is clear that the ship owner must decide the ship shape based on the speed and load condition that is more frequently used. However, the ship with a bulbous bow seems to be the best option at least for calm water. From the numerical results, we can also obtain a force decomposition, in viscous and pressure force, that will indicate the dominance of one or other forces for each velocity. This will help in the design process because it will clearly identify the area for shape optimization.

For instance, Figure 15 presents the total force decomposed in viscous (FV) and pressure (FP) force for LC light and the original BBM ship model. From this graph, we can appreciate that the viscous resistance grows linear, dominating for Fn < 0.25. On the other side, pressure resistance grows linear until Fn = 0.25. From this point, the pressure resistance becomes dominant, and it grows almost exponentially. It is clear then that for this ship, the critical point, as mentioned before, is Fn = 0.25. From this point onwards, the pressure resistance that is related to wave making plays a significant role.

Figure 16 presents the viscous force for the three hulls and the three load conditions. From these results, we can conclude that the viscous forces grow linear for all the cases. The BBM model is expected to have a higher viscous resistance due to its larger wetted surface area. This effect may be slightly noticeable for Fn values greater than 0.25. For lower Fn values, the forces are too small to draw any conclusions.

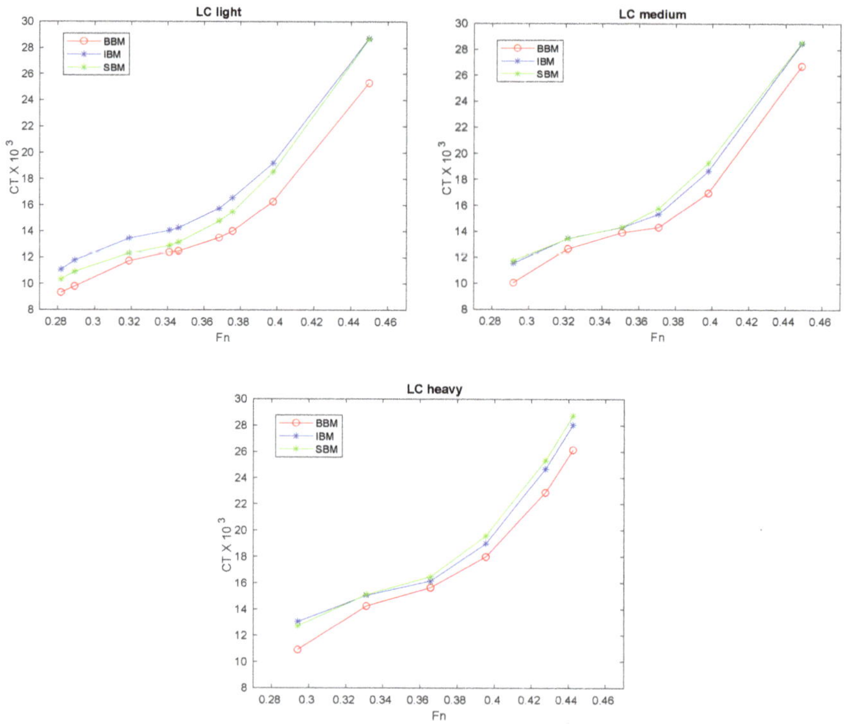

Figure 14. C_T vs. Fn comparison for the three hulls and for Fn > 0.3.

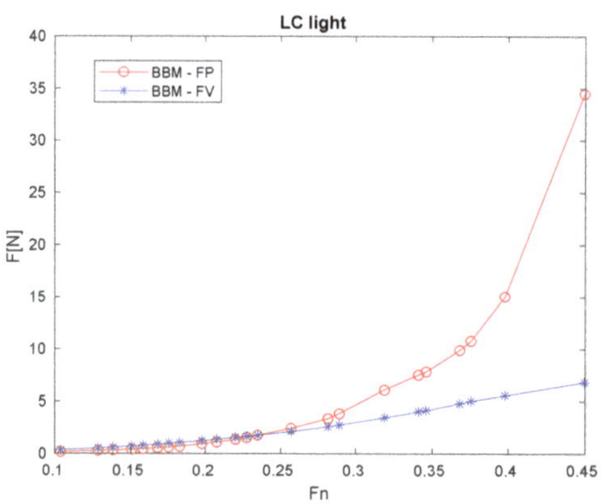

Figure 15. Pressure (FP) and viscous (FV) force for LC light and BBM.

Figure 17 presents the pressure forces for the three load conditions and ship hulls. The behavior of this force is similar for the different conditions. There is a linear growth until Fn = 0.25. Before this point, the pressure resistance was relatively low in comparison with the produced for higher Fn numbers. From Fn = 0.25 onwards, the growth in pressure resistance is almost exponential, becoming part of the resistance and the dominant one when compared with Figure 16. Besides, we can appreciate that the difference in total resistance between hulls is mainly due to the pressure resistance.

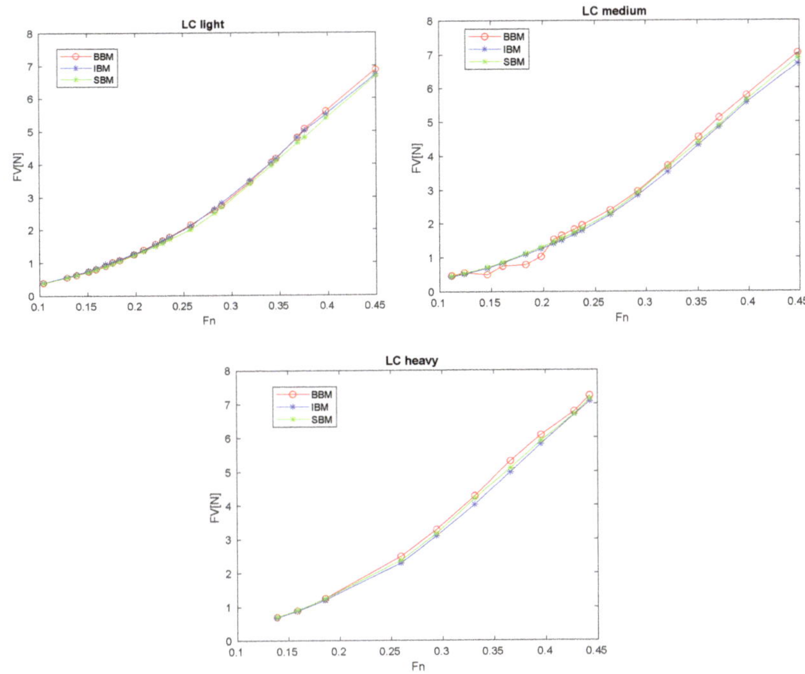

Figure 16. Viscous force comparison for the three hulls.

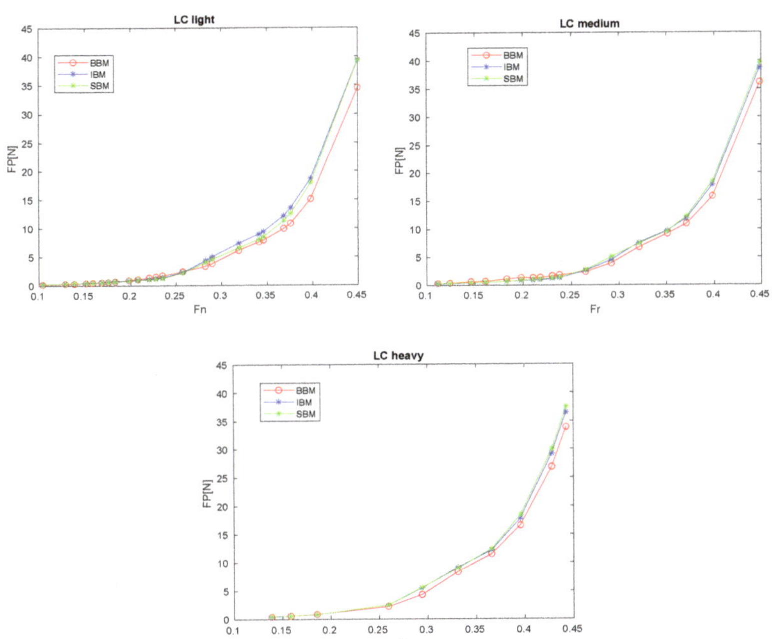

Figure 17. Pressure force comparison for the three hulls.

5. Conclusions

The conducted research tries to give more details about the hydrodynamic influence of three different bow configurations in fishing vessels. These configurations are the typical ones that can be found in this type of vessel. The presented studies were conducted at the model scale, either experimentally or numerically. The main conclusions are as follows:

- There is a critical point at Fn = 0.25 from where the resistance grows exponentially due to the high influence of the pressure resistance that is linked with wave resistance.
- For low velocities, the hulls with no bulbous bow in general show good efficiency until Fn 0.25, from where the hulls do not seem to have the correct shape.
- For the different load conditions and speeds after Fn = 0.25, the case of BBM shows higher efficiency. It is clear then that the bulbous bow works better, reducing pressure resistance.
- The viscous resistance grows linearly for all the cases. This resistance becomes more significant in the total resistance for Fn < 0.25, although the BBM does not seem to change the total amount of viscous force that much.
- The pressure resistance is predominant from Fn 0.25 onwards. This resistance is the main source of resistance, which means ship designers should focus on reducing this component of resistance.
- The pressure distribution around the hull demonstrates how it changes with the Fn number, revealing a larger hollow wake as Fn increases.
- The pressure distribution is practically the same for the two cases without a bulbous bow. When compared with a bulbous bow hull, it is noticeable that the hollows in the ship's shoulders are less pronounced, sometimes resulting in a nearly linear wake generation around the hull.

Finally, this research brings new information that is the first step to fully understanding the hydrodynamics of these kinds of ships and their design. Future research should focus on sailing in waves, which will guide us to decide if the conclusions for calm water are also supported in waves.

Author Contributions: Validation, S.O., R.S. and A.D.O.; Formal analysis, F.P.A.; Investigation, H.R.D.O. and F.P.A. All authors have read and agreed to the published version of the manuscript.

Funding: The Ministry of Science, Technology and Innovation of the Argentine Republic (MINCyT—Proyecto de Investigación y Desarrollo Tecnológico de la Iniciativa Pampa Azul—B2), the Universidad de Buenos Aires, Facultad de Ingeniería (Grant Peruilh).

Data Availability Statement: Data are contained within the article.

Acknowledgments: This work was supported by the Spanish Ministry of Ciencia e Innovación through grant PID2022-140481OB-I00. We thank the Argentine shipyard TecnoPesca Argentina for providing the fishing vessel lines.

Conflicts of Interest: The authors declare no conflicts of interest.

Nomenclature

List of variables:

L	Ship total length (m)
L_{WL}	Ship waterline length (m)
B	Ship beam (m)
T	Ship draught (m)
∇	Volumetric displacement of ship (m^3)
S	Wetted surface area (m^2)
C_B	Block coefficient
λ	Scale
C_M	Midship section coefficient
Fn	Froude number (Fn = $V\sqrt{gL_{WL}}$)

g	gravitational constant (m/s^2)
F	Total drag resistance (N)
FV	Viscous resistance (N)
FP	Pressure resistance (N)
ρ_∞	Water density
C_P	Pressure coefficient
C_T	Drag force coefficient coefficient
WH	Wave height (m)

References

1. Guldhammer, H.E.; Harvald, S. *SHIP RESISTANCE—Effect of Form and Principal Dimensions*; Danish Technical Press: Copenhagen, Danmark, 1965.
2. Holtrop, J. A statistical re-analysis of resistance and propulsion data. *Int. Shipbuild. Prog.* **1984**, *31*, 272–276.
3. Bilec, M.; Obreja, C.D. Ship resistance and powering prediction of a fishing vessel. *IOP Conf. Ser. Mater. Sci. Eng.* **2020**, *916*, 012011. [CrossRef]
4. Matínez-López, A.; Díaz Ojeda, H.R.; Míguez González, M.; Marrero, Á. Environmental Inefficiencies of Short Sea Shipping Vessels by Optimization Processes Based on Resistance Prediction Methods. *J. Mar. Sci. Eng.* **2022**, *10*, 1457. [CrossRef]
5. Larsson, L.; Raven, H.; Paulling, J. *Ship Resistance and Flow*; Principles of Naval Architecture; Society of Naval Architects and Marine Engineers: Alexandria, VR, USA, 2010.
6. Ferreiro, L.D. The social history of the bulbous bow. *Technol. Cult.* **2011**, *52*, 335–359. [CrossRef]
7. Barrass, B. *Ship Design and Performance for Masters and Mates*; Elsevier: Amsterdam, The Netherlands, 2004.
8. Díaz-Ojeda, H.; Pérez-Arribas, F.; Turnock, S.R. The influence of dihedral bulbous bows on the resistance of small fishing vessels: A numerical study. *Ocean. Eng.* **2023**, *281*, 114661. [CrossRef]
9. Pérez-Arribas, F.; Silva-Campillo, A.; Díaz-Ojeda, H.R. Design of Dihedral Bows: A New Type of Developable Added Bulbous Bows—Experimental Results. *J. Mar. Sci. Eng.* **2022**, *10*, 1691. [CrossRef]
10. Ren, Z.; Wang, J.; Wan, D. Numerical Simulations of Ship Bow and Shoulder Wave Breaking in Different Advancing Speeds. In Proceedings of the International Conference on Offshore Mechanics and Arctic Engineering, Madrid, Spain, 17–22 June 2018; American Society of Mechanical Engineers: New York City, NY, USA, 2018; p. V07AT06A010. [CrossRef]
11. Chakraborty, S. What's Importance of Bulbous Bow of Ships. *Marine Insight*. 21 May 2017. Available online: https://www.marineinsight.com/naval-architecture/why-do-ships-have-bulbous-bow/ (accessed on 1 February 2024).
12. Kracht, A. Design of bulbous bows. In Proceedings of the SNAME Annual Meeting, New York, NY, USA, 16–18 November 1978; Volume 86.
13. Hoyle, J.W.; Cheng, B.H.; Hays, B. *A Bulbous Bow Design Methodology for High-Speed Ships*; Society of Naval Architects and Marine Engineers-Transactions; Society of Naval Architects and Marine Engineers: Alexandria, VR, USA, 1986; Volume 94.
14. Hagen, G.E.A. *A Guide for Integrating Bow Bulb Selection and Design into the U.S. Navy's Surface Ship Hull Form Development Process, Naval Sea Systems Command*; Technical Note No. 885-55W-TN0001; NAVSEA: Washington, DC, USA, 1983.
15. Peri, D.; Rossetti, M.; Campana, E.F. Design optimization of ship hulls via CFD techniques. *J. Ship Res.* **2001**, *45*, 140–149. [CrossRef]
16. Sharma, R.; Sha, O.P. Practical hydrodynamic design of bulbous bows for ships. *Nav. Eng. J.* **2005**, *117*, 57–76. [CrossRef]
17. Watle, A. Flexible Bulbous Bow Design-A Hydrodynamic Study. Master's Thesis, NTNU, Trondheim, Norway, 2017.
18. Bertolotti, M.; Verazay, G.; Pagani, A.; Errazti, E.; Buono, J. Flota pesquera Argentina. Evolucion 1960–1998 con una actualizacion al 2000. In *El Mar Argentino y Sus Recursos Pesqueros. Evolución de la Flota Pesquera, Artes de Pesca y Dispositivos Selectivos*; INIDEP: Mar del Plata, Argentina, 2001.
19. Oyuela, S.; Sosa, R.; Otero, A.D.; Arribas, F.P.; Diaz-Ojeda, H.R. An experimental and numerical hydrodynamic study on the Argentinian fishing vessels. *IOP Conf. Ser. Mater. Sci. Eng.* **2023**, *1288*, 012047. [CrossRef]
20. Arribas, F.P.; Oyuela, S.; Otero, A.D.; Sosa, R.; Diaz-Ojeda, H.R. The use of Ctrl+Z in ship design: Removing a bulbous bow. *IOP Conf. Ser. Mater. Sci. Eng.* **2023**, *1288*, 012044. [CrossRef]
21. Specialist Committee: Procedures for Resistance, Propulsion and Propeller Open Water Tests of 23rd ITTC 2002. *ITTC—Recommended Procedures and Guidelines: Model Manufacture, Ship Models*; ITTC: Nairobi, Kenya, 2002
22. 28th ITTC Quality Systems Group. *ITTC—General Guidelines for Uncertainty Analisys in Resistance Tests*; ITTC: Nairobi, Kenya, 2017.

Journal of
Marine Science and Engineering

MDPI

Article

Parametric Design of a New Float-Type Wave Energy Generator and Numerical Simulation of Its Hydrodynamic Performance

Yu Zhang * and Dongqin Li *

School of Naval Architecture and Ocean Engineering, Jiangsu University of Science and Technology, Zhenjiang 212100, China
* Correspondence: zhangyu_1003@163.com (Y.Z.); mandy_ldq@163.com (D.L.)

Abstract: A novel float-type device for wave energy power generation, designed specifically for offshore wave environments, is introduced as an innovative technology in wave energy utilization. Herein, we present the design concept, structural composition, and energy conversion process of the device, and conduct mathematical modeling and theoretical research on its kinematic and dynamic characteristics. At the same time, we use a numerical wave pool based on the STAR-CCM+ boundary wave making method and damping dissipation method to analyze the motion response and output power of the wave energy generator in a five-order Stokes wave environment within one wave cycle and the entire operating cycle. Finally, in order to develop the best design strategy, we study the effect of changing the structural parameters of the power generation device on the hydrodynamic performance of the device.

Keywords: wave power generation; oscillating float; computational fluid dynamics; hydrodynamic performance

1. Introduction

The growth of the global economy and the excessive usage of fossil fuels have severely harmed and contaminated our ecosystem. Traditional energy sources are also becoming harder to obtain at the same time. As a result of this tendency, people are looking into new clean energy solutions to displace fossil fuels and numerous nations are gradually turning their focus to the large sector of ocean energy. Since waves have a tremendous amount of energy compared with wind and solar energy, developing and using wave energy offers several benefits, making coastal nations the center of interest for this technology today [1–3].

The design and analysis of hydrodynamic performance for wave energy power production devices represent a crucial field of study within marine renewable energy. Various types of power generation devices have been developed in this process. Due to its adaptable placement in the marine environment, oscillating float-type wave power generators [4] are one of the hotspots in the field of wave energy utilization and have high power generation efficiency.

Stephen Salter [5] created Salter's duck as a wave energy device. During the energy crisis, he was the first person to use the idea of oscillating floats to harness wave energy. This iconic tool is a springboard for further research on wave energy fields. In 2003, Korde [6] used a linked wave energy generation system with regular waves to numerically calculate the system's capture width ratio. In 2005, Eriksson et al. [7] performed some comparisons of the motion problem of a cylindrical float-type wave energy generator in regular and irregular waves, and they replaced the Power Take Off (PTO) mechanism in a wave energy power generator with linear springs and linear dampers. Babarit et al. [8] addressed the issue of energy capture efficiency in 2006 and offered two control approaches for the PTO system. The proposed method can adjust the device's energy capture efficiency in real time. In 2010, Griet et al. [9] used the Computational Fluid Dynamics (CFD) approach to simulate the absorbed power, impact velocity, and corresponding forces of the float under varied slamming loads. They discovered that, although it is also related to power loss, the risk of the float destabilizing is highly dependent on the draught depth of the float and the ocean environment. However, the likelihood of the risk occurring can be significantly decreased by adjusting the relevant control parameters of the device. The hydrodynamic performance of a pendulous float in shallow nearshore water compared with a float in an offshore surge was examined by Negri et al. in 2018 [10]. They discovered that the efficiency of a wave energy system's power generation dramatically rises with increasing sea depth. Eliseo Marchesi [11] devised a plan for a two–float PTO system in 2020, and while PTO damping in irregular waves was evaluated, the numerical simulation findings and testing results were in accord. The same year, Christophe Vogel [12] researched the effects of wave attenuation on the hydrodynamic performance of an asymmetric float wave energy generator and a hybrid energy collection system.

It is crucial to acknowledge that the energy industry is highly intricate and encompasses the administration of various interrelated elements. As an illustration, apart from the aim of generating energy, it is essential to account for the endurance of the machinery, which needs to withstand the challenging oceanic setting. While studies have advanced and yielded outcomes, it remains conceivable to aspire for greater achievements [13]. To meet our energy requirements and diminish our reliance on non-renewable resources, our community is increasingly integrating more eco-friendly energy approaches. The utilization of the abundant ocean energy requires the creation of innovative technologies that will lower global CO_2 emissions from the burning of fossil fuels and turn ocean waves and tidal energy into a sustainable and economical energy source [14].

Numerical studies, simulation analyses, and experimental testing are common analytical approaches for oscillating float-type wave energy devices. Many scholars have conducted research in this field with the goal of improving the stability and reliability of the devices for power generation [15]. Based on the existing oscillating float wave power generation device, this paper intends to further simplify the structure and energy conversion process, with the design of a new wave power generation device. We use theoretical analysis and numerical simulation to study the hydrodynamic performance of the device and the captured energy characteristics. The final analysis conclusion drawn from the above aims to provide a certain technical basis and reference for future research.

2. Mechanical Theory

Accurately and scientifically predicting the energy conversion performance of power generation equipment is paramount in the realm of wave energy generation technology utilization. The mechanical behavior of these devices in the ocean is closely related to the efficiency of their energy conversion and structural reliability. As a result, the development and application of mechanical formulas will provide a critical theoretical foundation for the design of diverse wave energy devices.

2.1. Basic Control Equations

Fluid motion should adhere to the physical laws, such as the law of mass conservation, the law of momentum conservation, and the law of energy conservation. The law of conservation of energy is not taken into account when analyzing the hydrodynamic performance of a new oscillating float wave energy generator in an incompressible viscous fluid, and only the conservation law is expressed by using the equations for mass and momentum conservation [16].

Equation (1) is used in both steady flow and unsteady flow for incompressible fluids:

$$\frac{\partial u}{\partial x} + \frac{\partial v}{\partial y} = 0. \tag{1}$$

The Navier–Stokes equations for incompressible fluids can be reduced using the continuity Equation (2):

$$\begin{aligned}&\frac{\partial(\rho\cdot u)}{\partial t}+\frac{\partial(\rho\cdot u^2)}{\partial x}+\frac{\partial(\rho\cdot u\cdot v)}{\partial y}+\frac{\partial(\rho\cdot u\cdot w)}{\partial z}\\&=F_x-\frac{\partial p}{\partial x}+\frac{\partial}{\partial x}\left(2\mu\frac{\partial u}{\partial x}\right)+\frac{\partial}{\partial y}\left[\mu\left(\frac{\partial v}{\partial x}+\frac{\partial u}{\partial y}\right)\right]+\frac{\partial}{\partial z}\left[\mu\left(\frac{\partial u}{\partial z}+\frac{\partial w}{\partial x}\right)\right].\end{aligned} \tag{2}$$

The following condensed version is produced when the coefficient of viscosity is assumed to be uniform:

$$\rho\left[\frac{\partial u}{\partial t}+u\frac{\partial u}{\partial x}+v\frac{\partial u}{\partial y}+w\frac{\partial u}{\partial z}\right]=F_x-\frac{\partial p}{\partial x}+\mu\left[\frac{\partial^2 u}{\partial x^2}+\frac{\partial^2 u}{\partial y^2}+\frac{\partial^2 u}{\partial z^2}\right]. \tag{3}$$

2.2. Turbulence Model

In nature, turbulence dominates most flows, its flow field is extremely stochastic and pulsing, and its instantaneous flow satisfies the N–S equation. The k–equation model, the most widely employed two-equation model up to this point, was proposed by Launder and Spalding in 1974. The turbulence model is characterized by the following characteristic scales, which are based on Taylor's hypotheses regarding the length of the turbulence:

$$l_T \sim k^{3/2}/\varepsilon. \tag{4}$$

In the equation, k is the kinetic energy of turbulence, and ε is the dissipation rate of kinetic energy.

Similar to this, turbulent kinetic energy controls the typical velocity. As a result, the fundamental premise of the two-equation model is established, namely that the vortex viscosity coefficient can be written as follows:

$$\nu_T = C_\mu k^2/\varepsilon. \tag{5}$$

C_μ is the coefficient.

The transport equation is more complicated than the k–equation, with many unknown second- and third-order moments and many terms that must be simulated using experiments and magnitude analysis.

The typical form of the k–equation is as follows:

$$\begin{aligned}\frac{D\varepsilon}{Dt}&=\frac{1}{\rho}\frac{\partial}{\partial x_k}\left(\frac{\mu_t}{\sigma_e}\frac{\partial\varepsilon}{\partial x_k}\right)+\frac{C_1\mu_t}{\rho}\frac{\varepsilon}{k}\left(\frac{\partial U_l}{\partial x_k}+\frac{\partial U_k}{\partial x_l}\right)\frac{\partial U_l}{\partial x_k}-C_1\frac{\varepsilon^2}{k},\\\frac{Dk}{Dt}&=\frac{1}{\rho}\frac{\partial}{\partial x_k}\left(\frac{\mu_t}{\sigma_k}\frac{\partial k}{\partial x_k}\right)+\frac{\mu_t}{\rho}\left(\frac{\partial U_l}{\partial x_k}+\frac{\partial U_k}{\partial x_l}\right)\frac{\partial U_l}{\partial x_k}-\varepsilon.\end{aligned} \tag{6}$$

Launder and Spalding provide the values of each individual constant:

$$C_\mu = 0.09,\ C_{\varepsilon 1} = 1.44,\ C_{\varepsilon 2} = 1.92,\ \sigma_e = 1.0,\ \sigma_\varepsilon = 1.3.$$

2.3. Wave Theory

There are many different types of waves in the water, and the dynamic equilibrium between inertial force, gravity, and surface tension defines their physical character. This paper adopts the fifth-order Stokes wave in the numerical pool wave generation because the wave height of a Stokes wave cannot be defined as infinitely small relative to the wavelength, this model can accurately respond to the nonlinear characteristics of shallow water waves, and it is a widely used wave theory in marine engineering calculations.

The potential wave velocity function of Stokes' fifth-order is as follows:

$$\Phi = \frac{c}{k}\sum_{n=1}^{5}\lambda_n \cosh\ nk(z+d)\sin\ n(kx-\omega t). \tag{7}$$

The wavefront equation is

$$\eta = \frac{1}{k}\sum_{n=1}^{5} \lambda_n \cos\ n(kx - \omega t). \quad (8)$$

The wave speed is

$$kc^2 = C_0^2(1 + \lambda^2 C_1 + \lambda^4 C_2). \quad (9)$$

The water quality point velocity and acceleration are

$$\begin{aligned} u_x &= \frac{\partial \phi}{\partial x} = c \sum_{n=1}^{5} n\lambda_n \cosh nk(z+d) \cos n(kx - \omega t), \\ u_z &= \frac{\partial \phi}{\partial z} = c \sum_{n=1}^{5} n\lambda_n \mathrm{sinh} nk(z+d) \sin n(kx - \omega t), \end{aligned} \quad (10)$$

$$\begin{aligned} a_x &= \frac{\partial u_x}{\partial t} = \omega c \sum_{n=1}^{5} n^2\lambda_n \cosh nk(z+d) \sin n(kx - \omega t), \\ a_z &= \frac{\partial u_z}{\partial t} = -\omega c \sum_{n=1}^{5} n^2\lambda_n \mathrm{sinh} nk(z+d) \cos n(kx - \omega t). \end{aligned} \quad (11)$$

Once the wave height, period, and water depth are known, it is possible to ascertain the coefficients and wavelengths of the fifth order wave of Stokes by utilizing the dispersion relation.

3. Design and Energy Conversion Mechanism

3.1. China's Offshore Wave Environment

In addition to adhering to the fundamental similarity requirement, the design of the wave energy generator model's parameters should take into account the actual ocean conditions. Figure 1 presents statistical representations of the hydrographic data of the sea region where the wave energy generator is situated in conjunction with earlier studies conducted by academics in China. The Luzon Strait and its western waters account for relatively large values of China's annual average effective wave height, with an annual average wave height of more than 1.8 m, according to the cloud diagram of the annual average effective wave height in China's offshore waters [17]. We referred to the conventional sea trial model and selected a certain scale ratio based on the simulation model. In conclusion, waves in China's offshore waters tend to be between 0.5 and 3 m tall, with periods between 4 and 8 s.

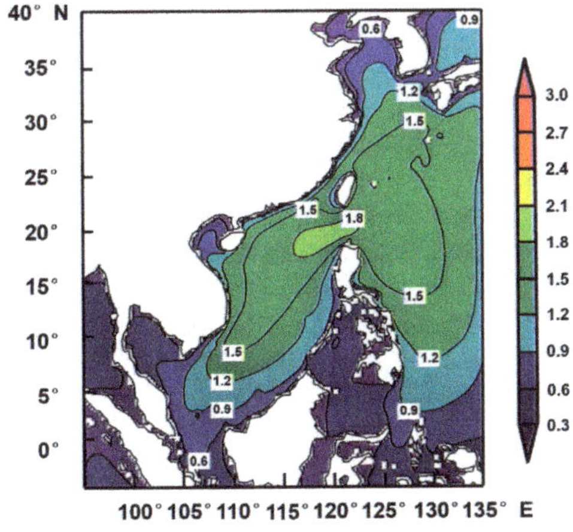

Figure 1. Annual average significant wave height in China's offshore waters (Unit/m).

Therefore, it can be inferred that the simulated wave heights of 0.04 m–0.12 m can fulfill the sea conditions of the real environment [18] when taking into account the probability of the occurrence of wave steepness in the Chinese offshore waters and the CFD numerical simulation circumstances. The wave height of the local wave field will increase sharply and the wavelength will decrease, which poses a significant challenge to the stability and safety of the wave energy generation device when the waves far from the shore propagate to the near shore and are influenced by topography, wind, and current.

In order to better match the more usual wave circumstances in China's actual marine environment, the wave height in this paper was chosen to be between 0.05 and 0.2 m, and the waveform used is the fifth-order Stokes wave.

3.2. Design Proposal

Several factors should be addressed while studying and designing wave energy power production devices, including proper size, the linear damping coefficient, and high mechanical transmission efficiency. A proper overall size can improve the device's application scenario. A suitable linear damping coefficient helps guarantee that wave energy is captured efficiently. The amount of the damping coefficient influences whether waves can readily drive the device. Higher mechanical efficiency can improve the device's energy transmission efficiency.

Because the complicated energy conversion of classic float wave energy power generating devices can result in low power generation efficiency, eliminating intermediate conversion links is critical for boosting device efficiency. Drawing on scientific research, the primary objective of this study is to streamline the composition and elucidate the process of energy conversion in the current oscillating float apparatus, ultimately establishing a novel form of wave energy power generator through oscillating float technology.

Figure 2 displays a three-dimensional depiction of the wave power generator. The Australian-developed "Poseidon" cylindrical float serves as an example of an oscillating float wave power generator. Alan Arthur Wells [19,20], the British engineer specializing in structural engineering, developed the innovative Wells turbine, which operates on low-pressure head air. A detailed representation of the turbine's structure can be observed on the lower left side of Figure 2.

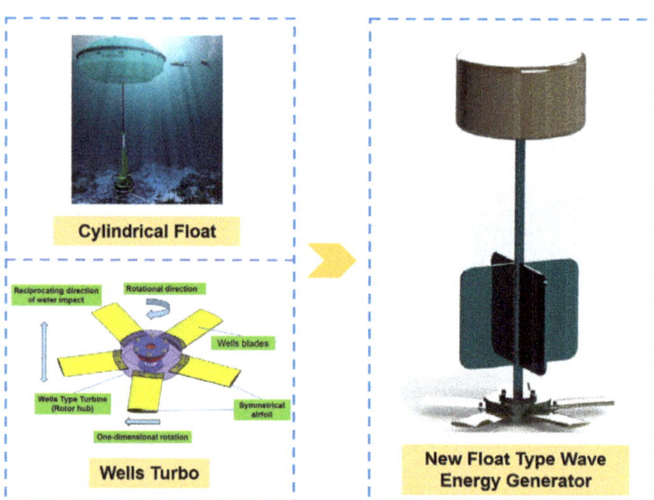

Figure 2. New float-type wave energy generator.

In Figure 2, bottom left, a drawing of the Wells turbine is displayed. The Wells blades are symmetrical airfoils that sustain unidirectional rotation in the reciprocating airflow. It consists of a symmetrical rotor with numerous sets of airfoil blades positioned along the hub with the chord normal to the rotating axis. It is frequently employed in air turbines for oscillating water column wave energy producers because of its inventive structural construction. It is also the most cost–effective option for wave energy generators due to its simple shape. In this study, a novel float-type wave power generator cleverly integrates the two.

A wave energy producing device must take into account a number of factors, including adequate size, linear damping coefficient, and high mechanical transmission efficiency. According to Figure 3, the design of this device fully incorporates the mechanism of the conventional cylindrical float and the Wells turbine, which can maintain unidirectional rotation under the influence of reciprocating airflow. Through the optimization of the transmission procedure, the device achieves the generation of torque by inducing unidirectional rotation of the blades during the upward and downward motion. This mechanism effectively propels the generator, resulting in power generation at its output.

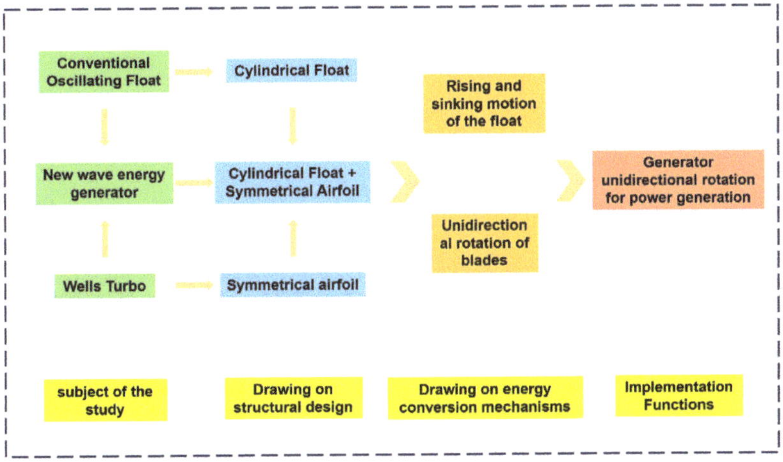

Figure 3. Design ideas for new float-type wave energy generator.

With an overall height of around 600 mm, the novel float-type wave energy generator is seen in Figure 4 below in an exploded view. The float is created in the shape of a hollow and sealed, with a diameter of 200 mm and a height of 120 mm. The overall weight of the device is about 2.8 kg. The float initially floats on the surface of water before moving like a pendulum as a result of the power of the waves. The transmission shaft is attached to the hub and the float through the bearing and the bearing seat, respectively, and the outside edge of the shaft is fitted with a sleeve. It houses a generator and a transmission mechanism.

A sleeve on the shaft's exterior acts as a connector and a protective way of securely connecting the turbofan and the float. The sleeve is fixed with four dampening plates that are 160 mm high but do not participate in the energy transfer. By increasing the efficiency of the device's energy harvesting in the direction of vertical oscillation, the dampening plates' presence can effectively minimize the amplitude of the device's deflection due to pitch and roll.

The novel device's design uses a suspension chain line mooring rope to secure the device's place in the water. Through a gear transmission mechanism, the entire device transfers energy. The power generator for the gadget is a permanent magnet brushless DC generator. The primary element of the energy conversion device and the structural innovation that sets it apart from other wave energy conversion devices is the turbofan section. As a result, the rotating blade is the primary research object in this paper, and its hydrodynamic properties and operational properties are thoroughly investigated.

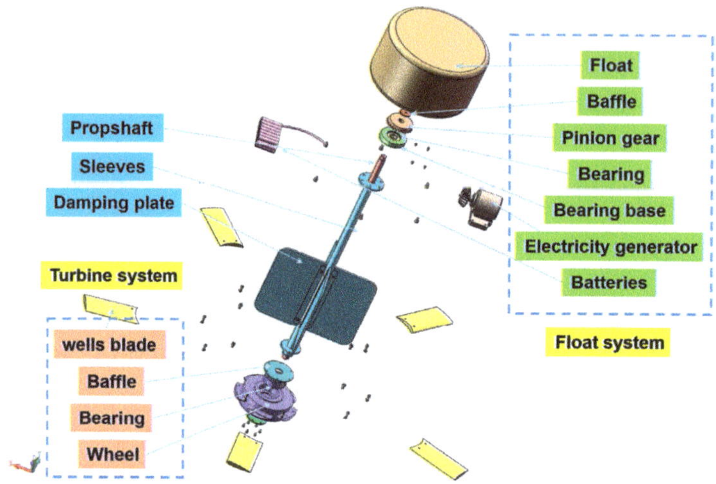

Figure 4. Exploded view of new float-type wave energy generator.

The rectangular bladed NACA0015 symmetric airfoil [21] is used in the turbofan. The blade is 100 mm long, the hub is 100 mm in diameter, and the thickness is 30 mm. The turbofan can successfully rotate unidirectionally in the reciprocating water flow thanks to the symmetric airfoil design. The specifications of the impeller system, which reacts to the float system's action by making a unidirectional rotating motion and driving the generator through the drive shaft to produce power, are shown in Table 1 below.

Table 1. Main parameters of rotating blades.

Parameters	Numerical Value
Number of leaves	5
Chord length/m	0.05
Wheel radius/m	0.05
Flange radius/m	0.05
Wheel thickness/m	0.03

3.3. Characteristics of the Energy Conversion

In a wave energy power generation device, the acquired energy is unidirectionally and irreversibly transferred and converted. The efficiency of energy conversion can be increased by reducing energy loss throughout the transfer process with the aid of a reasonable understanding of the device's energy conversion law. Through its distinctive structural design and transmission mechanism, the new float-type wave energy generator realizes the absorption and conversion of wave energy. Figure 5 depicts the transmission mechanism.

During wave transmission, the float reacts to the wave by oscillating, converting it into linear motion as mechanical energy. It is subjected to the effect of saltwater in the process of pendulum motion and makes a unidirectional rotating motion to transfer mechanical energy because it is rigidly coupled to the turbofan through a sleeve. By means of the drive train, the final generator facilitates the transformation of rotational mechanical energy into electrical energy.

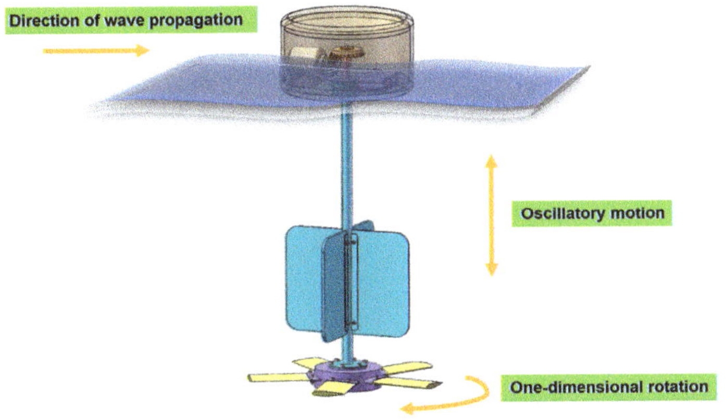

Figure 5. Transmission of the device.

The new wave energy generator, operating as a float-type system, undergoes an energy conversion process. This process is dependent on the efficiency of three components: the float's wave energy capture efficiency η_{cap}, the turbofan's hydrodynamic efficiency η_{hyd}, and the generator's power generation efficiency η_{gen}. The efficiency of the overall wave energy conversion device η can be mathematically represented as follows:

$$\eta = \eta_{cap} \cdot \eta_{hyd} \cdot \eta_{gen}. \tag{12}$$

The three are strongly influenced by nonlinear factors, and the mechanical and electromagnetic relationship is a complex interaction process.

The power input of a vortex fan is typically represented as the kinetic energy of the fluid that flows through the horizontal cross-section of the blades of the fan within a given period.

$$P_{input} = \frac{E_{input}}{t} = \frac{\frac{1}{2}\rho(\pi R^2)hV_A^2}{t} = \frac{1}{2}\rho\pi R^2 V_A^3, \tag{13}$$

At time t, the kinetic energy of a fluid cylinder with height h and radius R traversing the horizontal section of the blade can be represented by E_{input}. V_A denotes the velocity of the fluid relative to the rotating blade's horizontal cross-section.

Output power of turbofan P_{output}:

$$P_{output} = Tw = NT_s w, \tag{14}$$

where T is the output torque of the turbofan, w is the rotational speed (rad/s), N is the number of blades, and T_s is the torque generated by a single blade on the drive shaft.

The brushless DC motor is selected as the generator in the wave energy power generation device in this article. By definition [22], the generator power P is

$$P = \eta \cdot C_l \cdot \dot{x}^2, \tag{15}$$

C_l is the linear damping coefficient, and x is the pendulum displacement of the device. Generator rotational damping coefficient C_r:

$$C_r = \frac{1.5K_r^2}{R_1 + R_2}, \tag{16}$$

K_r is the torque constant of the generator, R_1 is the generator resistance, and R_2 is the external resistance of the circuit.

C_l can be expressed as

$$C_l = \frac{1.5K_r^2}{\eta(R_1 + R_2)e^2}, \tag{17}$$

The slope, denoted as e, represents the non-linear correlation between the rotational velocity and the velocity at which the device droops. The linear coefficient of damping is associated with the external load and may be regulated to attain the optimal damping ratio and wave capture ratio, which depend on the wave properties, by adjusting the effective value of the external load.

The output power P_{output} of the new float-type wave energy generator involves more variables, and its hydrodynamic characteristics are proposed to be investigated by a combination of theoretical analysis and numerical simulation.

3.4. Kinematic and Dynamic Characterizations

In this paper, the main structural component of energy conversion is the rotating turbofan, so the theoretical analysis focuses on the rotating turbofan part. The rotating blades of the power generation device do not have any change in shape along the direction of the drive shaft and each blade is exactly the same, so the section parallel to the rotating axis and parallel to the airfoil is taken to analyze the motion and force of individual blades. The blade motion and force are shown in Figure 6, and its horizontal cross-section is taken as the reference plane. When seawater impinges on the impeller motion with velocity V_A, it causes a current angle a, which generates a lift force F_L perpendicular to the current direction and a drag force F_D parallel to the current direction.

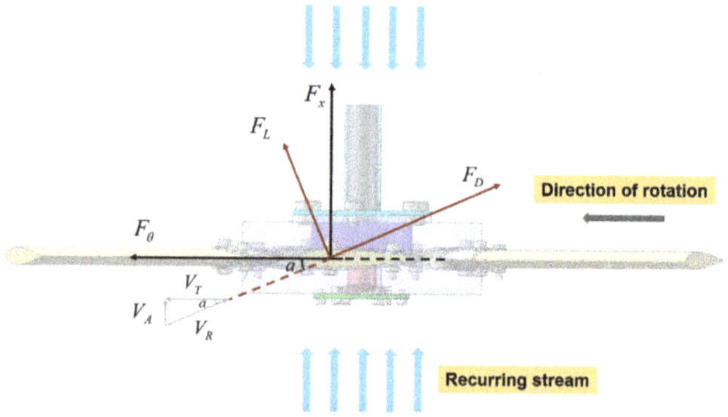

Figure 6. Velocity and force schematic of rotating blade.

As shown in Figure 6, V_T is the circumferential velocity of the fan blades, which is

$$V_T = w \cdot l, \tag{18}$$

where w is the rotational angular velocity of the blade and l is its radial diameter.

Then the combined velocity V_R of the blades relative to the incoming flow is

$$V_R = V_A - V_T. \tag{19}$$

Since V_A and V_T are perpendicular to each other, the mode $|V_R|$ of V_R can be computed in this respect:

$$|V_R| = \sqrt{|V_A|^2 + w^2l^2}. \tag{20}$$

The water flow angle a is

$$a = \arctan\frac{V_A}{w \cdot l}. \tag{21}$$

The blade lift F_L and drag F_D are

$$F_L = \frac{1}{2}\rho A V_R^2 C_L, \tag{22}$$

$$F_D = \frac{1}{2}\rho A V_R^2 C_D, \tag{23}$$

where ρ is the fluid density, A is the characteristic area of the blade, C_L is the lift coefficient, and C_D is the drag coefficient.

The above forces can be decomposed into an axial force F_x and a tangential force F_θ:

$$F_x = F_L \cos\alpha + F_D \sin\alpha, \tag{24}$$

$$F_\theta = F_L \sin\alpha - F_D \cos\alpha. \tag{25}$$

From the above analysis, it can also be found that for a given value of a, the direction of the tangential force F_θ is independent of the positive or negative value of a. Regardless of the direction of the water flow, the fan blades rotate in one direction.

4. CFD Simulation Analysis of a New Float-Type Wave Energy Device

4.1. Setup of CFD Simulation

In this chapter, a physical model must be chosen in order to calculate the STAR–CCM+. The main goal of the hydrodynamic simulation is to ascertain the dynamic reaction of the novel float-type wave energy generator under various wave circumstances. Table 2 displays the different parameter values used to simulate this paper.

Table 2. Setting of simulation parameters.

Combined Parameters	Physical Models
Space	Three-dimensional
Time	Implicit invariant (math.)
Eulerian multiphase flow model	Volume of fluid domain
Turbulence model	$k-\varepsilon$
Other models	Gravity, VOF waves

The Stokes fifth order waves are used using the boundary condition wave-making approach in this paper. The border satisfies the waves:

Velocity in x direction:

$$u_x = c\sum_{n=1}^{5} n\lambda_n cosh[nk(z+d)] \times cos[n(kx-\omega t)]. \tag{26}$$

Velocity in z direction:

$$u_z = c\sum_{n=1}^{5} n\lambda_n sinh[nk(z+d)] \times sin[n(kx-\omega t)]. \tag{27}$$

The instantaneous rise of the wavefront is

$$\eta = \frac{1}{k}\sum_{n=1}^{5} \lambda_n cos[n(kx-\omega t)], \tag{28}$$

where ω, d, and k are the circular frequency, water depth, and wave number, respectively.

Setting up a wave height meter, as seen in Figure 7, allows for monitoring the time course of the wavefront change on the free surface. It is clear that the waves simulated in this research are essentially consistent with the theoretical values of the fifth-order Stokes waves, laying the groundwork for an accurate numerical simulation of the subsequent device.

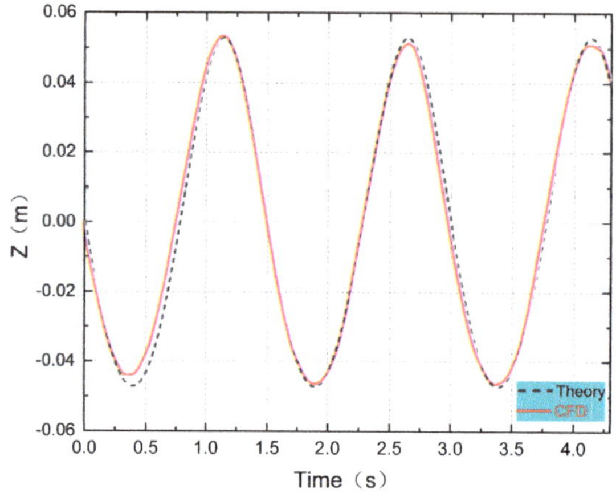

Figure 7. Time history curve of wave height.

To avoid the modification of waveforms and computations in the calculation area caused by wave reflection, the creation of a wave dissipation zone in the outflow region is necessary. This study establishes a length as a damped wave cancellation zone at the pool's outlet end using the damped wave cancellation method [23].

4.2. Computational Domain and Meshing

This article sets up the calculation domain for the fluid simulation computation when the wave period T = 1.5 s, in accordance with the guidelines of the International Towing Tank Conference (ITTC). The distance from the starting point to the boundary where the substance enters is 1λ, while the distance from the starting point to the boundary where the substance leaves is 2λ. The entire extent of the space is 3λ. The depth of the liquid is determined as 1 m, whereas it surpasses the water's surface by 1.08 m. The computational space possesses a width of 2.56 m and a height of 2.08 m. Around 1 million computational grids are secured within the water's surface, background, and the overlapping region of motion [24]. Figure 8 depicts the computational domain case configuration.

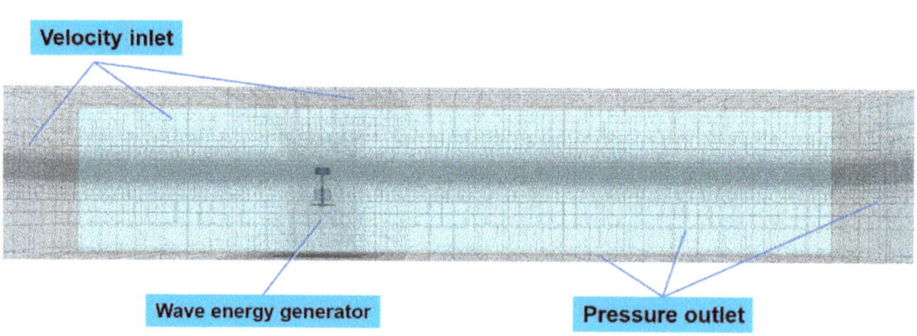

Figure 8. Computational domain setup and meshing.

Overlapping grids are more frequently employed in the CFD simulation of the coupling effect between waves and wave energy producers. The wavefront encrypted area is set up in the backdrop area, and the grid division of the wavefront encrypted area is built up in accordance with the wave parameters' wavelength and wave height H. This increases computation accuracy and wavefront stability. Figure 9 depicts the situation after encryption. Additionally, the encrypted area is built close to the revolving blade in order to better collect the device's motion state information. The primary encryption approach is to establish surface control and volume control.

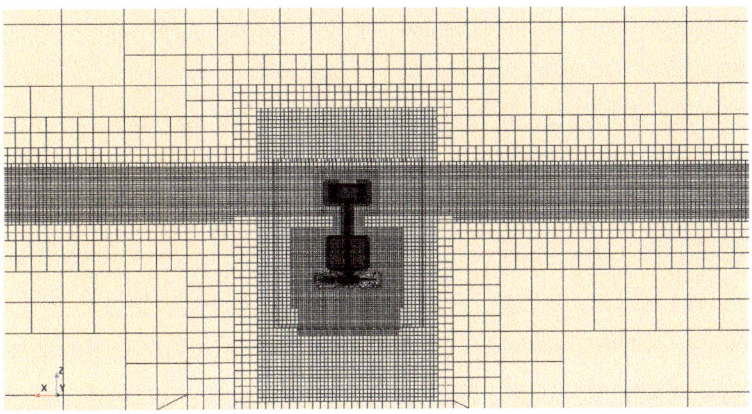

Figure 9. Encryption of meshes.

4.3. Analysis of Device Simulations during One Wave Cycle

The simulation analyzes the movement of the device in the wave with the wave height H = 0.1 m and the period T = 1.5 s in one cycle, and the dynamic simulation interface based on STAR–CCM+ is shown in Figure 10 in order to intuitively understand the law of movement of the power generation device in the wave.

These figures depict the device's status during one operation cycle. Within a cycle, the device goes from the horizontal plane to the peak, then to the trough, then back to the horizontal plane. Because the device's motion is periodic, analyzing the device's motion within one cycle is significant.

The device's overall pendulum motion speed is determined, among other things, by how the floats react to the waves, the inertial forces of the vortex fan, and the fluid resistance. It is clear from Figures 10–12 that the object is moving pendantly. In Figure 10a, orientation is shown as horizontal with minimal vertical displacement. The device is seen moving upwards along the wave surface, reaching its maximum velocity in the positive direction of the Z, measuring 0.151 m/s. Moving on to Figure 10b, the illustration demonstrates the device's upward motion along the wave face, ascending towards the peak of the wave. At this point, the vertical velocity of the pendant oscillation is close to zero at 0.021 m/s, and the greatest displacement is at 0.042 m. The floating body descends down the wave face to the horizontal plane, as shown in Figure 10c, at which point the vertical velocity is close to zero at 0.021 m/s and the displacement of the pendent oscillation achieves its maximum value of 0.042 m. The floating body goes upward along the wave face to meet the wave, as shown in Figure 10c. The device's highest vertical velocity, or 0.146 m/s, is along the Z-axis's negative direction. The floating body has a vertical displacement of 0.042 m. Figure 10e depicts the device making a climbing motion along the wave surface and re–entering at the position of the time plane. Figure 10d depicts the device reaching the trough of the wave, where the vertical displacement of the device in the opposite direction of the extended Z reaches a maximum. The motion of the power generating unit throughout one cycle was described by the comprehensive simulation of Figures 10–12.

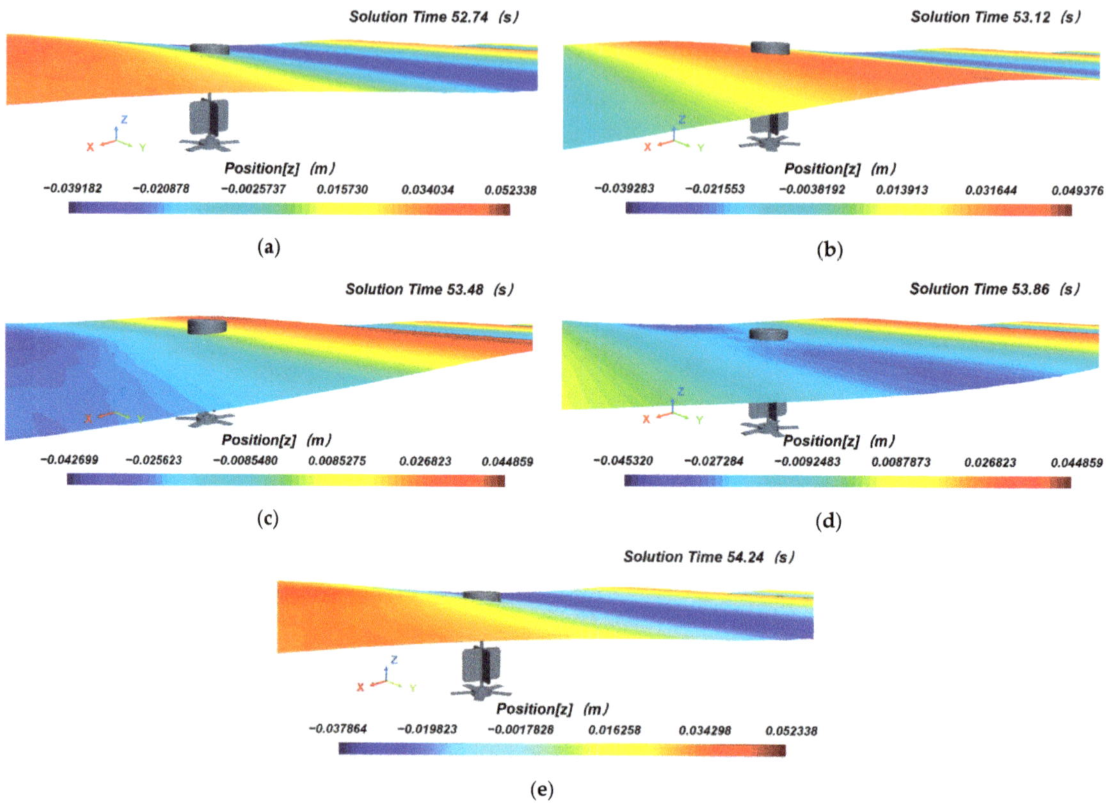

Figure 10. T = 1.5 s, H = 0.1 m diagram of the motion of the power generator in one λ. (**a**) t = 52.74 s; (**b**) t = 53.12 s; (**c**) t = 53.48 s; (**d**) t = 53.86 s; (**e**) t = 54.24 s.

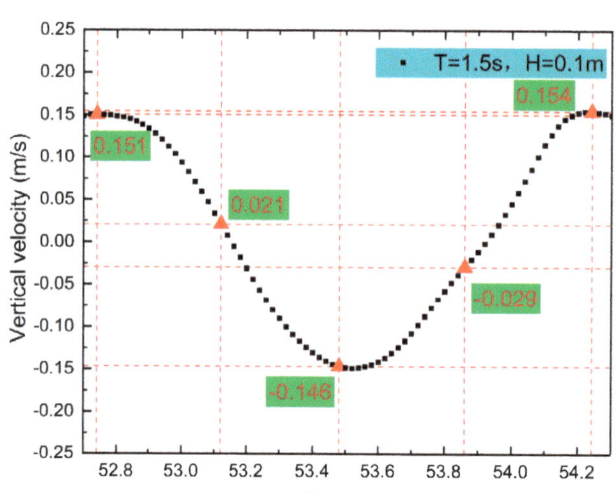

Figure 11. Time history curve of device droop velocity.

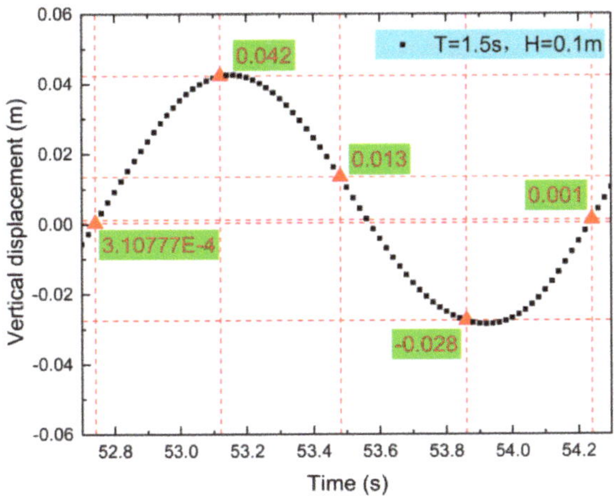

Figure 12. Time history curve of device droop displacement.

The results of the numerical simulation are accurately visualized in Figures 11 and 12 by the use of scatter plots, with a time interval of 0.02 s between each node. Further investigation reveals that the device's movement velocity is decreasing and the velocity decreases more quickly at the point when the device is moving from the horizontal direction towards the crest of the wave, or when the device is moving from 52.74 s to 53.12 s. The device's pendulum velocity in the opposite direction increases slowly and then decreases quickly as it transitions from the peak to the trough, or from 53.12 to 53.86 s. The device's motion velocity in the pendent direction gradually increases as it goes from the wave valley to the horizontal plane, or from 53.86 to 54.24 s, and achieves its maximum at the wave peak. The gadget is subject to both the action of the wave and the action of its own gravity, which results in a certain regularity in the change in speed of motion.

The dynamic pressure cloud of the blades is illustrated in Figure 13. The leading edge of the blade exhibits the highest concentration of pressure, with a maximum pressure difference of 446.18 Pa [25]. The pressure side of the blade is predominantly controlled by positive pressure, while the suction side goes through negative pressure. The output torque of the turbofan is mainly determined by the pressure disparity found at the blade's leading edge [26,27]. Additionally, this analysis acts as an extensive point of reference for a subsequent examination of the device's hydrodynamic efficiency throughout a complete operational cycle.

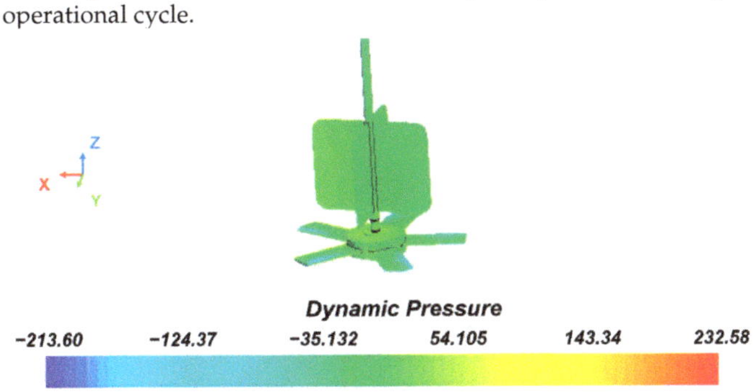

Figure 13. Turbofan dynamic pressure cloud diagram at T = 1.5 s.

A perfect balance between the rotational speed and rotational torque can maximize the power output of the device. The rotational speed of the blades is a significant factor determining the output power and operating parameters of the device. The rotating angular velocity of the blades exhibits regularity, as shown in Figure 14. The turbofan continues to revolve due to inertia at 53.12 s, when the float is at the wave crest position, and at this moment the angular velocity is the least, at roughly 3.419 rad/s.

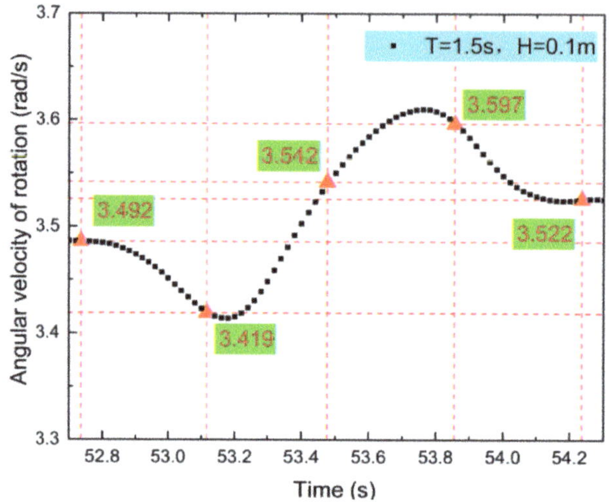

Figure 14. Time history curve of fan blade rotation angular velocity.

The float reaches the wave valley at roughly 53.86 s, and the rotational angular velocity is likewise rapid, at about 3.597 rad/s. The float's angular velocity decreases as it moves from the wave valley to the wave crest, but it considerably increases when it moves from the wave crest to the wave valley position. The angular velocity considerably increases through position.

The turbofan's cycle of instantaneous output power variations is depicted in Figure 15 and is governed by the blade torque and rotational angular velocity. The greatest output power is around 0.05 W during the blade's transition from the crest to the horizontal plane.

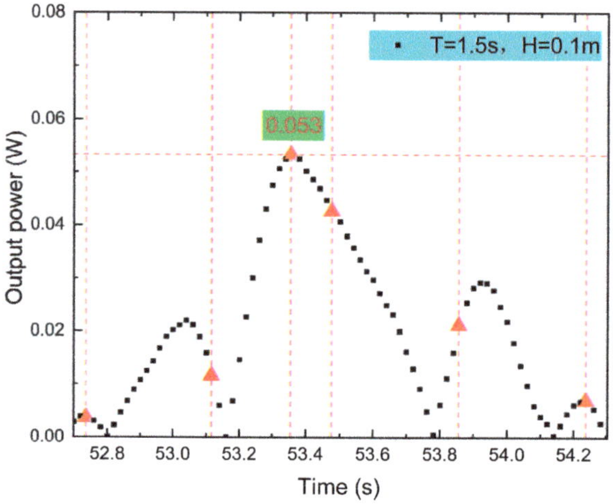

Figure 15. Output power.

The transfer function of the device's motion may be intuitively determined by analyzing how it operates and collects energy in one cycle. Additionally, this analysis acts as an extensive point of reference for a subsequent examination of the device's hydrodynamic efficiency throughout a complete operational cycle.

4.4. Numerical Study of the Effect of Hydrodynamic Performance of Power Generating Units over the Full Operating Cycle

The new float-type wave energy power generating gadget creates periodic motions in the water, and the law may be thought of as the superposition of innumerable periodic motions when the simulation results tend to converge. This part examines the device's force, motion, and power production after it has been operating steadily. Additionally, this portion offers an analytical framework for later optimization of the device's structural properties.

Figures 16 and 17 indicate that after the device has been operating steadily, the pendant velocity and displacement both exhibit periodic fluctuations, with the maximum pendant velocity being around 0.15 m/s. The maximum displacement amplitude of the device, measured in reference to the horizontal plane, is roughly 0.04 m. The device is subjected to buoyancy and gravity during movement, and the unidirectional rotation of the blades can cause disturbance to the water flow, resulting in uneven displacement of the device.

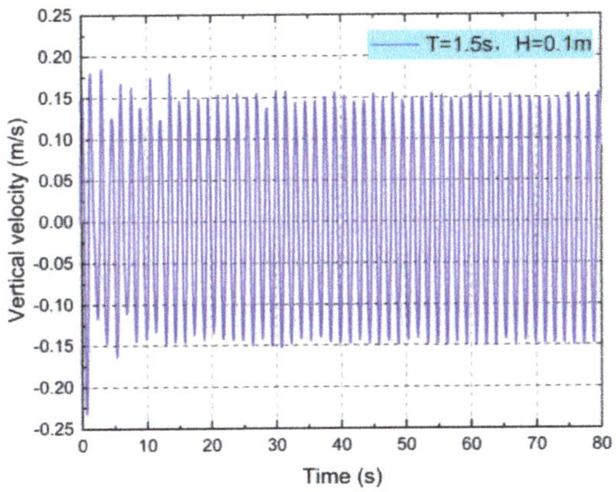

Figure 16. Time–calendar curve of pendulum velocity for full-cycle operation of the device.

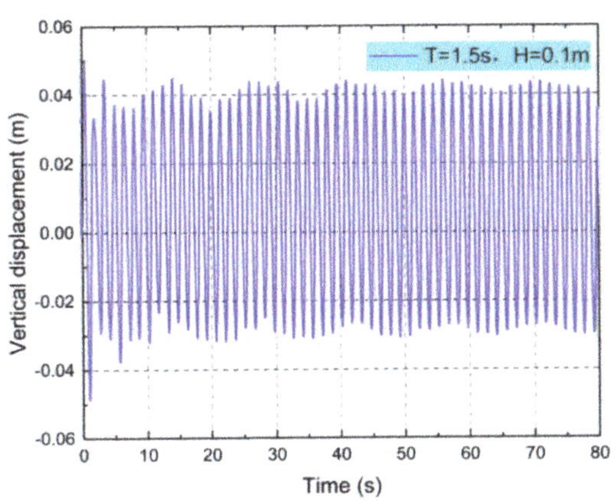

Figure 17. Time–history curves of pendulum displacements for full-cycle operation of the device.

Figure 18 shows that the device's response to the wave moment is periodic, with a larger, approximately 0.015 Nm, blade moment in the valley of the fan blade subjected to reverse rotation and a 0 Nm blade moment in the peak moment. However, due to the turbofan's inertia, the blades continue to rotate even in the case where the blade moment is 0 Nm.

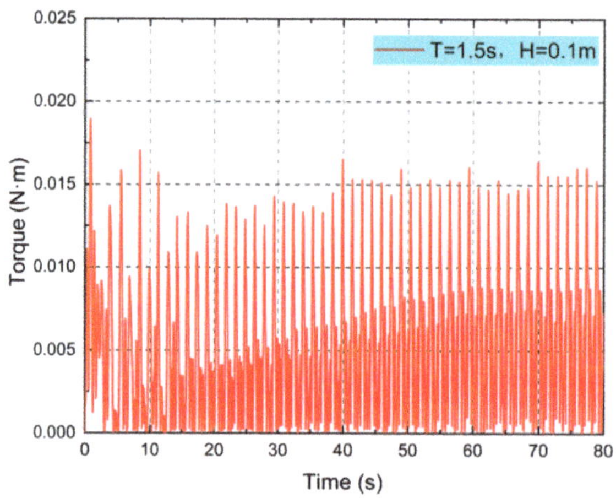

Figure 18. Time–calendar curve of fan blade torque variation for full-cycle operation of the device.

According to Figure 19, the device's rotational angular velocity increases sporadically over time. After 55 s, the blades' rotational angular velocity tends to stabilize at about 3.613 rad/s, which also suggests that at this point the device's motion has reached a converged state.

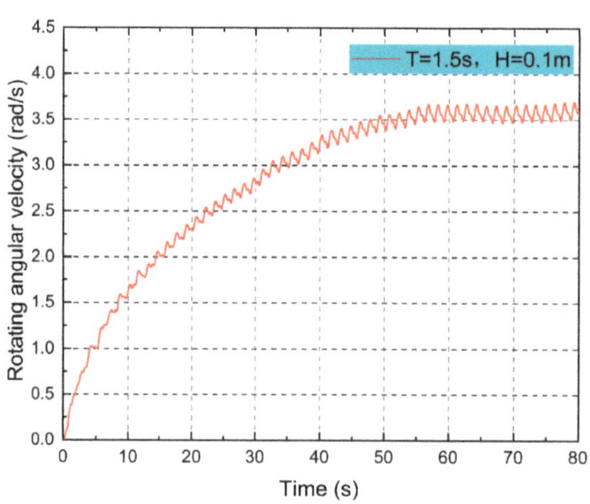

Figure 19. Time–calendar curve of rotational angular velocity variation for full-cycle operation of the device.

The device's output power is influenced by the blade's rotating torque, rotational speed, and transmission efficiency. The output power of the device similarly enters a constant state of change after 55 s and the maximum instantaneous output power is around 0.053 W (as shown in Figure 20).

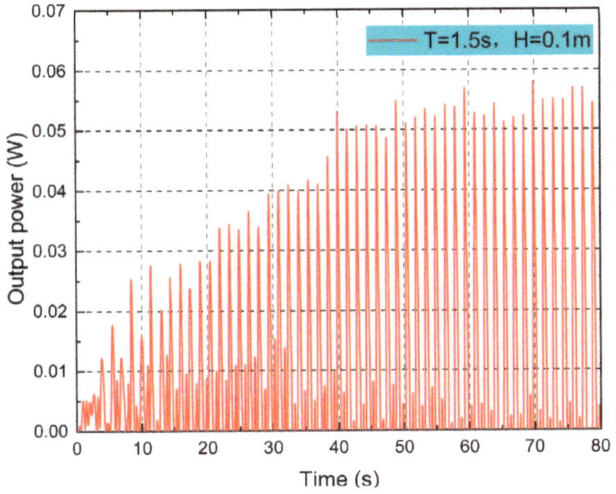

Figure 20. Time–calendar curve of output power change for full-cycle operation of the device.

This part expands from a local analysis to a stable convergence study, concentrating on the hydrodynamic and energy outputs of the device in a single cycle as well as in a whole cycle with a blade number of five. Because changes to the device's structural parameters have no effect on the device's overall general functioning in a wave environment, the research in this part also acts as a comparative reference for the optimization of the structural parameters in the following section.

4.5. Numerical Study of the Effect of Structural Parameters on the Hydrodynamic Performance of Power Generation Units

One of the crucial factors in the structural design of a rotating turbofan is the number of its blades; too many or too few moving elements on the impeller will influence the device's dependability and performance qualities. This section analyzes the bottom rotating turbofan force and motion response in waves using four different devices with three, four, five, and six blades. Table 3 displays the chosen operating conditions.

Table 3. Main parameters for different number of blades.

Working Condition	Number of Leaves	Period T(s)	Wave Height (m)
A_1	3	1.5	0.1
A_2	4	1.5	0.1
A_3	5	1.5	0.1
A_4	6	1.5	0.1

The analysis of the device's motion in the wave primarily utilizes major factors including the velocity of motion along the direction of the overall pendulum oscillation, displacement, moment exerted by the blade, rotational speed of the vortex fan, dynamic pressure, and power generated as output [28]. The device's oscillatory displacement and velocity exhibit periodic changes, as shown in Figures 21 and 22. The number of blades has less of an impact on the pendulum displacement and pendulum velocity of the device.

The time–domain curve of the moment of the turbofan blades is depicted in Figure 23. The rotational moment is significantly impacted by the periodic fluctuations in the blade moment caused by the action of the waves. When there are six blades, the rotational torque is greatest when compared with other circumstances. The greatest moment amplitude is around 0.012 Nm, whereas the smallest moment amplitude is about 0.008 Nm. With an increase in blade count, the spinning moment of the blades increases.

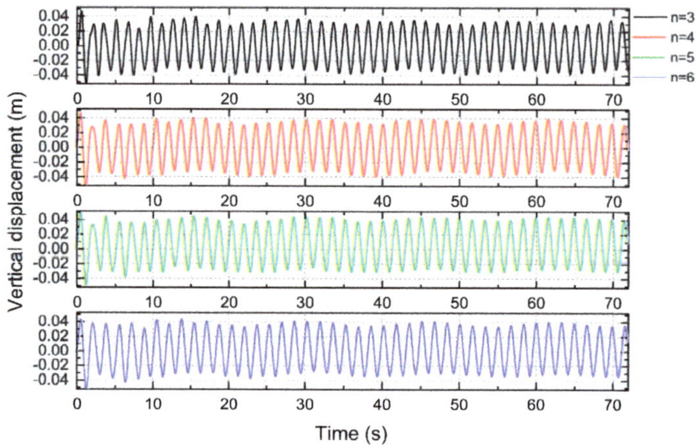

Figure 21. Encrypt vertical displacement.

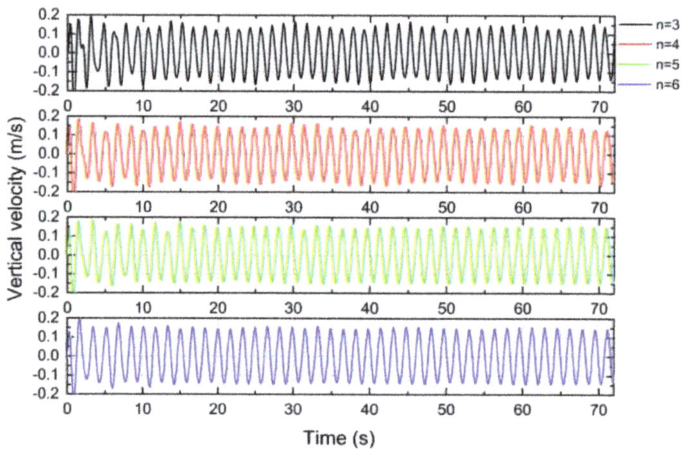

Figure 22. Vertical velocity.

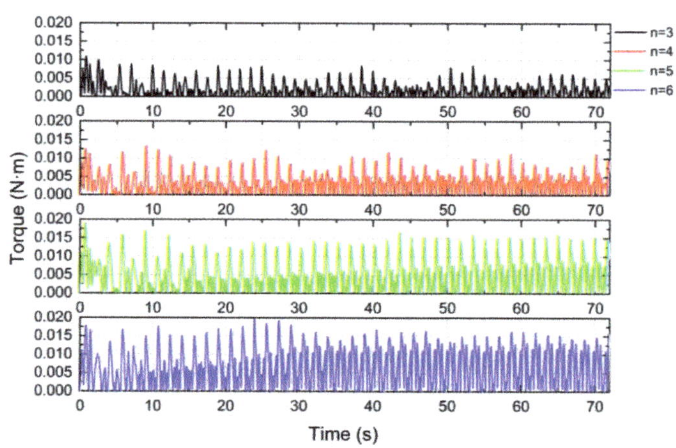

Figure 23. Time history curve of fan blade torque.

The time–calendar curve of the turbofan's angular velocity is depicted in Figure 24. As seen in the image, the turbofan's rotating angular velocity steadily rises with the number of blades and, over time, tends to stabilize for all of them. The convergence of the turbofan rotational speed is very similar when there are five and six blades. The rotational speed of a turbofan with five blades is approximately 3.613 rad/s and a turbofan with six blades is approximately 3.707 rad/s, which is somewhat higher than the constant rotational speed of these two devices. The turbofan speed converges steadily to 2.11 rad/s, the minimum speed, when there are three blades. As can be seen, the turbofan speed increases steadily as the number of blades grows and will eventually converge to a steady value.

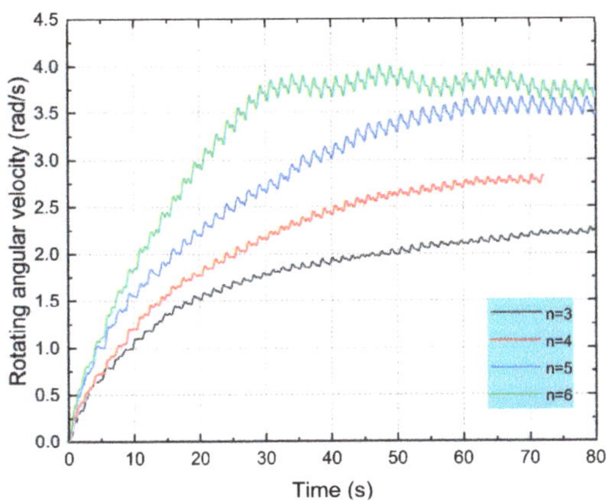

Figure 24. Time history curve of blade rotation speed.

The comparative analysis of Figures 13 and 25 unveils the similarities in the cloud diagrams illustrating the dynamic pressure of the four distinct device types. It becomes evident that the distribution pattern of blade pressure exhibits resemblances. Specifically, the leading edge predominantly houses the high-pressure region, while the differential pressure on the blade surface gradually diminishes along the chord length, eventually approaching zero near the trailing edge section [29].

When there are three blades, the maximum differential pressure of the blade is 139.556 Pa; when there are four blades, the maximum differential pressure is 338.257 Pa; and when there are six blades, the highest differential pressure is 424.52 Pa.

With an increase in the number of blades, the differential pressure at the leading edge of the blade surface gradually rises. In other words, a sufficient increase in the number of blades can increase the pressure differential between the two sides of the leading edge of the blades, increase the size of the high-pressure differential zone, and consequently enable the turbofan to produce a greater rotating torque. This study also found that increasing the velocity relative to the blade's leading edge can enhance the pressure disparity on both sides. This expansion of the high-pressure difference region leads to a greater rotational torque within the absorber. The blade pressure is negative, and the blade can spin in one direction as long as the pressure differential on both sides of the blade is bigger than the back chord.

Figure 26 displays a graph of the device's instantaneous output power. It is evident that, after some time in the same wave environment, the output power of the power production equipment tends towards a stable amplitude of change. The turbofan's highest instantaneous output power is 0.059 W when it has six blades, while its lowest instantaneous output power is approximately 0.012 W when it has three blades. It was discovered that the instantaneous output power rises with the number of blades when the other three conditions are combined.

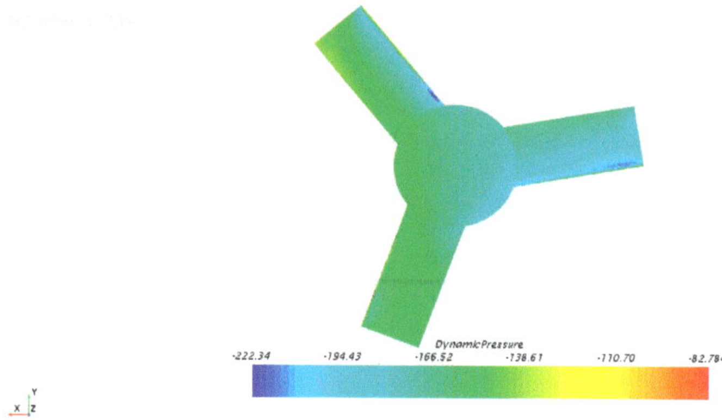

(a)

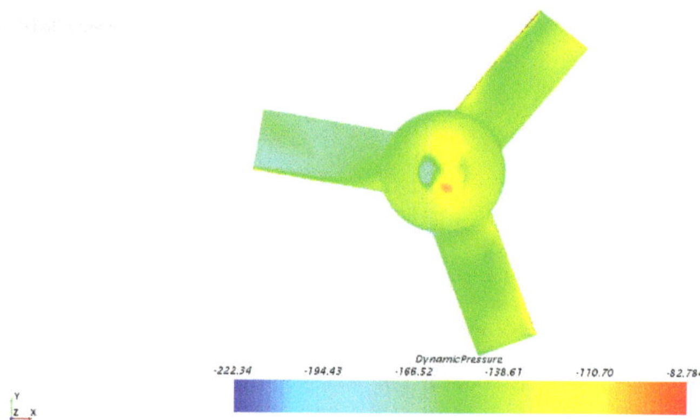

(b)

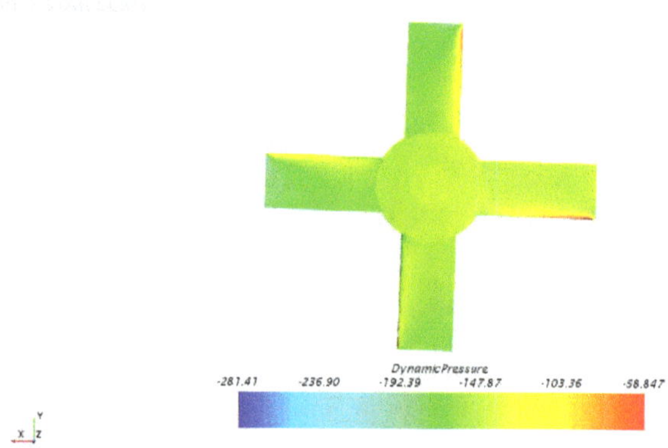

(c)

Figure 25. *Cont.*

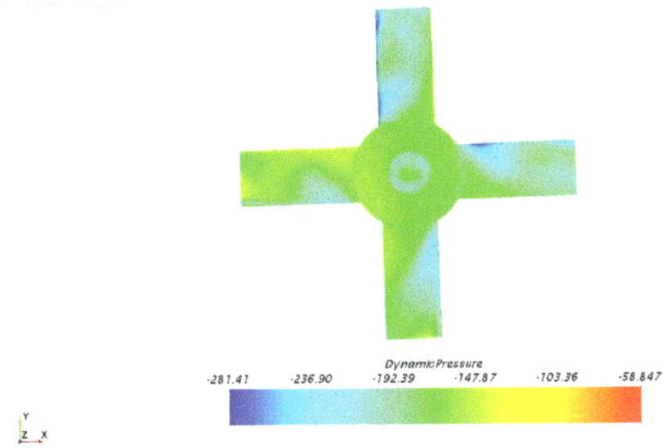

(d)

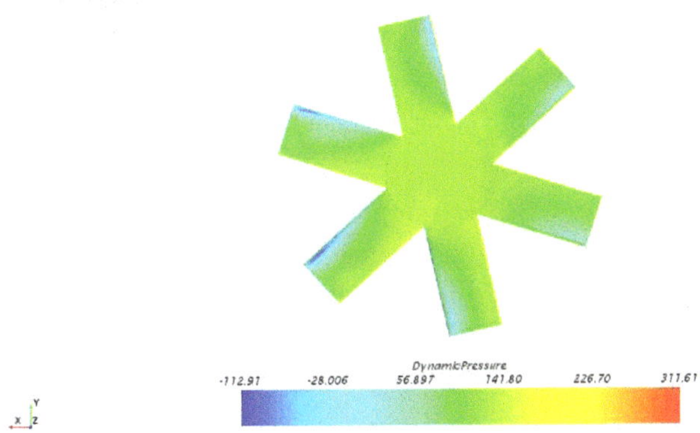

(e)

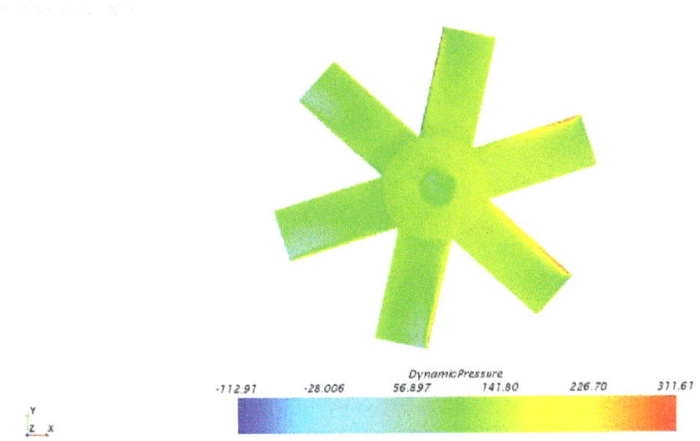

(f)

Figure 25. Turbofan dynamic pressure clouds for devices with different blade numbers. (**a**) Pressure side of turbofan with three-bladed unit; (**b**) Suction side of turbofan with three-bladed unit; (**c**) Pressure side of turbofan with four-bladed unit; (**d**) Suction side of turbofan with four-bladed unit; (**e**) Pressure side of turbofan with six-bladed unit; (**f**) Suction side of turbofan with six-bladed unit.

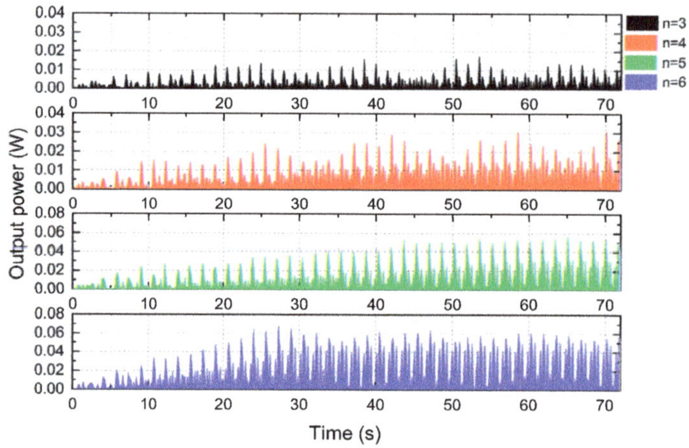

Figure 26. Output power of units with different blades.

Table 4 summarizes the findings of the calculations made by adding up all the calculated working conditions:

Table 4. Comparison of calculation results.

Working Condition	Amplitude of Pendulum Displacement (m)	Amplitude of Pendulum Speed (m/s)	Amplitude of Rotational Torque (N·m)	Rotational Speed (rad/s)	Power (W)
A_1	0.372	0.158	0.004	2.352	0.012
A_2	0.381	0.157	0.009	2.831	0.023
A_3	0.407	0.161	0.014	3.613	0.053
A_4	0.411	0.159	0.015	3.707	0.059

As shown in Table 4, when there are five blades, the device's pendulum oscillation speed is higher, averaging 0.161 milliseconds, and the instantaneous output power is comparatively high at this time, second only to the six-bladed device. The number of blades has less of an impact on the generator's overall pitching displacement and pitching speed [30].

As the number of blades increases, the turbofan's rotational speed gradually rises, as does the rotational torque, which is caused by an increase in the ratio of the high-pressure differential area to the area of a single blade. In terms of rotation speed, as the number of blades grows, the device's power and rotation speed steadily increase. We also referred to the option of floating fans for the setting of the five blades for the five and six blades. At the moment, cost is a major impediment to the development of innovative energy gadgets. On the one hand, we examined the structure's rotating speed and output power. Based on this, we believe that fewer blades are preferable. In other ways, it is to cut costs. As a result, the device's overall torque rises, as does the power output, and the five-bladed turbofan's overall performance is better from the standpoint of efficiency.

5. Conclusions

This paper presents a novel float-type wave energy power generation device specifically designed for the offshore wave environment in Asia. The objective of this study is to develop a wave energy power generation device that incorporates a simple transmission method, allows for flexible layout, and achieves high conversion efficiency. The study begins by analyzing the operational behavior of the device within a single cycle under optimal wave conditions (simulation period: 1.5 s, wave height: 0.1 m). Subsequently, the hydrodynamic performance of the crucial impeller system is examined, with a particular focus on the impact of varying blade numbers on device performance, in order to identify the optimal design solution. The following specific conclusions are drawn:

1. We constructed a numerical wave pool using STAR-CCM+ and applied overlapping grid technology to simulate the movement and power generation efficiency of a new float-type wave energy power generation device under Stokes fifth-order waves. Our findings indicate that this numerical method accurately captures changes in complex flow fields, and the simulated waves closely match the theoretical waves. Hence, it can serve as a reliable tool for numerical simulation of this device.
2. In operation following the stability of the law, the new type of float wave power generating device in saltwater to perform the cycle movement may be viewed as the superposition of the movement of countless cycles. The device's speed drops as it moves from the horizontal to the crest of the movement of the moment, and the speed declines ever faster, according to the examination of the device's movement in a cycle. The pendulum velocity of the device increases slowly and then lowers quickly as it transitions from the crest to the trough. The device's motion speed in the vertical direction gradually increases until it reaches its maximum speed at the wave peak as it passes from the wave valley to the horizontal plane. The torque and speed of the rotating blades have a significant impact on the device's output power.
3. According to the structural design and optimization of the device, it was found that the overall movement of the device is less affected by changes in the number of blades, and regardless of the changes in the number of blades, the vertical displacement and velocity of the device exhibit periodic changes. Under the influence of waves, the moment of the blades varies on a regular basis, and the moment of the blades rises as the number of blades grows. A suitable increase in the number of blades can cause the turbofan to create a greater rotating moment because the pressure differential at the leading edge of the blade surface steadily increases with the number of blades. The output power of the device increases with the number of blades, and the five blades turbofan performs better overall in terms of equipment cost, operation, and power generation efficiency.

Author Contributions: Writing—original draft, Y.Z.; Writing—review & editing, D.L. All authors have read and agreed to the published version of the manuscript.

Funding: This research received no external funding.

Data Availability Statement: Data are contained within the article.

Acknowledgments: We appreciate the journal's editors and reviewers for their insightful criticism of the research.

Conflicts of Interest: The authors declare no conflict of interest.

References

1. Side, J.; Gallego, A.; James, M.; Davies, I.; Heath, M.; Karunarathna, H.; Venugopal, V.; Vögler, A.; Burrows, M. Developing methodologies for large scale wave and tidal stream marine renewable energy extraction and its environmental impact: An overview of the Tera Watt project. *Ocean Coast. Manag.* **2017**, *147*, 1–5. [CrossRef]
2. Leeney, R.H.; Greaves, D.; Conley, D.; O'Hagan, A.M. Environmental Impact Assessments for wave energy developments —Learning from existing activities and informing future research priorities. *Ocean Coast. Manag.* **2014**, *99*, 14–22. [CrossRef]
3. Venugopal, V.; Nemalidinne, R.; Vögler, A. Numerical modelling of wave energy resources and assessment of wave energy extraction by large scale wave farms. *Ocean Coast. Manag.* **2017**, *147*, 374–378. [CrossRef]
4. Wang, X.; Liang, H.; Qiao, D.; Yan, J.; Ning, D.; Ou, J. Hardware-In-the-Loop test for the optimal damping identification of oscillating buoy wave energy converters. *Ocean Eng.* **2023**, *287*, 115730. [CrossRef]
5. Salter, S.H. Wave power. *Nature* **1974**, *249*, 720–724. [CrossRef]
6. Korde, U.A. Phase control of floating bodies from an on-board reference. *Appl. Ocean Res.* **2001**, *23*, 2512–2562. [CrossRef]
7. Eriksson, M.; Isberg, J.; Leijon, M. Hydrodynamic modelling of a direct drive wave energy converter. *Int. J. Eng. Sci.* **2005**, *43*, 1377–1387. [CrossRef]

8. Babarit, A.; Clément, A.H. Optimal latching control of a wave energy device in regular and irregular waves. *Appl. Ocean Res.* **2006**, *28*, 77–91. [CrossRef]
9. De Backer, G.; Vantorre, M.; Frigaard, P.; Beels, C.; De Rouck, J. Bottom slamming on heaving point absorber wave energy devices. *J. Mar. Sci. Technol.* **2010**, *15*, 119–130. [CrossRef]
10. Negri, M.; Malavasi, S. Wave Energy Harnessing in Shallow Water through Oscillating Bodies. *Energies* **2018**, *11*, 2730. [CrossRef]
11. Marchesi, E.; Negri, M.; Malavasi, S. Development and analysis of a numerical model for a two-oscillating-body wave energy converter in shallow water. *Ocean Eng.* **2020**, *214*, 107765. [CrossRef]
12. Zhang, H.; Zhou, B.; Vogel, C.; Willden, R.; Zang, J.; Geng, J. Hydrodynamic performance of a dual-floater hybrid system combining a floating breakwater and an oscillating-buoy type wave energy converter. *Appl. Energy* **2020**, *259*, 114212. [CrossRef]
13. Curto, D.; Franzitta, V.; Guercio, A. Sea Wave Energy. A Review of the Current Technologies and Perspectives. *Energies* **2021**, *14*, 6604. [CrossRef]
14. Khan, M.Z.A.; Khan, H.A.; Aziz, M. Harvesting Energy from Ocean: Technologies and Perspectives. *Energies* **2022**, *15*, 3456. [CrossRef]
15. Ahn, S.; Haas, K.A.; Neary, V.S. Wave energy resource characterization and assessment for coastal waters of the United States. *Appl. Energy* **2020**, *267*, 114922. [CrossRef]
16. Plummer, M.L.; Feist, B.E. Capturing energy from the motion of the ocean in a crowded sea. *Coast. Manag.* **2016**, *44*, 4644–4685. [CrossRef]
17. Zheng, C. Theoretical Research on the Utilization of Renewable Energy (Wave Energy, Wind Energy) Resources at Sea. Ph.D. Thesis, National University of Defense Technology, Changsha, China, 2020.
18. Chen, W.; Dolguntseva, I.; Savin, A.; Zhang, Y.; Li, W.; Svensson, O.; Leijon, M. Numerical modelling of a point-absorbing wave energy converter in irregular and extreme waves. *Appl. Ocean Res.* **2017**, *63*, 901–905. [CrossRef]
19. Wang, X.S.; Wang, L.H.; Song, X.F.; Ning, B. 3D Wave Simulation Basing on VOF Method and Dynamic Grid Technology. *Adv. Mater. Res.* **2013**, *774*, 3443–3446. [CrossRef]
20. Kotb, A.T.M.; Nawar, M.A.A.; Abd El Maksoud, R.M.; Mohamed, M.H. Comprehensive and synergistic analysis of geometry effect on an axial turbine performance for wave energy conversion. *Ocean Eng.* **2021**, *233*, 109212. [CrossRef]
21. Noel, A.B.; Abdaoui, A.; Elfouly, T.; Ahmed, M.H.; Badawy, A.; Shehata, M.S. Structural health monitoring using wireless sensor networks: A comprehensive survey. *IEEE Commun. Surv. Tutor.* **2017**, *19*, 1403–1423. [CrossRef]
22. Cambuli, F.; Ghisu, T.; Virdis, I.; Puddu, P. Dynamic interaction between OWC system and Wells turbine: A comparison between CFD and lumped parameter model approaches. *Ocean Eng.* **2019**, *191*, 106459. [CrossRef]
23. Setoguchi, T.; Takao, M. Current status of self rectifying air turbines for wave energy conversion. *Energy Convers. Manag.* **2006**, *47*, 2382–2396. [CrossRef]
24. Zhang, Y.; Wu, P.; Duan, M. A mesh-independent technique to evaluate stress singularities in adhesive joints. *Int. J. Adhes. Adhes.* **2015**, *57*, 105–117. [CrossRef]
25. Zhang, Y.; Li, D.Q.; Hong, S.H.; Zhang, M. Design of a new oscillating-buoy type wave energy converter and numerical study on its hydrodynamic performance. *Hrčak* **2023**, *74*, 145–168. [CrossRef]
26. Falnes, J. Wave-energy conversion through relative motion between two single-mode oscillating bodies. *J. Offshore Mech. Arct. Eng.* **1999**, *121*, 32–38. [CrossRef]
27. Zhang, Y.; Duan, M.; Wang, Y.; Chu, G. Analytical study of the strength of adhesive joints of riser pipes. *Ships Offshore Struct.* **2015**, *10*, 545–553. [CrossRef]
28. Cui, Y.; Hyun, B.-S. Numerical study on Wells turbine with penetrating blade tip treatments for wave energy conversion. *Int. J. Nav. Archit. Ocean Eng.* **2016**, *8*, 4564–4565. [CrossRef]
29. Zhao, X.; Xue, F.; Chen, L.; Göteman, M.; Han, D.; Geng, J.; Sun, S. Hydrodynamic analysis of a floating platform coupled with an array of oscillating bodies. *Ocean Eng.* **2023**, *287*, 115439. [CrossRef]
30. Cui, Y.; Zhao, H. Marine renewable energy project: The environmental implication and sustainable technology. *Ocean Coast. Manag.* **2022**, *232*, 106415. [CrossRef]

Journal of
Marine Science and Engineering

MDPI

Article

High-Order Spectral Irregular Wave Generation Procedure in Experimental and Computational Fluid Dynamics Numerical Wave Tanks, with Application in a Physical Wave Tank and in Open-Source Field Operation and Manipulation

Young Jun Kim [1,2,3], Maxime Canard [1], Benjamin Bouscasse [1], Guillaume Ducrozet [1], David Le Touzé [1] and Young-Myung Choi [2,3,*]

1 Nantes Université, École Centrale Nantes, CNRS, LHEEA, UMR 6598, F-44000 Nantes, France; david.letouze@ec-nantes.fr (D.L.T.)
2 Department of Naval Architecture and Ocean Engineering, Pusan National University, Busan 46241, Republic of Korea
3 Global Core Research Center for Ships and Offshore Structures (GCRC-SOP), Pusan National University, Busan 46241, Republic of Korea
* Correspondence: youngmyung.choi@pusan.ac.kr

Abstract: The accurate generation of a target sea state in numerical or experimental wave tanks is a fundamental line of research for the ocean engineering community. It guarantees the quality and relevance of wave–structure interaction tests. This study presents a reproducible irregular wave generation and qualification procedure, accounting for the nonlinear aspects of wave propagation. It can be used for both numerical simulation and experiments. The presented numerical and experimental results are obtained from the OpenFOAM solver and the Ecole Centrale Nantes wave tank facilities, respectively. The procedure comprises two steps: First, the wavemaker motion is calibrated numerically to generate the target wave spectrum at the position of interest. This is achieved with a wavemaker-equipped nonlinear potential flow solver. The open-source HOS-NWT solver, based on the high-order spectral method, was employed in this study. Then, the corrected wavemaker motion is used directly in the experimental wave tank. OpenFOAM simulations were performed to generate waves with the relaxation method, using wave elevation and velocity field data from HOS-NWT. The procedure was finally tested for mild and extreme breaking sea states. The waves generated by the HOS-NWT solver, the experiment, and the OpenFOAM simulation were compared from both stochastic and deterministic perspectives.

Keywords: irregular wave; breaking wave; experimental wave tank; numerical wave tank; high-order spectral; OpenFOAM

1. Introduction

Floating bodies operating in the ocean encounter various environmental conditions throughout their lifetime, including wind, currents, and waves. The primary load on the floating bodies is induced by waves in the ocean. Wave–structure interaction (WSI) is a major concern in the design of ship and marine structures. Many academic and industrial actors have conducted wave basin experiments or computational fluid dynamics (CFD) simulations to test WSI performances [1–4]. The waves generated for these tests must accurately represent specific in situ wave conditions and be easily reproduced in both an experimental wave tank (EWT) and CFD-based numerical wave tank (CFD-NWT). Therefore, it is essential for WSI tests to establish a uniform procedure to reproduce accurately irregular wave fields for both EWTs and NWTs. This procedure should accurately generate the target sea state at the location of interest within the domain and ensure the quality of the wave corresponding waves.

In the EWT wave generation procedure, the wavemaker motions are typically constructed using a linear transfer function applied to the target wave spectrum with random phases. However, the natural nonlinear characteristics of waves lead to changes in their statistical properties as they propagate from the wavemaker. Furthermore, replicating a specific target wave, such as a wave field generated by a numerical code, using a linear transfer function is a complex challenge.

In the case of a CFD-NWT simulation, mainly three wave generation approaches can be considered. The first approach is the modeling of a moving wavemaker, where the CFD solver generates and propagates waves by manipulating the wavemaker's stroke. While this method can replicate experimental waves in an NWT, it requires significant computational resources to model the whole wave tank, including the wavemaker. The second approach is a inner-domain wave-making method [5], which can model the propagation of the intermediate and shallow water waves using a Boussinesq-type model [6,7]. The third approach imposes the velocity and elevation of the incident wave on a specific boundary or region of a smaller-size CFD domain, facilitating the efficient generation of a target incident wave. This approach is widely used in various NWT applications, such as NWT simulations of semisubmersible [8]. This approach allows the simulation of nonlinear wave interaction, such as wave breaking, in the CFD domain while maintaining efficiency by minimizing the computational domain.

As the waves propagate along the tank, the shape and parameters of the wave spectrum are affected by nonlinear wave interaction and dissipation phenomena. These effects have been extensively studied experimentally [9,10] and theoretically [11,12] in recent years. Therefore, for both EWTs and NWTs, to ensure the accurate generation of a specific sea state at the position of interest, a wave qualification procedure is essential. The latter mainly relies on two features/parameters/measures: the wave spectrum at the target location is compared with the design wave spectrum, and the wave crest height distribution, which represents the severity of the wave field, is compared with semiempirical references. On this ground, the irregular wave generation procedure includes a wave calibration step that ensures the quality of the spectrum at the target position. This calibration consists of iterative correction of the wave input (such as the wavemaker motion) to balance the wave propagation effects from the wavemaker to the target position. For the EWT, this wave calibration step is highly time-consuming and may require up to a quarter of the total experimental time. For the CFD-NWT, this procedure is almost impossible to apply because of its prohibitive computational cost.

Therefore, there is a need for alternative methods for wave calibration. In the recommended procedures and guidelines of the International Towing Tank Conference (ITTC) for the laboratory modeling of waves [13], various nonlinear irregular wave modeling methods were reviewed. From those methods, the high-order spectral (HOS) method was chosen to be employed in this study. The HOS method is a nonlinear wave generation method that can be extended up to an arbitrary order [14,15], with the advantage of being computationally efficient. In this study, we used the open source HOS-NWT program [16], which is an extension of the HOS method designed to replicate waves in an NWT using a wavemaker. The HOS-NWT solver includes all the necessary nonlinear wave propagation features and is relatively cost-effective. Furthermore, the iterative correction of the wave spectrum within such an NWT was successfully validated in previous studies [17].

The present study aimed to build and validate an irregular wave generation and qualification procedure for irregular wave experiments and CFD simulations based on an iterative HOS-NWT calibration. This procedure enables the generation and propagation of the same qualified nonlinear irregular waves in both EWTs and CFD-NWTs without a calibration process. It was expected to not only reduce the time required for experimental wave calibration but also to facilitate easy synchronization of the experimental results with the CFD-NWT results. Detailed information about and the validation of the suggested procedure in the present paper are organized as follows:

- Section 2 presents the target wave conditions used in the wave generation and qualification procedure. A strong wave breaking condition is included to assess the limitations of the suggested procedure;
- Section 3 presents the wave generation and qualification procedure and workflow of the present study. The deterministic and stochastic comparison approaches are also introduced;
- Section 4 presents details on the iterative wave calibration procedure using the HOS-NWT solver. The iteration with the error criterion is applied to obtain an appropriate wave spectra in the numerics;
- Section 5 describes the experimental setup in the École Centrale Nantes (ECN) Ocean Engineering tank;
- Section 6 details the numerical features of the OpenFOAM simulations, the wave generation methodology, and validation with spatial and temporal refinement;
- Section 7 gives a comparative analysis between the HOS-NWT, the experiment, and the CFD simulation. Deterministic analyses were performed by comparing the wave elevation and the wave profile during wave breaking events. Stochastic quantities such as the wave spectrum and the crest probability of exceedance were also compared;
- Section 8 provides a summary of the wave generation and qualification procedures, the limitations of the study, and our future plans.

2. Wave Conditions

In this study, the suggested procedure was applied to two design sea states, named `Case1` and `Case3`. These sea states are the same as the wave conditions considered during the "Reproducible Offshore CFD" joint industrial project [18], which has been conducted by various academic and industrial groups. For both cases, the design wave spectra follow the unidirectional Joint North Sea Wave Project (JONSWAP) model [19].

Table 1 presents the JONSWAP parameters for `Case1` and `Case3`. The significant wave height is denoted as H_s, the peak period as T_p, the peak wavelength as λ_p, the magnification factor as γ, and the steepness parameter as $\epsilon = H_s/\lambda_p$. The Benjamin–Feir Index (BFI) parameter, characterizing the nonlinear wave interactions (also known as Benjamin-Feir instability), was estimated using the formula suggested by Serio et al. [20].

Table 1. Irregular wave parameters based on the JONSWAP spectrum.

Case	H_s (m)	T_p (s)	h (m)	λ_p (m)	γ	ϵ (%)	BFI
`Case1`	6	12.25	500	234	1	2.5	0.318
`Case3`	17	15.5	500	375	2.6	4.5	0.677

The normalized spectrum shapes of `Case1` and `Case3` are depicted in Figure 1. Both wave conditions can be assumed to be deep-water conditions. `Case3` corresponds to the Gulf of Mexico 1000-year return period sea state, where frequent and intense wave breaking events are expected, and `Case1` represents a relatively mild sea state, where wave breaking barely occurs.

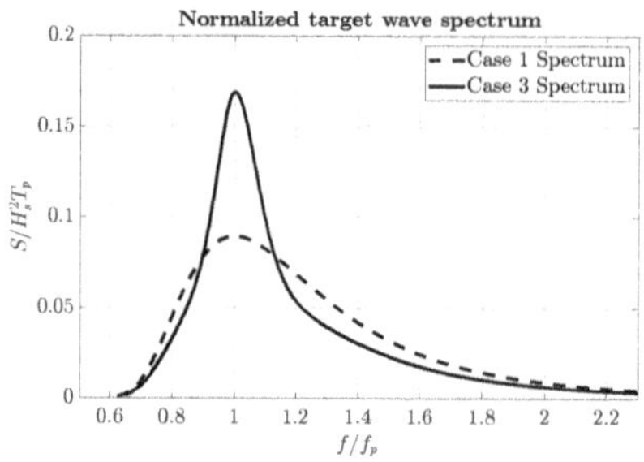

Figure 1. Normalized target wave spectra for `Case1` and `Case3`.

3. Wave Generation and Qualification

Generating consistent irregular waves for WSI tests, in both experimental and numerical studies, presents challenges owing to the nonlinearity of the waves. The target sea state has to be generated at a given distance from the wavemaker denoted X_t and denoted the target location. In severe sea states, the stochastic properties of the wave field evolve with propagating distance, primarily due to nonlinear wave interactions and energy dissipation. Therefore, a procedure to accurately reproduce a designated wave field, satisfying both stochastic and deterministic criteria, at the target location is essential. The following sections introduce a data analysis methodology and an algorithm designed to reproduce nonlinear waves in both the experimental wave tank and the CFD-NWT.

3.1. Metrics for the Data Analysis

The metrics for the wave qualification procedure fulfill two distinct objectives: first, to evaluate the similarity between two temporal wave measurements for a deterministic comparison; and second, to compare the stochastic characteristics of different sea states.

3.1.1. Deterministic Comparison

When the free surface elevation is provided as spatio-temporal information, a relative difference function $D(t)$ in Equation (1) can be introduced to observe the time-varying difference.

$$D(t) = \frac{1}{H_{ref} \cdot N} \sum_{i=1}^{N} |\eta_A(t, x_i) - \eta_B(t, x_i)| \tag{1}$$

where H_{ref} is the reference wave height that is chosen to be the significant wave height for irregular waves, N is the number of measurement locations x_i along the propagation, and $\eta_A(t, x_i)$ and $\eta_B(t, x_i)$ are the wave elevations compared. Furthermore, the averaged Pearson correlation coefficient $\bar{r}(t)$ can be considered with a moving time window [21].

$$\bar{r}(t) = \frac{1}{N} \sum_{i=1}^{N} r_i(t) \tag{2}$$

where $r_i(t)$ represents the local Pearson correlation coefficient of each elevation measurement is given as:

$$r_i(t) = \frac{\text{cov}[\eta(\mathcal{T}(t), x_i), \eta_{ref}(\mathcal{T}(t), x_i)]}{\sigma[\eta(\mathcal{T}(t), x_i)]\sigma[\eta_{ref}(\mathcal{T}(t), x_i)]}, \qquad \mathcal{T}(t) = \{\tau | \tau \in (t - T_w/2, t + T_w/2)\} \tag{3}$$

where $\text{cov}[p, q]$ represents the covariance of data p and q, $\sigma[q]$ is the standard deviation of data q, and T_w represents the time window duration. In the present study, $T_w = 5T_p$ was chosen following Choi et al. [22]. With this approach, the time shift (δt) from the reference run (maximizing the cross correlation) is computed to estimate a time-dependent deterministic error. Therefore, the time shift (δt) and the averaged Pearson correlation coefficient ($\bar{r}(t)$) are compared.

3.1.2. Stochastic Comparison

The wave spectrum is evaluated using Welch's spectral density estimation method with a smoothing window length of $15T_p$ for each realization [23]. Each spectrum can be bounded by a 95% confidence interval, which accounts for the sample variability of the data. The spectra are then averaged over the N realizations to obtain the averaged spectrum $S(f, x = X_t)$.

The stochastic properties of waves, such as the significant wave height H_s, are evaluated from the wave spectrum. The wave peak frequency $\hat{f}_p = 1/\hat{T}_p$ is evaluated using the Young [24] technique.

$$H_s = 4\sqrt{m_0}, \qquad m_0 = \int_0^\infty S(f)df, \qquad \hat{f}_p = \frac{\int_0^\infty fS^4(f)df}{\int_0^\infty S^4(f)df} \tag{4}$$

The probability distribution of the wave crest heights from single realizations (PDSR) and the ensemble realization (PDER) is also considered. Since different wave realizations give different wave crest PDSR curves, Huang and Zhang [25] proposed the PDSR range with a confidence level. The PDER is evaluated for N realizations to include all events. For the analysis of PDER, the Jeffrey confidence intervals [26] are also utilized to estimate the sampling variability error. For this analysis, the uncertainty limit was set at the probability of the 20th most extreme event.

3.2. *Wave Qualification Criteria*

Criteria for qualifying the spectrum and crest height distributions were established in the 'Reproducible CFD modeling Practice for Offshore Applications' Joint Industrial Project [18].

According to these criteria, a generated sea state is considered qualified if the relative spectrum error ($E(f)$), as defined in Equation (5), remains within ±5% of the target frequency range $f \in \left[\frac{3}{4}f_p, \frac{3}{2}f_p\right]$, where f_p is the wave peak frequency of the sea state.

$$E(f) = \frac{S(f, x = X_t) - S_{target}(f)}{S_{target}(f)} \quad , \quad f \in \left[\frac{3}{4}f_p, \frac{3}{2}f_p\right] \tag{5}$$

where $S(f, x = X_t)$ is the measured wave spectrum from EWT, HOS-NWT, and CFD-NWT, and $S_{target}(f)$ is the target wave spectrum.

The probability distribution of the wave crest height is also used for wave qualification. The measured PDSR and PDER are compared with reference wave crest height distributions. Longuet-Higgins [27] showed that the wave crest distribution of linear waves with assumptions of a Gaussian distribution and a narrow-band spectrum follows the Rayleigh distribution. However, the Rayleigh distribution is known to underestimate the height of the wave crests, particularly in severe sea states. Forristall [28] suggested the wave crest distribution based on second-order theory. Later, Huang and Zhang [25] proposed a semiempirical formula for the PDER curve and the mean and 99% upper and lower bounds of the PDSR curves, derived from the numerical wave simulation of 305 sea states. Therefore, the Rayleigh distribution, Forristall distribution, Huang's PDER distribution, and the mean and 99% bounds of Huang's distributions were used as references for comparing the measured PDSR and PDER curves. However, studies such as Onorato et al. [29] and Canard et al. [30] showed that, in a wave tank environment, the distribution of the wave crest heights depends mainly on the distance from the wavemaker due to modulational instabilities that develop during the propagation of unidirectional irregular sea states. The references mentioned above are dedicated to open ocean description and consequently do not consider these effects and, thus, should be used carefully.

3.3. Available Tools and Adopted Procedure

Obtaining converged wave stochastic properties in experiments or CFD simulations is extremely time-consuming because a significant number of realizations of irregular waves are required. Additionally, proposing a calibration procedure in EWTs or CFD-NWTs will be highly time-consuming, making it practically unusable. Therefore, a dedicated and efficient numerical wave tank tool based on potential flow theory can be used in a preliminary step.

To this end, this study proposes a novel wave calibration and generation procedure that employs the HOS-NWT solver as an iterative wave calibration tool. This will provide the adequate wavemaker motion for EWTs and/or the nonlinear incident wave field for CFD-NWTs. Figure 2 illustrates the workflow of irregular wave calibration and generation.

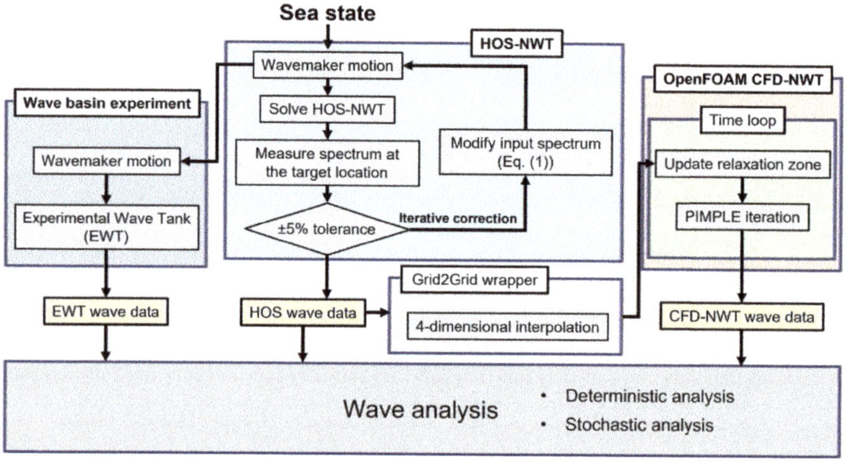

Figure 2. Wave generation and qualification workflow for the HOS-NWT, the wave basin experiment, and the OpenFOAM simulations.

When the target sea state is chosen, the iterative wave calibration process is performed with the HOS-NWT solver (see Section 4 for details). The wavemaker motion in the HOS model is adjusted iteratively, continuing until the relative difference in the wave spectrum ($E(f)$ in Equation (5)) falls within a ±5% tolerance. After the iteration procedure is completed, the wavemaker motions for the EWT and the HOS wave data for the CFD-NWT are available to conduct the WSI study of interest. One objective of the present study was, thus, to perform the wave qualification process with the waves measured experimentally and the ones computed with the CFD-NWT (see Section 7). Finally, the wave qualification process is performed for the waves measured from the experiment and the CFD simulation.

4. Iterative Wave Calibration with HOS-NWT

The HOS-NWT solver, which is an extension of the HOS method, incorporates specific features of a wave tank environment, such as a wavemaker, lateral reflective walls, and an absorbing beach. It has been extensively validated and used as a digital twin of the ECN wave basin.

The nonlinear potential flow models naturally neglect vorticity and viscosity effects by employing irrotational and inviscid assumptions, which dramatically facilitate the wave modeling. However, these assumptions prevent the use of potential flow solvers for phenomena involving strong viscous effects or high vorticity such as breaking waves.

To overcome this limitation, the Tian–Barthelemy wave breaking model was employed. This model uses Tian's eddy viscosity model with Barthelemy's breaking criterion [31,32]. The wave breaking model predicts the onset of wave breaking events and estimates the amount of dissipation using the wave profile.

Calibration at the Target Position Using HOS-NWT

The preliminary step of generating calibrated HOS-NWT waves for sea states `Case1` and `Case3` was performed using an iterative procedure. As the detailed process of this iterative wave calibration using HOS-NWT is thoroughly described in Canard et al. [17], this section will briefly outline the numerical setup of the HOS-NWT solver.

The 2D numerical domain (length, depth, and wavemaker geometry) was set to mimic the ECN Ocean Engineering wave tank to allow consistent comparisons with the experiment. The dimensions of the 2D numerical wave tank were $L_x = 5000$ m in length and $h = 500$ m in depth. The target location from the mean wavemaker was set to 708 m (i.e., $X_t = 3\lambda_p$) for `Case1` and 750 m (i.e., $X_t = 2\lambda_p$) for `Case3`. Note that the numerical absorbing beach was tuned to minimize wave reflection.

Table 2 presents the HOS-NWT numerical setup adopted for `Case1` and `Case3`. Spatial discretization was specified using $k_{max} = N_x\pi/L_x$, i.e., the largest wave number solved, which is related to the number of nodes in the x direction (N_x). N_z is the number of HOS nodes in the vertical direction. The HOS simulation time (T_{simul}) for each seed was 10,800 s, excluding the transitory stage for wave propagation. The wave breaking model was activated only for `Case3`. The intensity of the eddy viscosity (α_e), introduced for the wave breaking model in Tian et al. [31], acted as the damping when the wave breaking event occurred. For this study, $\alpha_e = 0.04$ was used. Finally, the HOS order was set to $M = 3$, which ensured an accurate and efficient solution of the nonlinear problem.

Table 2. HOS-NWT numerical setup for `Case1` and `Case3`.

Wave Case	N_x	N_z	α_e	M	T_{simul}
`Case1`	1024 ($k_{max} = 25k_p$)	17	0	3	10,800 s ($880T_p$)
`Case3`	400 ($k_{max} = 15k_p$)	17	0.04	3	10,800 s ($700T_p$)

The iterative spectrum correction procedure was applied to `Case1` and `Case3` at their respective target locations. Following the calibration procedure in Canard et al. [17], after 1 and 4 iterations for `Case1` and `Case3`, respectively, the mean wave spectrum at X_t satisfied the $\pm5\%$ criterion. In Canard et al. [17], 100 realizations per sea state were calibrated. Since such a large number of realizations cannot be performed in CFD or in experiments, for the present study, we used selected 10 realizations for each sea state.

Figure 3 presents the `Case3` wave spectrum and the crest ensemble distribution computed using the 100 seeds (left plots), and the 10 seeds selected for CFD and experimental reproduction (right plots). The thick solid black line, which is the mean wave spectrum, satisfies the $\pm5\%$ criterion. The gray lines in the spectrum plots correspond to the spectra computed from the individual realizations. The confidence intervals are also displayed to quantify the increase in the sampling variability error that follows the selection of the 10 seeds.

The calibrated HOS-NWT results were used to generate the equivalent EWT (Section 5) and the corresponding CFD-NWT simulation (Section 6). Finally, the associated stochastic quantities (spectrum, crest distribution, significant wave height, peak period) are presented and compared in Section 7.

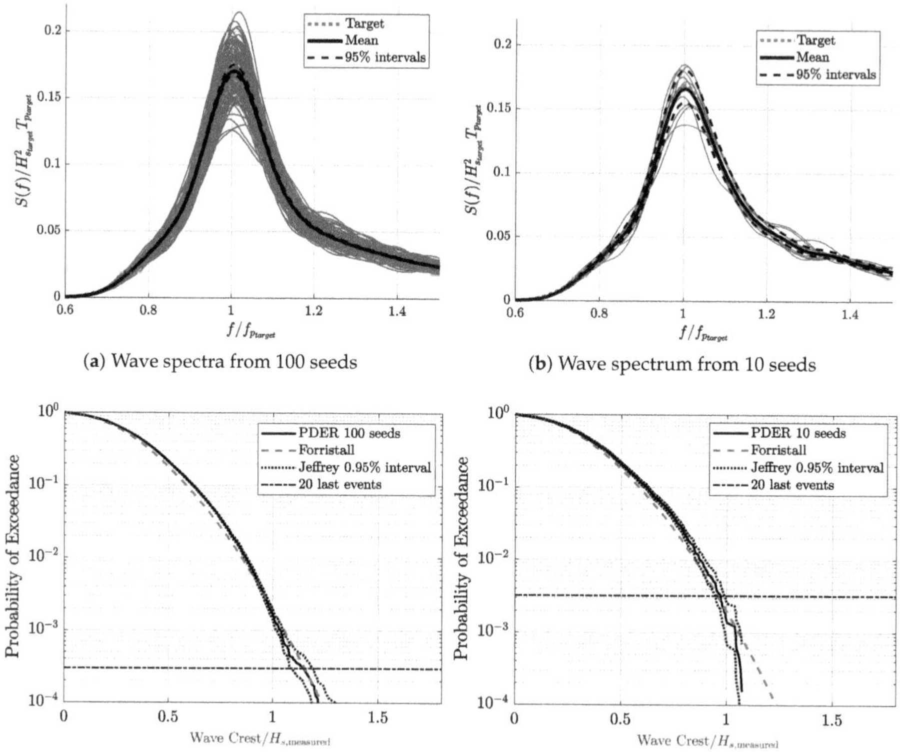

(**a**) Wave spectra from 100 seeds

(**b**) Wave spectrum from 10 seeds

(**c**) Ensemble crest distribution from 100 seeds

(**d**) Ensemble crest distribution from 10 seeds

Figure 3. `Case3` HOS-NWT wave spectra and ensemble crest distribution at X_t after iteration, using the 100 seeds (**a**,**c**), and the 10 seeds selected for CFD and experimental reproduction (**b**,**d**). The grey lines in the spectrum plots correspond to the spectra computed from the individual realizations.

5. Experiments in ECN Ocean Wave Tank

The wave generation experiments, following the workflow in Figure 2, were conducted in the ECN Ocean Engineering tank. In these experiments, the calibrated wavemaker motion derived from the iterative procedure using the HOS-NWT solver was applied to the experimental wavemaker. For both sea states `Case1` and `Case3`, 10 realizations were reproduced in the wave tank.

This section briefly describes the experimental setup, and the experimental results are discussed in Section 7. Further details on the wave basin experiments can be found in Canard et al. [33].

Experimental Setup

Figure 4 presents the dimensions of the ECN Ocean Engineering tank, which measures 46.4 m in length, 29.74 m in width, and 5 m in depth. The tank features a segmented wavemaker at one end, consisting of 48 independent flaps with hinges positioned 2.147 m above the bottom. At the opposite end of the tank, a parabolic absorbing beach is installed to reduce wave reflections, limiting the reflection to under 5% of the wave amplitude for the range of wave frequencies considered for the present study.

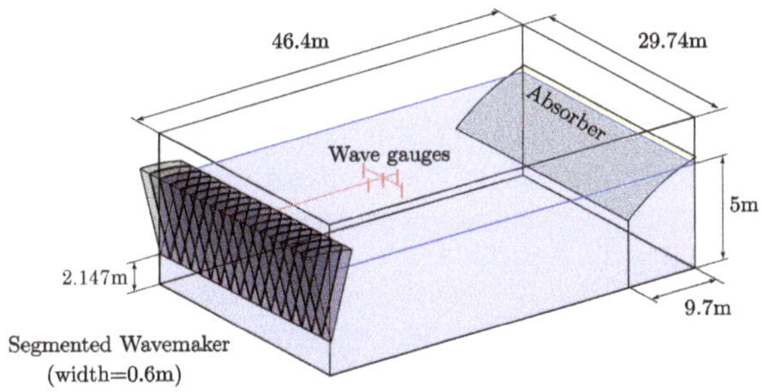

Figure 4. ECN ocean engineering tank dimensions.

To use optimized wavemaker capabilities and position the target near the middle of the wave tank, the waves for `Case1` and `Case3` were generated on a scale of 1:40 and 1:50, respectively. The tank was equipped with resistive wave gauges measuring the free surface elevation time series with a sampling frequency of 100 Hz. The uncertainty tests for the same wave measurement system were conducted in Canard et al. [30], and the uncertainty of corrected calibration factor of the wave gauge was about 2.5% in the calibration stage. The positions of these gauges, both in the model and full-scale, are listed in Table 3. WG1, positioned at the target location, was the reference wave gauge for the experiment. WG4 and WG5, installed for validation purposes, were placed at the same distance from the wavemaker along a transversal line.

Table 3. The position of wave gauges in the x-direction from the wavemaker in the ocean engineering tank. WG1 was the reference wave gauge.

Sea State	WG1, WG4, WG5	WG2	WG3
`Case1` (Scale: 1/40)	17.7 m	13.95 m	21.45 m
`Case3` (Scale: 1/50)	15 m	11.25 m	18.75 m

6. CFD Simulations with the OpenFOAM `foamStar` Solver

This study employed Open-source Field Operation And Manipulation (OpenFOAM) [34] as a CFD tool. OpenFOAM provides a C++ toolbox for the development of user-customized numerical solvers with pre-/postprocessing utilities, mainly for fluid mechanics solutions.

In this study, the OpenFOAM simulation was performed with an additional in-house code package, `foamStar`, co-developed by Bureau Veritas and ECN, which supports specific functionalities for naval applications such as wave generation and floating body controls. It also interacts with an external library, `Grid2Grid`, which is a wrapper program that reconstructs and interpolates the HOS-NWT wave elevations to any point where it is needed in the CFD domain following the protocol given by Choi et al. [22]. Details of the OpenFOAM formulation can be found in Seng [35], and the numerical configurations for the CFD simulation are briefly discussed.

6.1. Numerical Configurations

The governing equations were solved with a finite volume method (FVM) using an unstructured polyhedral volume mesh with appropriate spatial and temporal schemes and boundary conditions. OpenFOAM supports numerous user-selectable spatial and temporal schemes. The simulation results were sensitive to the selection of numerical schemes, especially convection and time-integration schemes. For all simulations in this study, the set of spatial discretization schemes outlined in Table 4 was used.

Table 4. Spatial discretization schemes.

Item	Equation	FV Scheme
Gradient	∇q	cellLimited Gauss linear 1.0
Surface Normal gradient	$\mathbf{n} \cdot (\nabla q)_f$	corrected 0.5
VOF convection	$\nabla \cdot (\mathbf{u}\alpha)$	Gauss vanLeer + MULES limiter
VOF relative flux	$\nabla \cdot (\mathbf{u}_r \alpha (1-\alpha))$	Gauss linear
Momentum convection	$\nabla \cdot (\rho \mathbf{u}\mathbf{u})$	Gauss linearUpwindV grad(U)
Momentum Laplacian	$\nabla \cdot (\mu \nabla (\mathbf{u}))$	Gauss linear corrected 0.5
Pressure Laplacian	$\nabla \cdot (\frac{1}{q_\rho} \nabla (p))$	Gauss linear corrected 0.5
k convection	$\nabla \cdot (\rho \mathbf{u} k)$	Gauss upwind
ω convection	$\nabla \cdot (\rho \mathbf{u} \omega)$	Gauss upwind
Under-relaxation	q	1.0

For the Reynolds-averaged Navier–Stokes equations (RANSE) simulation based on the FVM in the naval engineering field, k-ω SST models are widely used. However, in the OpenFOAM community, it is known that the k-ω SST model generates a nonphysical high viscosity beneath the free surface in two-phase flow simulations. This excessive turbulence viscosity dramatically dampens the waves within a few wavelengths, which is unrealistic in wave propagation. To avoid nonphysical damping due to the turbulence model, Larsen and Fuhrman [36] introduced a buoyancy production term for two-phase flow applications. The current work adopted this concept to address the variation in density of the two-phase flow near the free surface, naming the adopted model the 'free surface k-ω SST'. In the application of the free surface k-ω SST model, the selection of the convection scheme is crucial for the stability and computational efficiency of the simulation, due to the mass flux $(\rho\mathbf{u})$ in the convection term. Therefore, the first-order `Gauss upwind` scheme was employed for both turbulence convection terms.

The simulations were conducted using the 0.99 off-centered Crank–Nicolson scheme, identified as the most efficient time-integration scheme for wave simulation in a previous study [37]. For turbulence transport, however, the implicit Euler time-integration scheme was employed to enhance numerical stability. Eight PIMPLE iterations and four PISO iterations were utilized to minimize the error from nonlinear iterations. Since the meshes were fully orthogonal, no nonorthogonal corrector was used.

6.2. Wave Relaxation Method

An explicit relaxation method [35,38] was adapted to generate and absorb waves. It smoothly blends the computed solution with the target solution in a relaxation zone. The computed solution (χ) is blended with the target solution (incident wave) (χ_I) as shown in Equation (6). An exponential blending weight function (w) was employed that varies between 0 and 1, with the normalized coordinate of $\xi_{relax} \in [0,1]$ changing in the relaxation zone.

$$\chi = (1-w)\chi + w\chi_I \quad , \quad w(\xi_{relax}) = \frac{e^{\xi_{relax}^{3.5}} - 1}{e-1} \tag{6}$$

6.3. CFD Numerical Domain and Spatial Discretization

The CFD numerical domain and the numerical configuration employed for wave simulations are detailed here. Figure 5 provides a full-scale representation of the 2D rectangular NWT for numerical simulations. The CFD simulation was conducted with the model scale (1/100). However, for consistency, all values are given in full-scale terms. For a clear definition of each solver's numerical setup, two distinct coordinate systems were introduced: one for the HOS domain and another for the CFD domain. The origin of the HOS coordinate system was established at the mean position of the wavemaker and the mean free surface. The length of the CFD domain was 1200 m and was positioned in the middle of the HOS computation domain from $x_{HOS} = 100$ m to $x_{HOS} = 1300$ m. In addition, the water depth was 500 m. Both inlet and outlet relaxation zone lengths were set to 300 m regions, and a 600 m long CFD zone was used for pure wave propagation.

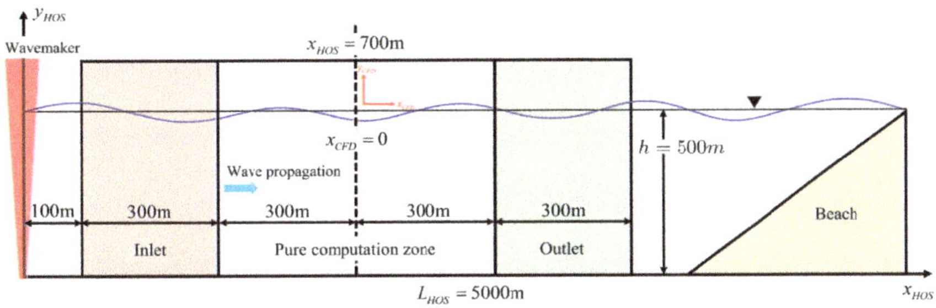

Figure 5. Schematic view of the HOS and CFD domains for irregular wave simulation.

At the initial time ($t = t_{start}$), the wave elevation and velocity were initialized across the entire CFD domain. The VOF (volume of fluid) field was set based on the HOS wave elevation, and the fluid velocity beneath the free surface was derived from the HOS wave velocity. To maximize efficiency on a multinode computer cluster system, the 3 h sequence was divided into seven or eight intervals. In each interval, 10 peak periods ($10T_p$) overlapped with the preceding and subsequent runs. Table 5 outlines the start time (t_{start}) and the end time (t_{end}) of each interval. Each interval corresponds to 110 peak periods: 1225 s for `Case1` and 1550 s for `Case3`. The total simulation time (T_{simu}) was 9922.5 s for `Case1` and 10,650 s for `Case3`. The validation of this approach is provided in Section 7.

Table 5. Time intervals of the irregular wave simulations for `Case1` and `Case3` in full-scale.

	`Case1` $H_s = 6$ m $T_p = 12.25$ s		`Case3` $H_s = 17$ m $T_p = 15.5$ s	
Run	t_{start} **(s)**	t_{end} **(s)**	t_{start} **(s)**	t_{end} **(s)**
Run1	1077.5	2425	1050	2700
Run2	2302.5	3650	2550	4200
Run3	3527.5	4875	4050	5700
Run4	4752.5	6100	5550	7200
Run5	5977.5	7325	7050	8700
Run6	7202.5	8550	8550	10,200
Run7	8427.5	9775	10,050	11,700
Run8	9652.5	11,000	-	-

Table 6 provides a summary of the mesh refinement parameters selected for each sea state. The cell sizes in the free surface refinement region are denoted as Δx and Δz in horizontal and vertical directions, respectively. A constant cell aspect ratio ($\Delta z/\Delta x = 0.5$) was used. The free surface refinement region was defined between $z = -H_s$ and $z = 1.5H_s$. The upper and lower refinement zones were linearly stretched from cell sizes comparable to those of the free surface refinement zone to large cells at the top and bottom domain boundaries. The first two refinement parameters in the table refer to the reference setup. Considering that the target frequency range was $f \in [\frac{3}{4}f_p, \frac{3}{2}f_p]$, the number of cells per shortest target wavelength (λ_{min}) was approximately $\lambda_{min}/\Delta x \approx 50$ for `Case1` and $\lambda_{min}/\Delta x \approx 80$ for `Case3`. The computational cost of single `Case3` simulation was about 120 CPU hours. The lower two mesh refinements setup in Table 6 were only used for the wave breaking event convergence test.

Table 6. Mesh refinement parameters in full-scale for the free surface refinement region ($-H_s < z < 1.5H_s$). The upper two refinements were the reference setup for the irregular wave simulations.

Wave Case	Δx (m)	Δz (m)	$H_s/\Delta z$	h (s)	Total N Cells
Case1	2	1	6	0.04	60,000
Case3	2	1	17	0.04	73,200
Case3 Dx4m	4	2	8.5	0.08	18,300
Case3 Dx1m	1	0.5	34	0.02	292,800

7. Results and Discussion

As presented in Figure 2, this section provides a comparative analysis of the results obtained from the HOS-NWT, OpenFOAM (`foamStar`), and experimental data. The experimental results are briefly described, and the results of the deterministic analysis and stochastic analysis are presented. The deterministic analysis emphasizes how the time series differed when comparing the three types of results. Specific attention was paid to the time and space where the wave breaking events were identified using the HOS-NWT wave breaking model and the possible differences that can be observed in the CFD and the experiments. The stochastic analysis focuses on a spectral analysis of wave signals corresponding to a calibrated wave condition and the probability distribution on the wave crest height (PDSR and PDER).

7.1. Experimental Result

In the analysis of the experimental data, considerable attention was paid to the evaluation of the accuracy and uncertainties of the experimental process and results. The following results briefly provide an estimate of errors that affect both deterministic and stochastic variables of interest. For more comprehensive details on this experiment, the readers are referred to the work of Canard et al. [33].

7.1.1. Repeatability

During the experimental campaign, the sea state `Case1 Seed1` experiment was conducted daily to test repeatability. This process resulted in eight runs and all runs started under calm water conditions, with the exceptions of `Test1` and `Test4`, which were started just after a previous run to assess the influence of preliminary noise in the tank.

Figure 6 offers a deterministic comparison of the eight runs. The first plot in the figure presents the comparison of the time series. The second and third graphs present the comparison of the time shift (δt) varying throughout the runs to maximize the Pearson correlation coefficient, and `Test6` was chosen as the reference data. For the runs that started in calm water and over the displayed time window, the correlation was around 0.99, and the time shift was below $\delta t < 0.05$ s $= 2.5\%T_p$. The errors increased for the other two runs (`Test1` and `Test4`), providing insight into the deterioration of the waves due to the influence of preliminary noise.

7.1.2. Transverse Measurements

The waves measured from WG1, WG4, and WG5 were compared to verify the unidirectionality of the generated waves and quantify the possible generation of spurious transverse waves. Figure 7 displays the mean spectrum and wave crest height PDER for `Case1`, as measured by the three wave gauges. It is observed that the PDERs were nearly indistinguishable up to the sampling variability limit. However, the spectra present notable differences. While WG1 and WG5 showed similar results, the WG4 measurements clearly deviated from the other two wave gauges, which reached 12% around $f = 1.4f_p$. This deviation was most likely due to the transverse waves, which are almost unavoidable in a wave basin experiment. Since WG1 was the reference wave gauge of the experiment, hereafter, only the experiment results of WG1 are discussed.

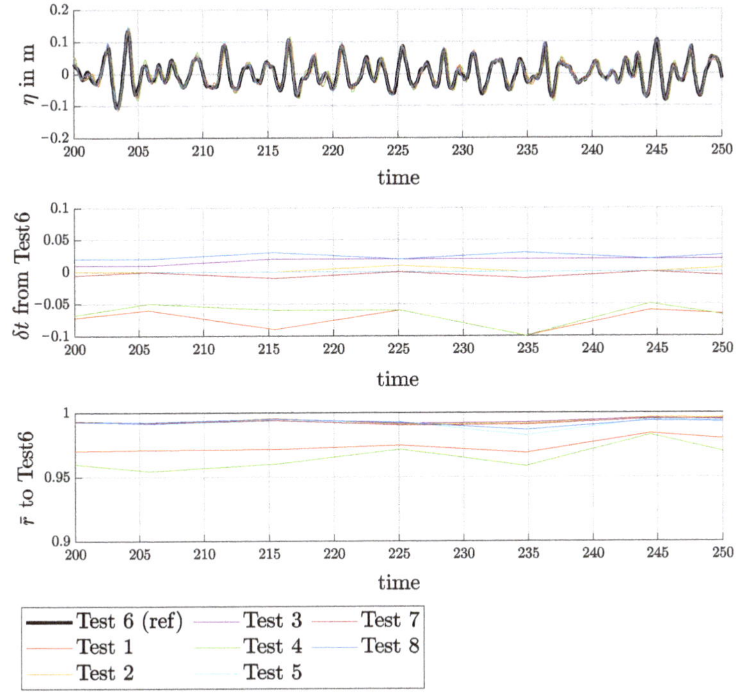

Figure 6. Deterministic comparison among the 8 runs of the repeatability data set (`Case1 Seed1`), WG1 measurements.

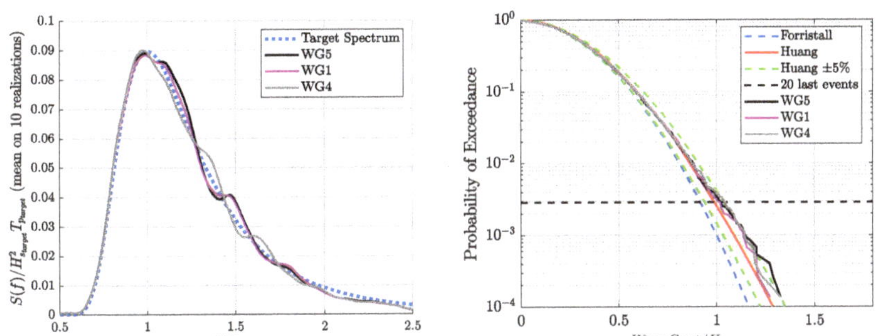

Figure 7. `Case1` wave spectrum and wave crest height PDER measured by the three wave gauges at $x = X_t$ with different lateral positions.

7.2. Deterministic Analysis

7.2.1. Influence of the Initial Time in the CFD-NWT Simulation

A notable advantage of this wave generation procedure for the CFD-NWT is that the user can choose the start time of the CFD simulation. Although performing a long CFD-NWT simulation is important for validating the proposed workflow, this approach remains beneficial since the appropriate length of the CFD simulation might be shorter than the typical 3 h realization.

This approach raises two key questions: First, how much transient time is needed for the CFD to evolve its own solution from the flow input? Second, are there accumulation errors or dysfunctions of the absorption capabilities that make the solution depend on the total simulation time? To answer these questions, the split simulations, as detailed in Table 5, were compared with a 3 h simulation.

A difference indicator function in Equation (1) was used to clearly observe the discrepancy of waves between the computational setups. Here, the reference wave elevation was the HOS-NWT, and the CFD-NWT waves were compared. The difference indicator function measures the normalized difference between the HOS-NWT wave elevation and the CFD wave elevation in the pure CFD computational (150 m $\leq x_{CFD} \leq$ 450 m).

In Figure 8, the wave elevation from the 3 h simulations is compared to the split simulation for the test cases `Case1 Seed1 Run8` and `Case3 Seed5 Run7`, which are the last time intervals, as described in Table 5. The computational setups for the split and 3 h simulations were identical. The first and third panels of Figure 8 display the normalized wave time series at $x_{CFD} = 0$ m for `Case1` and `Case3`, while the second and fourth panels illustrate the difference indicator function over time for the same cases. In the graphs, solid green lines represent the results of the split simulations, while orange dashed lines depict the results from the 3 h simulations. The solid blue line in the first and third panels indicates the wave elevation discrepancy between the `foamStar` 3 h simulations and the split simulation. For both `Case1` and `Case3`, the time series of the wave elevation between the split and the 3 h simulations generally showed good agreement.

The solid black line shows the discrepancy ratio between two respective difference functions. For `Case1`, the discrepancy ratio remained below 2.5% and gradually decreased as the simulation progressed. For `Case3`, the discrepancy ratio between two respective difference functions slowly decreased after 20 wave peak periods. These results suggest that, while CFD-NWT and HOS-NWT may produce locally varying time series due to different approaches in modeling nonlinearities, particularly wave breaking, the impact of memory effects in CFD computations appeared to be minimal. In conclusion, both the split and the 3 h simulation methods are effective for long irregular wave simulations, although further investigation is required to clarify the minor differences observed between these two approaches, particularly in `Case1`.

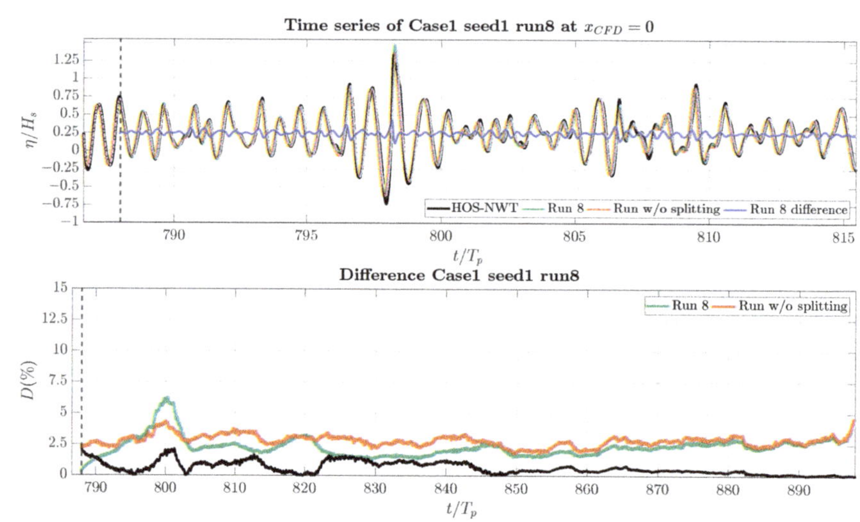

Figure 8. *Cont.*

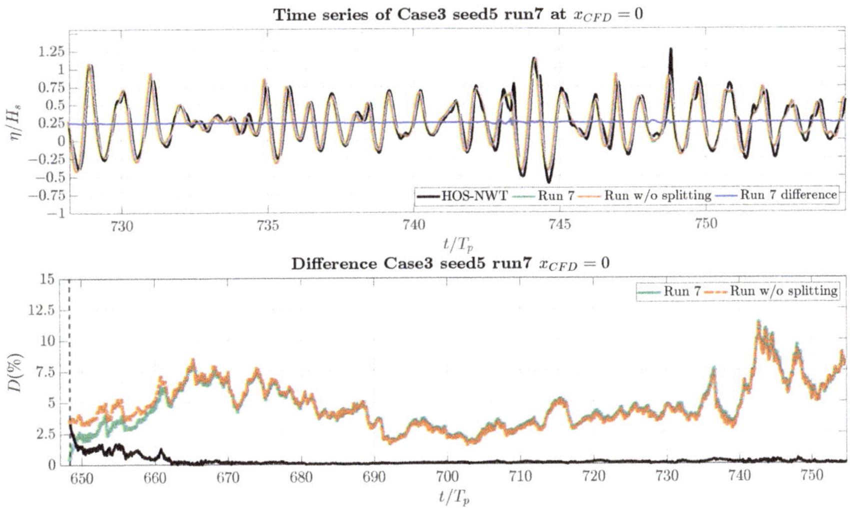

Figure 8. Comparison between split and simulation result for `Case1 Seed1 run8` (upper two plots) and `Case3 Seed5 Run7` (lower two plots). The black solid line in the second and fourth panels present the discrepancy ratio between the two respective difference functions.

7.2.2. Comparative Analysis of HOS-NWT, CFD-NWT, and EWT Focusing on Wave Breaking Events

A deterministic analysis of the wave elevation time series during wave breaking events was performed using both experimental and numerical results. The analysis was performed for the test case `Case3 Seed6 Run4`.

The timing and locations of wave breaking events were initially checked to facilitate further analysis. Figure 9 shows the wave breaking events identified using the Tian–Barthelemy model during HOS-NWT simulations. The x-axis represents the longitudinal position of the computational domain of HOS-NWT, and the y-axis indicates time. The thick black vertical dotted lines are the boundary between the pure CFD computational zone and the relaxation zone. Horizontal dotted lines indicate the time interval between simulations. Each colored box represents a breaking event, where the color of the box corresponds to its wave crest height (H_c). The dimensions of each box represent the specific position and duration of the wave breaking event.

Figure 10 presents a comparison of the wave signal from the ECN experiment, the HOS-NWT simulation, and the OpenFOAM (`foamStar`) simulation of the test case `Case3 Seed6`. As in Figure 6, the wave time series, the time shift, and the correlations were compared using the HOS-NWT wave as a reference.

In general, the wave signals from both `foamStar` and the experimental data demonstrated good agreement with the HOS-NWT results. Specifically, `foamStar` exhibited a larger correlation, while the experimental data showed a smaller time shift. During this event, both the `foamStar` and the experimental waves showed good agreement. However, after the wave breaking event, around 6400 s, a decrease in correlation was observed in the experimental waves. This result implies that the Tian–Barthelemy wave breaking model effectively replicates the dissipation of waves during the wave breaking event. However, the memory effect of the wave breaking event shown in the experiment was not fully reproduced in the HOS-NWT and `foamStar` simulations.

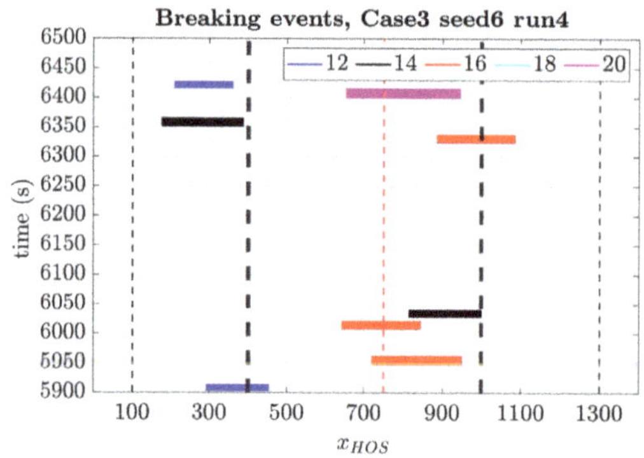

Figure 9. Examples of wave breaking events identified using the Tian–Barthelemy breaking model for HOS-NWT waves. Each box represents space and time where and when the breaking events occur, and the color of the box corresponds to its wave crest height (H_c). The thin black vertical lines represent the boundary of the CFD zone, whereas the thick black vertical dotted lines are the boundary between the pure CFD computational zone and the relaxation zone. The red vertical line is the reference position.

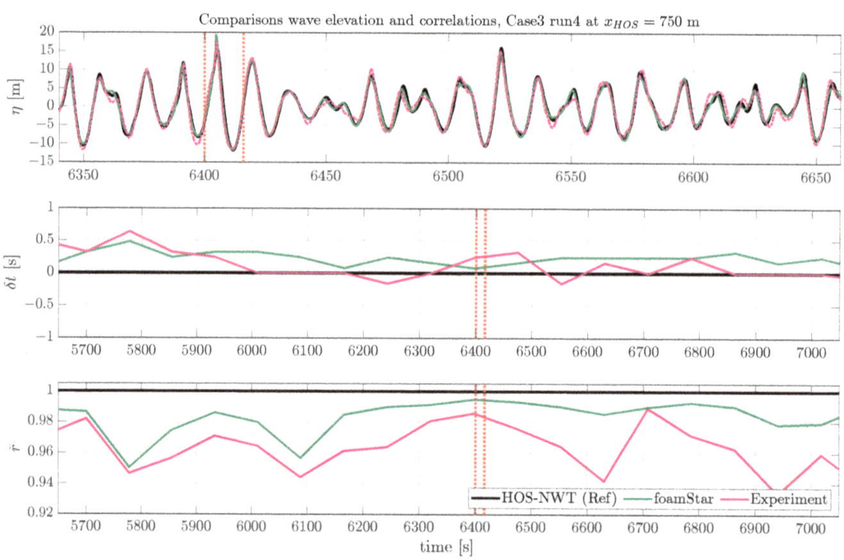

Figure 10. Comparison of the wave elevation time series from the experiment `foamStar` and HOS-NWT at x_{CFD} = 750 m `Case3 Seed6`. The red vertical lines represent the wave breaking events.

Another deterministic analysis was conducted that mainly aimed to compare the difference in the spatial distribution of the wave elevation during the wave breaking events. The spatial distribution of the wave elevation from the HOS-NWT, `foamStar`, and point measurement from the experiment were compared.

In Figure 11, the five plots present the spatial distributions of the wave elevation of the wave breaking event measured within the time interval of [5940 s, 5980 s] and the space range of $x = [700, 1000]$ m. The vertical dotted lines are the boundaries between the pure computational zone and the relaxation zones in the `foamStar` simulation. The results of the CFD simulation were obtained from three different mesh refinements, which are listed in Table 6. Before the wave broke, the CFD waves globally followed the spatial distribution of HOS-NWT. When the wave breaking event occurred, the CFD waves deviated from

HOS-NWT and their values were closer to the experimental measurements. This result was expected since the wave breaking model implemented in the HOS-NWT did not represent the exact physical process during the breaking phase. It was designed to be representative of the real physics in terms of postbreaking properties (energy dissipation and redistribution). This was validated in the observation that, after breaking, the waves quickly recovered the excellent agreement between the HOS-NWT, CFD-NWT, and EWT. Also, after the wave breaking event, the CFD results exhibited different patterns of local wiggling, depending on the mesh refinement level. This implies that the CFD mesh refinements used in the present study are sufficient to reproduce the wave breaking event.

7.3. Stochastic Analysis

A stochastic analysis on the irregular waves was performed to estimate statistical properties and check the quality of the generated waves. The analysis was carried out using 10 realizations of waves measured at the target location. Each realization of a wave corresponded to approximately 3 h of measurements at full-scale.

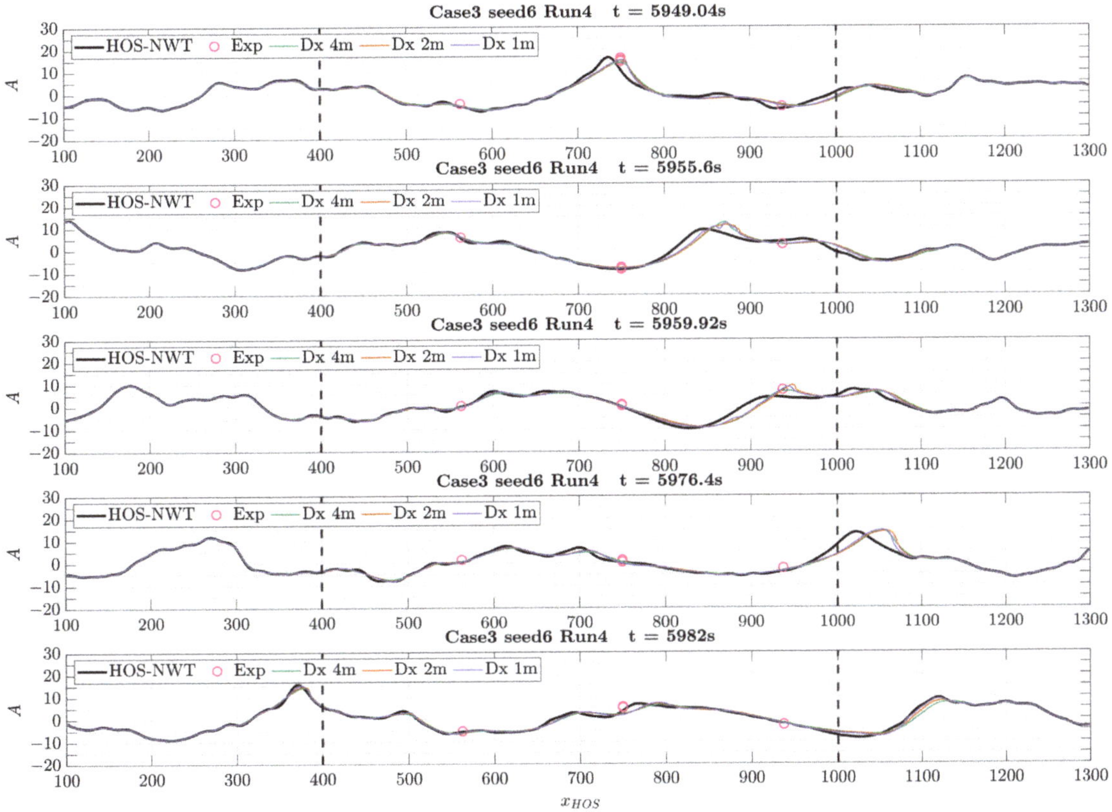

Figure 11. Comparison of the wave profile from the HOS-NWT solver and `foamStar` solver with three resolutions during the wave breaking event of `Case3 Seed6 Run4`.

7.3.1. Spectral Analysis

For spectral analysis, the averaged wave spectrum of 10 realizations was calculated and compared. The following analysis compares four different wave spectra, which are the target wave spectrum and the averaged wave spectrum obtained from the HOS-NWT, `foamStar`, and experiment. Figure 12 compares the mean wave spectra of `Case1` and `Case3` with the target spectrum. The x-axis is normalized to its peak wave frequency, and the y-axis is normalized to its significant wave height and peak period. The mean wave spectrum is defined as the average of all wave spectra from the 10 three-hour realizations. The following analysis compares four different wave spectra, which are the target wave spectrum and the averaged wave spectrum obtained from the HOS-NWT, `foamStar`, and experiment. Since the HOS-NWT waves were already corrected to the target spectrum, the HOS-NWT spectrum showed very good agreement. The `foamStar` results generally showed slightly underestimated results, and the experiment showed small oscillations around the target, depending on the wave frequency.

Figure 13 presents a quantitative comparison of the mean wave spectra by assessing the difference ratio between the wave spectra. The measured wave spectra are compared with the target spectrum on the left graph and also with the HOS-NWT spectrum on the right graph of Figure 13. A comparison with the HOS-NWT wave spectrum was considered important since the HOS-NWT was the input of the experiment and the CFD simulation. In this graph, the black dotted lines denote the bounds of the ±5% target criterion region, while the red dotted lines indicate the target frequency range of $[0.75f_p\ 1.5f_p]$.

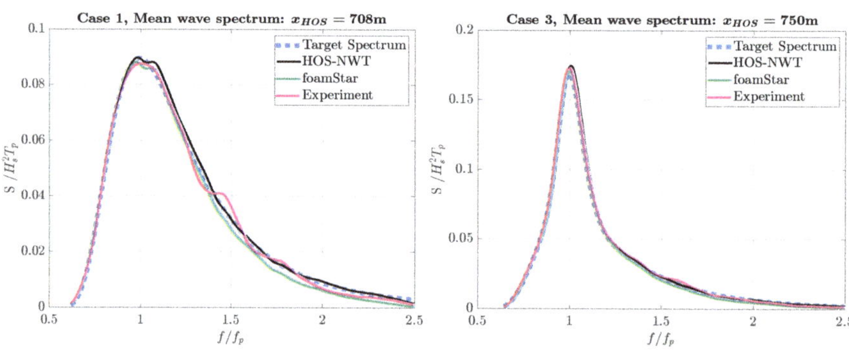

Figure 12. Comparison of the irregular wave spectrum from the HOS-NWT, experiment, and `foamStar` for `Case1` and `Case3` at the target location $x_{CFD} = 0$ m.

When the difference ratio of the wave spectrum from the target spectrum was compared, only the HOS-NWT spectrum satisfied the ±5% criterion in the target frequency range. On the other hand, comparing the experiment and the `foamStar` wave spectra using the HOS-NWT wave spectrum as a basis, neither of the two wave spectra satisfied the ±5% criterion in the target frequency range. The experimental result for both `Case1` and `Case3` generally aligned well with the HOS-NWT, while a larger discrepancy wave was measured for the high-frequency components. Looking at the variation in the wave spectrum in Figure 7, it suggests that the variation in the high-frequency components was related to the transverse effects in the wave basin. The `foamStar` simulation result generally showed a smaller wave spectrum than the HOS-NWT, and the wave spectrum gradually decreased as the wave frequency increased. Considering that the damping of a regular wave was highly related to the mesh resolution per wavelength, the underestimation of the spectrum, along with the frequency, was presumably due to the insufficient mesh resolution. The difference ratio of `Case3` showed a more oscillatory reduction along the frequency than that of `Case1`.

Table 7 summarizes the wave parameters obtained at the target locations for both `Case1` and `Case3`. It presents the significant wave height (H_s), the wave peak period (T_p, maximum of spectrum), and the true wave peak period ($\hat{T_p}$) defined in Equation (4). The ITTC recommendation for wave modeling [13] proposed a tolerance of ±5% on both the significant wave height and the peak period. For both sea states, the significant wave heights and the wave peak periods of all wave spectra remained within the range ±5%.

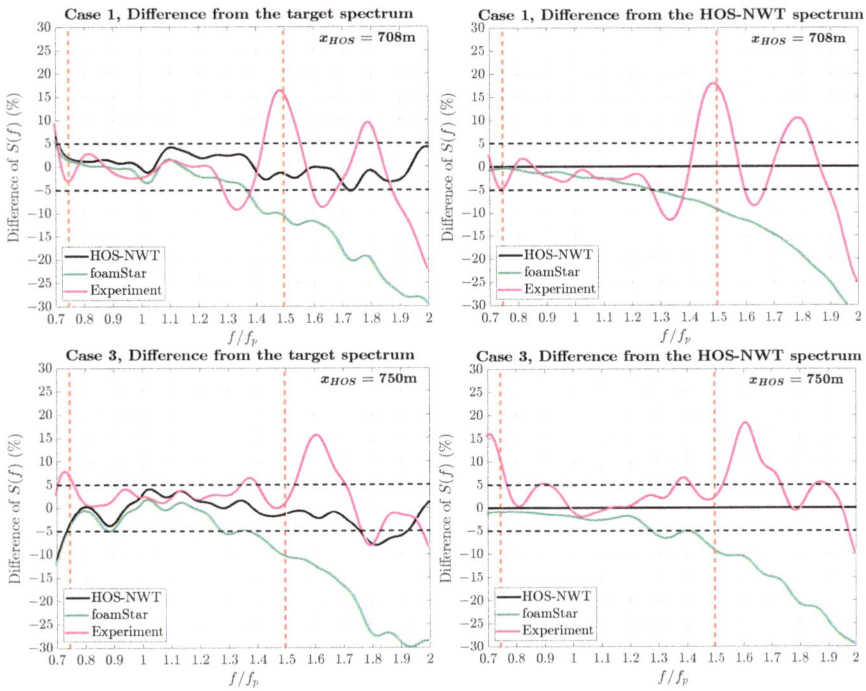

Figure 13. Comparison of the irregular wave spectrum from the HOS-NWT, experiment, and `foamStar` for `Case1` and `Case3` at the target positions.

Table 7. Significant wave height, peak periods, and maximum difference ratio from the HOS-NWT spectrum of the irregular waves measured at x_{HOS} = 708 m for `Case1` and x_{HOS} = 750 m for `Case3`.

`Case1`	**Target**	**HOS-NWT**	`foamStar`	**Experiment**
H_s [m]	6	5.96 (−0.64%)	5.74 (−4.31%)	5.91 (−1.57%)
T_p [s]	12.25	12.49 (1.94%)	12.55 (2.41%)	12.17 (−0.63%)
$\hat{T}_p$ [s]	12.25	11.65 (−4.87%)	11.73 (−4.23%)	11.67 (−4.76%)
Max diff. $S(f)$	-	-	9.17%	18.04%
`Case3`	**Target**	**HOS-NWT**	`foamStar`	**Experiment**
H_s [m]	17	17.02 (0.14%)	16.52 (−2.81%)	17.12 (0.72%)
T_p [s]	15.5	15.43 (−0.47%)	15.43 (−0.47%)	15.49 (−0.09%)
$\hat{T}_p$ [s]	15.5	15.31 (−1.25%)	15.32 (−1.16%)	15.34 (−1.01%)
Max diff. $S(f)$	-	-	9.55%	9.87%

In summary, the wave spectra of the `foamStar` simulations and the experiment generally showed good agreement with the HOS wave spectrum, while they did not satisfy the ±5% criterion in the entire target frequency range. Also, the wave parameters satisfied the ITTC recommendation.

7.3.2. Wave Crest Height Distribution

The PDSR (single realization) and PDER (ensemble realization) of the wave crest height were compared with the reference distributions. Note that the previous study of Canard et al. [30] showed that, depending on the target location from the wavemaker, even if the spectrum was qualified, the PDER curve of wave crest height could be significantly different. However, in this study, the target location from the wavemaker was only two and three wavelengths, and the evolution of the wave statistic properties along the domain was not expected to be significant. Figure 14 compares the probability of exceedance (POE) of the wave crest height measured at the target locations. The upper graphs show PDSR markers for `Case1` and `Case3`, while the lower graphs show the PDER curves. Each PDSR marker presents a measured wave crest height. The y-axis of the graph presents the POE of the wave crest height, and the x-axis of the graph presents the normalized wave crest height. Note that the wave crest is normalized by the measured significant wave height tabulated in Table 7.

The PDSR markers of the experiments and numerical simulations were compared with a Rayleigh distribution, the Forristall distribution [28], and 99% of the upper and lower bounds of the wave crest distribution built by Huang and Zhang [25]. For `Case1`, the PDSR markers of the experiment, HOS-NWT, and `foamStar` mostly fell between the 99% lower and upper bounds of Huang's distribution. Moreover, a few smaller wave crest heights of `foamStar` and larger wave crest heights of the experiment and HOS-NWT were captured. In the case of `Case3`, relatively smaller PDSR markers were captured compared to the Huang mean distribution.

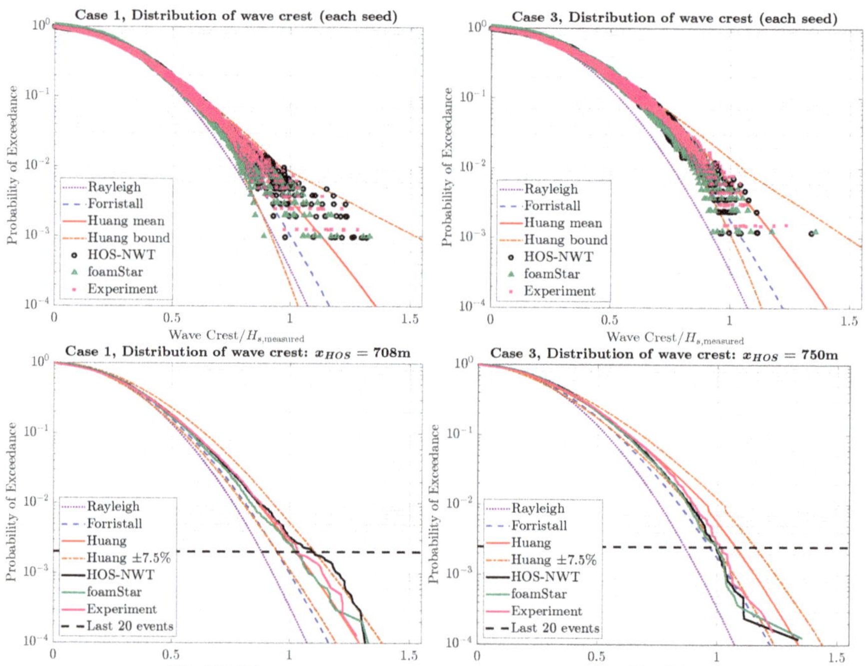

Figure 14. Probability distribution of wave crests height at the target locations. The upper plots show the probability distribution of single realizations and the lower plots show the probability distribution of wave crest for an ensemble of realizations.

The PDER curves were compared with the Huang ensemble distribution and the Forristall distribution. Additional ±7.5% bounds of the Huang ensemble distribution were also used to estimate the difference. The horizontal dotted line represents the probability of the last 20 events of the HOS-NWT simulation. In the PEDR analysis, the results under this dotted line are considered unreliable due to the lack of wave crest height data.

For `Case1`, overall, the three curves were in good agreement within the range of statistical convergence. Small discrepancies were observed for`foamStar`, which predicted slightly underpredicted values. This might be related to the lower energy in the high-frequency range of the spectrum. Then, these curves compared quite well with the Huang

PDER curve. For `Case3`, all three curves agreed well, excluding the last 20 events. This demonstrates that the wave breaking model in the HOS-NWT was able to correctly model the wave breaking observed in the CFD-NWT and the experiment. All three measured PDERs were underestimated compared to the Huang ensemble distribution.

8. Conclusions

The main objective of this paper was to propose and validate a wave generation and qualification procedure for irregular wave experiments and an OpenFOAM CFD simulation based on the HOS-NWT method. A wave generation and calibration procedure based on the HOS-NWT solver is proposed, and each part of the workflow is described. The details of the HOS-NWT wave solver are presented, and the setup of the wave basin experiment is outlined. Then, the numerical setup of the OpenFOAM simulation is introduced.

The proposed irregular wave generation process was tested for two long-crested irregular wave conditions: `Case1` and `Case3`. One representing mild wave conditions and the other including frequent wave breaking events. For each wave condition, 10 three-hour realizations were performed to obtain the stochastic properties of the irregular waves. Finally, comparisons of the HOS-NWT, OpenFOAM (`foamStar`), and experiment were performed. Both deterministic and stochastic analyses were performed to assess the quality of the irregular waves, and the wave qualification criteria proposed for the stochastic analysis were applied. Finally, from the analysis, the following remarks can be made:

- Using the OpenFOAM solver `foamStar`, a split simulation with multiple time intervals was performed. The simulation was compared with the full 3 h simulation. Considering that the difference between the two simulations almost disappeared within $20T_p$ for `Case3`, it was found that the influence of splitting is minimal and the memory effect from the nonlinearity of irregular waves is small. This result confirms that the split simulation of irregular waves is applicable. However, regarding its application to the WSI simulation, further investigation is required;
- In the deterministic analysis, a specific wave breaking event was selected for examination, focusing on comparing its time-series and spatial-wave profiles. The difference between the HOS-NWT, OpenFOAM, and experimental wave was small until they reached the wave breaking limit. When the HOS-NWT and OpenFOAM wave profiles were compared, a clear phase difference was captured, whereas a comparable wave dissipation was measured due to wave breaking. This finding suggests that the Tian–Barthelemy wave breaking model integrated into the HOS-NWT solver effectively replicates wave dissipation during the wave breaking event, while it does not induce a phase change associated with wave breaking events. The resultant waves of OpenFOAM generally followed the HOS-NWT, but they deviated from the HOS-NWT when wave breaking occurred and matched well with those in the experiment. Therefore, the CFD simulation with the suggested workflow can be used to model wave breaking events;
- A stochastic analysis of the wave spectrum was performed, comparing the measured wave spectrum with the target wave spectrum and the HOS-NWT wave spectrum. The wave parameters satisfied the ITTC $\pm 5\%$ criterion. The wave spectra of the OpenFOAM simulation and the experiment generally agreed well with the HOS-NWT wave spectrum, but they did not stay within the $\pm 5\%$ region in the entire target frequency range. The resultant OpenFOAM wave spectrum showed an underestimated wave spectrum, especially for the high-frequency components, due to an insufficient mesh resolution;
- A stochastic analysis was performed on the probability distribution of wave crest height. The measured PDSR and PDER curves were compared with the reference distributions. For both sea states, the PDSR markers fell mostly between the $\pm 99\%$ bounds of Huang's empirical formula. For both `Case1` and `Case3`, the three curves generally agreed well, excluding the last 20 events. This validates that, for `Case3`, the wave breaking model in the HOS-NWT can correctly model the wave breaking

observed in the CFD-NWT and the experiment. Small underpredictions of the wave crest height were observed for `foamStar`, which might be related to the lower energy in the high-frequency range of the spectrum. For `Case1`, the PDER curves compared quite well with the Huang ensemble distribution, while smaller PDER curves were observed for the wave breaking condition. This result confirms that the suggested procedure effectively reproduces wave stochastic properties in both CFD simulations and experiments.

With these remarks, this study confirms that the HOS-NWT-based nonlinear irregular wave calibration and generation procedure, including the wave breaking model, can effectively reproduce nonlinear waves for both experiments and CFD-NWT simulations, saving experimental and computational time in wave calibration. However, the wave spectra of the experiment and CFD-NWT did not stay within the ±5% region in the target frequency range. The variation in the wave spectra of the experiment is considered to be an unavoidable error due to the wave basin characteristics and experimental limitations. In the case of the CFD-NWT, further studies on the CFD mesh resolution for irregular wave simulations are required to explore an adequate numerical setup for certain sea states.

Author Contributions: Conceptualization, Y.J.K., M.C. and B.B.; methodology, Y.J.K. and M.C.; validation, Y.J.K. and M.C.; formal analysis, Y.J.K. and M.C.; investigation, Y.J.K. and M.C.; software, Y.-M.C. and G.D.; writing—original draft preparation, Y.J.K. and M.C.; writing—review and editing, Y.-M.C., B.B., G.D. and D.L.T.; visualization, Y.J.K. and M.C.; supervision, Y.-M.C., B.B., G.D. and D.L.T.; project administration, B.B., G.D. and D.L.T.; funding acquisition, Y.-M.C., B.B., G.D. and D.L.T. All authors have read and agreed to the published version of the manuscript.

Funding: This research received no external funding.

Institutional Review Board Statement: Not applicable.

Informed Consent Statement: Informed consent was obtained from all subjects involved in the study.

Data Availability Statement: Data are contained within the article.

Acknowledgments: This work was performed in the framework of the Chair Centrale Nantes—Bureau Veritas on the Hydrodynamics of Marine Structures and was supported by the WASANO project, funded by the French National Research Agency (ANR) as part of the "Investissements d'Avenir" Programme (ANR-16-IDEX-0007). We also acknowledge funding from the Joint Industry Project *Reproducible Offshore CFD JIP* [18]. This research was supported by Basic Science Research Program through the National Research Foundation of Korea (NRF) funded by the Ministry of Education (RS-2023-00243370).

Conflicts of Interest: The authors declare no conflicts of interest.

Abbreviations

WSI	Wave–Structure Interaction
OpenFOAM	Open-source Field Operation And Manipulation
CFD	Computational Fluid Dynamics
EWT	Experimental Wave Tank
NWT	Numerical Wave Tank
HOS	High-Order Spectral
FVM	Finite Volume Method
JONSWAP	JOint North Sea WAve Project
PDSR	Probability Distribution of the wave crest heights from Single Realizations
PDER	Probability Distribution of the wave crest heights from Ensemble Realizations
VOF	Volume Of Fluid

References

1. ITTC. *Recommended Procedures and Guidelines: 7.5-02-07-02.1 Seakeeping Experiment*; ITTC Association: Zürich, Switzerland, 2021; Revision 07.
2. Chuang, T.C.; Yang, W.H.; Yang, R.Y. Experimental and numerical study of a barge-type FOWT platform under wind and wave load. *Ocean Eng.* **2021**, *230*, 109015. [CrossRef]
3. Mohapatra, S.C.; Islam, H.; Hallak, T.S.; Soares, C.G. Solitary Wave Interaction with a Floating Pontoon Based on Boussinesq Model and CFD-Based Simulations. *J. Mar. Sci. Eng.* **2022**, *10*, 1251. [CrossRef]
4. Gao, J.l.; Lyu, J.; Wang, J.h.; Zhang, J.; Liu, Q.; Zang, J.; Zou, T. Study on Transient Gap Resonance with Consideration of the Motion of Floating Body. *China Ocean Eng.* **2022**, *36*, 994–1006. [CrossRef]
5. Chawla, A.; Kirby, J.T. A source function method for generation of waves on currents in Boussinesq models. *Appl. Ocean Res.* **2000**, *22*, 75–83. [CrossRef]
6. Gao, J.; Ma, X.; Dong, G.; Chen, H.; Liu, Q.; Zang, J. Investigation on the effects of Bragg reflection on harbor oscillations. *Coast. Eng.* **2021**, *170*, 103977. [CrossRef]
7. Gao, J.; Shi, H.; Zang, J.; Liu, Y. Mechanism analysis on the mitigation of harbor resonance by periodic undulating topography. *Ocean Eng.* **2023**, *281*, 114923. [CrossRef]
8. Wang, L.; Robertson, A.; Kim, J.; Jang, H.; Shen, Z.R.; Koop, A.; Bunnik, T.; Yu, K. Validation of CFD simulations of the moored DeepCwind offshore wind semisubmersible in irregular waves. *Ocean Eng.* **2022**, *260*, 112028. [CrossRef]
9. Shemer, L.; Sergeeva, A.; Liberzon, D. Effect of the initial spectrum on the spatial evolution of statistics of unidirectional nonlinear random waves. *J. Geophys. Res. Ocean.* **2010**, *115, C12039*. [CrossRef]
10. Latheef, M.; Swan, C. A laboratory study of wave crest statistics and the role of directional spreading. *Proc. R. Soc. A Math. Phys. Eng. Sci.* **2013**, *469*, 20120696. [CrossRef]
11. Janssen, P.A. Nonlinear four-wave interactions and freak waves. *J. Phys. Oceanogr.* **2003**, *33*, 863–884. [CrossRef]
12. Fedele, F.; Cherneva, Z.; Tayfun, M.A.; Guedes Soares, C. Nonlinear Schrödinger invariants wave statistics. *Phys. Fluids* **2010**, *22*, 1–9. [CrossRef]
13. ITTC. *Recommended Procedures and Guidelines: 7.5-02-07-01.2 Laboratory Modelling of Waves*; ITTC Association: Zürich, Switzerland, 2021; Revision 01.
14. West, B.J.; Brueckner, K.A.; Janda, R.S.; Milder, D.M.; Milton, R.L. A new numerical method for surface hydrodynamics. *J. Geophys. Res. Ocean.* **1987**, *92*, 11803–11824. [CrossRef]
15. Dommermuth, D.G.; Yue, D.K. A high-order spectral method for the study of nonlinear gravity waves. *J. Fluid Mech.* **1987**, *184*, 267–288. [CrossRef]
16. Ducrozet, G.; Bonnefoy, F.; Le Touzé, D.; Ferrant, P. A modified High-Order Spectral method for wavemaker modeling in a numerical wave tank. *Eur. J. Mech. B/Fluids* **2012**, *34*, 19–34. [CrossRef]
17. Canard, M.; Ducrozet, G.; Bouscasse, B. Generation of 3-hr long-crested waves of extreme sea states with HOS-NWT solver. In Proceedings of the 39th International Conference on Ocean, Offshore and Arctic Engineering (OMAE 2020), Online, 3–7 August 2020.
18. Fouques, S.; Croonenborghs, E.; Koop, A.; Lim, H.J.; Kim, J.; Zhao, B.; Canard, M.; Ducrozet, G.; Bouscasse, B.; Wang, W.; et al. Qualification Criteria for the Verification of Numerical Waves—Part 1: Potential-Based Numerical Wave Tank (PNWT). In Proceedings of the ASME 2021 40th International Conference on Ocean, Offshore, and Arctic Engineering, Virtual, 21–30 June 2021.
19. Hasselmann, K.; Barnett, T.P.; Bouws, E.; Carlson, H.; Cartwright, D.E.; Eake, K.; Euring, J.A.; Gicnapp, A.; Hasselmann, D.E.; Kruseman, P.; et al. Measurements of wind-wave growth and swell decay during the joint North Sea wave project (JONSWAP). *Ergnzungsh. Zur Dtsch. Hydrogr. Z. R.* **1973**, *8*, 1–95.
20. Serio, M.; Onorato, M.; Osborne, A.; Janssen, P. On the computation of the Benjamin-Feir Index. *Nuovo Cim. Della Soc. Ital. Di Fis. C* **2005**, *28*, 893–903. [CrossRef]
21. Papadimitriou, S.; Sun, J.; Yu, P.S. Local correlation tracking in time series. In Proceedings of the Sixth International Conference on Data Mining (ICDM'06), Hong Kong, China, 18–22 December 2006; pp. 1–10. [CrossRef]
22. Choi, Y.M.; Bouscasse, B.; Ducrozet, G.; Seng, S.; Ferrant, P.; Kim, E.S.; Kim, Y.J. An efficient methodology for the simulation of nonlinear irregular waves in computational fluid dynamics solvers based on the high order spectral method with an application with OpenFOAM. *Int. J. Nav. Archit. Ocean Eng.* **2023**, *15*, 100510. [CrossRef]
23. Welch, P.D. The Use of Fast Fourier Transform for the Estimation of Power Spectra: A Method Based on Time Averaging Over Short, Modified Periodograms. *IEEE Trans. Audio Electroacoust.* **1967**, *15*, 70–73. [CrossRef]
24. Young, I.R. The determination of confidence limits associated with estimates of the spectral peak frequency. *Ocean Eng.* **1995**, *22*, 669–686. [CrossRef]
25. Huang, Z.; Zhang, Y. Semi-empirical single realization and ensemble crest distributions of long-crest nonlinear waves. In Proceedings of the ASME 2018 37th International Conference on Ocean, Offshore and Arctic Engineering, Madrid, Spain, 17–22 June 2018; Volume 1. [CrossRef]
26. Brown, L.D.; Cai, T.T.; Das Gupta, A. Interval estimation for a binomial proportion. *Stat. Sci.* **2001**, *16*, 101–117. [CrossRef]
27. Longuet-Higgins, M.S. On the Statistical Distribution of the Heights of Sea Waves. *J. Mar. Res.* **1952**, *11*, 245–266.
28. Forristall, G.Z. Wave crest distributions: Observations and second-order theory. *J. Phys. Oceanogr.* **2000**, *30*, 1931–1943. [CrossRef]

29. Onorato, M.; Osborne, A.R.; Serio, M.; Cavaleri, L.; Brandini, C.; Stansberg, C.T. Extreme waves, modulational instability and second order theory: wave flume experiments on irregular waves. *Eur. J. Mech. B/Fluids* **2006**, *25*, 586–601. [CrossRef]
30. Canard, M.; Ducrozet, G.; Bouscasse, B. Varying ocean wave statistics emerging from a single energy spectrum in an experimental wave tank. *Ocean Eng.* **2022**, *246*, 110375. [CrossRef]
31. Tian, Z.; Perlin, M.; Choi, W. An eddy viscosity model for two-dimensional breaking waves and its validation with laboratory experiments. *Phys. Fluids* **2012**, *24*, 036601. [CrossRef]
32. Barthelemy, X.; Banner, M.L.; Peirson, W.L.; Fedele, F.; Allis, M.; Dias, F. On a unified breaking onset threshold for gravity waves in deep and intermediate depth water. *J. Fluid Mech.* **2018**, *841*, 463–488. . [CrossRef]
33. Canard, M.; Ducrozet, G.; Bouscasse, B. Experimental reproduction of an extreme sea state in two wave tanks at various generation scales. In Proceedings of the OCEANS 2022—Chennai, Chennai, India, 21–24 February 2022.
34. Jasak, H. Error Analysis and Estimation for the Finite Volume Method with Applications to Fluid Flows. Ph.D. Thesis, Imperial College of Science, Technology and Medicine, London, UK, 1996.
35. Seng, S. Slamming and Whipping Analysis of Ships. Ph.D. Thesis, Technical University of Denmark, Kongens Lyngby, Denmark, 2012.
36. Larsen, B.E.; Fuhrman, D.R. On the over-production of turbulence beneath surface waves in Reynolds-averaged Navier-Stokes models. *J. Fluid Mech.* **2018**, *853*, 419–460. [CrossRef]
37. Kim, Y.J.; Bouscasse, B.; Seng, S.; Le Touzé, D. Efficiency of diagonally implicit Runge-Kutta time integration schemes in incompressible two-phase flow simulations. *Comput. Phys. Commun.* **2022**, *278*, 108415. [CrossRef]
38. Jacobsen, N.G. *waves2Foam Manual*; Technical Report August; Deltares: Delft, The Netherlands, 2017.

MDPI AG
Grosspeteranlage 5
4052 Basel
Switzerland
Tel.: +41 61 683 77 34

Journal of Marine Science and Engineering Editorial Office
E-mail: jmse@mdpi.com
www.mdpi.com/journal/jmse

www.ingramcontent.com/pod-product-compliance
Lightning Source LLC
Chambersburg PA
CBHW061517210926
51850CB00008B/1945
* 9 7 8 3 7 2 5 8 5 8 9 3 4 *